Agricultural Research for Sustainable Food Systems in Sri Lanka

Buddhi Marambe • Jeevika Weerahewa •
Warshi S. Dandeniya
Editors

Agricultural Research for Sustainable Food Systems in Sri Lanka

Volume 1: A Historical Perspective

Springer

Editors
Buddhi Marambe
Department of Crop Science
Faculty of Agriculture
University of Peradeniya
Peradeniya, Sri Lanka

Jeevika Weerahewa
Department of Agricultural Economics and Business Management
Faculty of Agriculture
University of Peradeniya
Peradeniya, Sri Lanka

Warshi S. Dandeniya
Department of Soil Science
Faculty of Agriculture
University of Peradeniya
Peradeniya, Sri Lanka

ISBN 978-981-15-2151-5 ISBN 978-981-15-2152-2 (eBook)
https://doi.org/10.1007/978-981-15-2152-2

This Springer imprint is published by the registered company Springer Nature Singapore Pte Ltd.
The registered company address is: 152 Beach Road, #21-01/04 Gateway East, Singapore 189721, Singapore

This book was possible owing to the live agricultural research culture that has continuously evolved for 70 years in the premier agriculture higher education institute of Sri Lanka, the Faculty of Agriculture of the University of Peradeniya. The faculty provided leadership to agriculture research and established strong collaborations with key stakeholders for making the food systems of the country more sustainable. The academia and young minds that were trained under the wings of the faculty have provided the leadership in the agriculture and allied sectors over the past seven decades, nationally and globally. This book is dedicated to the past, present and the future agricultural scientists in the quest for making food systems more sustainable.

Acknowledgements

The Faculty of Agriculture at University of Peradeniya, the pioneer institute on agriculture higher education in Sri Lanka, reached its 70 years of Agriculture Higher Education and Research in 2018. Since its inception in 1948, the Faculty of Agriculture has made a commendable contribution to uplift the agriculture sector of Sri Lanka through education, research and outreach activities. To commemorate this important milestone, the faculty decided to issue a publication bringing together the views of expert agricultural scientists of the country on the evolution of agricultural research in Sri Lanka and challenges they foresee in making the food systems of the country more sustainable.

The launch of the two volumes titled *Agricultural Research for Sustainable Food* Systems *in Sri Lanka* is an outcome of this effort. Volume I of the publication is devoted to describe the historical perspective and Volume II is a pursuit of advancement.

The editors wish to record a word of appreciation for the following individuals and committees. First of all, the Vice Chancellor, University of Peradeniya and the Dean, Faculty of Agriculture are acknowledged for providing leadership to the activities carried out. The editors also would like to thank the members of the Faculty of Agriculture of University of Peradeniya and the Government Department of Agriculture, Sri Lanka, for pioneering and supporting teaching and research activities in agriculture in Sri Lanka, which formed the basis for this book. Authors of the book chapters represent aforementioned two institutions. Their immense contribution to compilation of this volume is greatly acknowledged. The initiatives taken by the Faculty Board of Faculty of Agriculture, University of Peradeniya, facilitating the publication of agricultural research for sustainable food systems are highly appreciated.

The Publications and Public Relations Committee of the Faculty of Agriculture—2018 is the body which initiated and coordinated the key activities in compiling the volume. Dr. Ramya Malkanthi Fonseka, Secretary, Publications and Public Relations Committee and her members are acknowledged and appreciated. Professor Nimal Gunawardena, Chairperson, Faculty Review and Editorial Committee and his members led the reviewing of the chapters. The reviewers of the book chapters (experts outside of the Faculty of Agriculture) are acknowledged for the services rendered. The anonymous reviewers of the book proposal are acknowledged for provision of constructive comments. The editors also wish to thank Springer Nature for identifying the value of this publication and considering for publishing.

The editors expect that Volume I will serve as a reference module for the future generation of agricultural scientists in the developing world and in Sri Lanka in particular.

August 27, 2019

Buddhi Marambe
Jeevika Weerahewa
Warshi S. Dandeniya

Contents

Editors and Contributors

About the Editors

Buddhi Marambe is a Senior Professor at the Department of Crop Science, Faculty of Agriculture, University of Peradeniya, Sri Lanka. He received his B.Sc. (Agriculture) from the University of Peradeniya, and his M.Agr. and D.Agr. degrees (Plant Environmental Sciences) from Hiroshima University, Japan. His main research focuses are on weed science, climate change, and food security. Prof. Marambe is currently the President of the Weed Science Society of Sri Lanka (WSSSL), Chairman of the National Invasive Species Specialist Group (NISSG), and Chairman of the National Experts Committee on Climate Change Adaptation (NECCCA). He has also served as a consultant on agriculture-related issues to the World Bank, United Nations Development Programme (UNDP), Food and Agriculture Organization (FAO), International Center for Tropical Agriculture (CIAT), and the World Agroforestry Center (ICRAF).

Jeevika Weerahewa is a Professor of Agricultural Economics at the Department of Agricultural Economics and Business Management, Faculty of Agriculture, University of Peradeniya, Sri Lanka. She holds B.Sc. and M.Phil. degrees in Agriculture from the University of Peradeniya, and a Ph.D. in Agricultural Economics from the University of Guelph, Canada. Her primary research area is agricultural policy analysis. She has served as President of the Sri Lankan Agricultural Economics Association, a Collaborator of the International Food Policy Research Institute (IFPRI), a Hewlett Fellow of the International Agricultural Trade Research Consortium, and a Fellow of the Canadian Agricultural Trade Policy Research Network.

Warshi Dandeniya is a Senior Lecturer and presently the Head of the Department of Soil Science, University of Peradeniya, Sri Lanka. She received her B.Sc. and M.Phil. degrees in Agriculture from the University of Peradeniya, and her Ph.D. in Soil Science from Cornell University, USA. Her expertise is in soil microbial ecology, and her research work focuses on investigating the impacts of agricultural practices on soil biodiversity, and on improving nutrient management in cropping systems using microorganisms. Dr. Dandeniya is a member of the FAO/WHO Roster of Experts for Antimicrobial Resistance and a Young Affiliate of The World

Academy of Science (TWAS). She is currently Vice President of the Soil Science Society of Sri Lanka.

Contributors

Saman Athauda Department of Animal Science, Faculty of Agriculture, University of Peradeniya, Peradeniya, Sri Lanka

Nimal Chandraratna National Aquaculture Development Authority (NAQDA), Battaramulla, Sri Lanka

Warshi S. Dandeniya Department of Soil Science, Faculty of Agriculture, University of Peradeniya, Peradeniya, Sri Lanka

C. M. B. Dematawewa Department of Animal Science, Faculty of Agriculture, University of Peradeniya, Peradeniya, Sri Lanka

M. P. Dhanapala Rice Research and Development Institute (RRDI), Department of Agriculture, Bathalagoda, Sri Lanka

P. B. Dharmasena Department of Agriculture, Rajarata University of Sri Lanka, Mihintale, Sri Lanka

H. Hemal Fonseka Onesh Agri Pvt Ltd, Colombo, Sri Lanka

K. S. Hemachandra Department of Agricultural Biology, Faculty of Agriculture, University of Peradeniya, Peradeniya, Sri Lanka

Venura Herath Department of Agricultural Biology, Faculty of Agriculture, University of Peradeniya, Peradeniya, Sri Lanka

Dimanthi Jayatilake Department of Agricultural Biology, Faculty of Agriculture, University of Peradeniya, Peradeniya, Sri Lanka

S. S. B. D. G. Jayawardena Sri Lanka Council for Agricultural Research Policy, Ministry of Agriculture, Colombo, Sri Lanka

Pradeepa Korale-Gedara Department of Agricultural Economics and Business Management, Faculty of Agriculture, University of Peradeniya, Peradeniya, Sri Lanka

Ranjith B. Mapa University of Peradeniya, Peradeniya, Sri Lanka

Buddhi Marambe Department of Crop Science, Faculty of Agriculture, University of Peradeniya, Peradeniya, Sri Lanka

S. M. M. R. Mawalagedera School of Education, Faculty of Arts, Deakin University, Geelong, VIC, Australia

S. A. C. N. Perera Department of Agricultural Biology, Faculty of Agriculture, University of Peradeniya, Peradeniya, Sri Lanka

D. V. Pahan Prasada Department of Agricultural Economics and Business Management, Faculty of Agriculture, University of Peradeniya, Peradeniya, Sri Lanka

S. Premarathne Department of Animal Science, Faculty of Agriculture, University of Peradniya, Peradniya, Sri Lanka

K. Samarasinghe Department of Animal Science, Faculty of Agriculture, University of Peradniya, Peradniya, Sri Lanka

G. L. L. P. Silva Department of Animal Science, Faculty of Agriculture, University of Peradeniya, Peradeniya, Sri Lanka

C. Sivayoganathan University of Peradeniya, Peradeniya, Sri Lanka

Udaya W. A. Vitharana Department of Soil Science, Faculty of Agriculture, University of Peradeniya, Peradeniya, Sri Lanka

Jeevika Weerahewa Department of Agricultural Economics and Business Management, Faculty of Agriculture, University of Peradeniya, Peradeniya, Sri Lanka

W. A. P. Weerakkody Department of Crop Science, Faculty of Agriculture, University of Peradeniya, Peradeniya, Sri Lanka

W. M. W. Weerakoon Department of Agriculture, Peradeniya, Sri Lanka

Chatura Sewwandi Wijetunga Faculty of Animal Science and Export Agriculture, Uva Wellassa University of Sri Lanka, Badulla, Sri Lanka

Hemantha Wijewardena Sri Lanka Council for Agricultural Research Policy, Ministry of Agriculture, Colombo, Sri Lanka

Abbreviations

AAIB	Agricultural and Agrarian Insurance Board
AEU	Agriculture Education Unit
AFS	Australian Friesian Sahiwal
AI	Artificial Insemination
AI	Agricultural Instructor
AICRP(VC)	All India Coordinated Research Project (Vegetable Crops)
AMZ	Australian Milking Zebu
AO	Agriculture Officer
ASC	Agrarian Service Centre
ASD	Advisory Services Department
ASF	Animal-sourced foods
ASMEC	Annual Symposium of Minor Export Crops
AT	Adaptability testing
AVRDC	Asian Vegetable Research and Development Center (Presently known as the World Vegetable Center)
BDG	Brewers dried grains
BLB	Bacterial Leaf Blight
BMI	Body Mass Index
BPH	Brown plant-hopper
CAGR	Compound Annual Growth Rate
CARI	Central Agricultural Research Institute
CAS	Ceylon Agricultural Society
CBSL	Central Bank of Sri Lanka
CCB	Coconut Cultivation Board
CDO	Coconut Development Officer
CEU	Cyber Extension Unit
CFC	Ceylon Fisheries Cooperation
CGRD	Coconut Genetic Resources Database
CICL	Ceylinco Insurance Company Limited
CIDA	Canadian International Development Agency
CIMMYT	International Maize and Wheat Improvement Center
CISL	Ceylinco Insurance and Securities (Pvt.) Limited
CKDu	Chronic Kidney Disease of uncertain etiology
COGENT	International Coconut Genetic Resources Network

CP	Central Province
CPRS	Central Poultry Research Station
CPUE	Catch Per Unit Effort
CRBS	Central Rice Breeding Station
CRI	Coconut Research Institute
CRVT	Coordinated Rice Varietal Testing program
CTVS	Cascaded Tank-Village Systems
DAEO	District Agricultural Extension Officer
DAPH	Department of Animal Production and Health
DArT	Diversity Array Technology
DC	Desiccated coconut
DCS	Department of Census and Statistics
DEA	Department of Export Agriculture
DEMs	Digital Elevation Models
DFP	Department of Food Production
DGWG	Dee-Geo-Woo-Gen
DHS	Demographic and Health Survey
DOA	Department of Agriculture
DZ	Dry Zone
EEZ	Exclusive Economic Zone
EMI	Electromagnetic Induction
ERP	Eppawala Rock Phosphate
EU	European Union
FAO	Food and Agriculture Organization of the United Nations
FCRDI	Field Crops Research and Development Institute
FHB	Family Health Bureau
FPO	Food Production Overseer
FRDI	Fruit Crops Research and Development Institute
GAP	Good Agricultural Practices
GAP	Granary Area Program
GDP	Gross Domestic Production
GE	Genetic Engineering
GFMI	Genetic, Feeding and Management Improvement
GIAHS	Globally Important Agricultural Heritage System
GIFT	Genetically Improved Farm Tilapia
GIZ	German International Cooperation Agency
GM	Gall midge
GM	Genetically modified
GMO	Genetically modified organisms
GN	*Grama Niladhari*
GPR	Ground Penetrating Radar
GPS	Global Positioning System
GSP	Generalised Scheme of Preferences
HI	Harvest Index
HIES	Household Income and Expenditure Surveys

HORDI	Horticultural Crop Research and Development Institute
HS	Hybridization and selection
HYV	High yielding varieties
HYW	Hungarian Yellow Wax
IASRI	Indian Agricultural Statistics Research Institute
ICPT	Integrated Crop Protection Technologies
ICT	Information and Communications Technology
IDD	Iodine Deficiency Disorders
IFC	International Finance Corporation
InDels	Insertion–Deletion Length Polymorphism
IPM	Integrated Pest Management
IPNS	Integrated Plant Nutrition System
IPS	Institute of Policy Studies
IR	Indian River
IRRI	International Rice Research Institute
ISE	Introduction Selection and Evaluation
IZ	Intermediate Zone
JICA	Japanese International Cooperation Agency
KVS	*Krushikarma Vyapthi Sevaka* (English translation: Agriculture Extension Officer)
LDI	Livestock Development Instructor
LOAM	Lanka Organic Agriculture Movement
MABB	Marker-Assisted Backcross Breeding
MAB	Molecular Accelerated Breeding
MAS	Marker Assisted Selection
MB	Maintenance Breeding
MI	Mahailluppallama
MILCO	Milk Industries of Lanka Company Limited
MPCI	Multi-Peril Crop Insurance
MRY	Meuse-Rhine-Yssel
MSY	Maximum Sustainable Yield
MYT	Major Yield Trial
NAEC	National Agricultural Extension Committee
NAIC	National Agricultural Information and Communication Centre
NAP	National Agriculture Policy
NAQDA	National Aquaculture Development Authority
NARA	National Aquatic Resources Research and Development Agency
NCD	Non-Communicable Disease
NCVT	National Coordinated Varietal Trial
NDVI	Normalized Difference Vegetation Index
NFS	National Fertilizer Secretariat
NGOs	Non-Governmental Organizations
NGS	Next-Generation Sequencing Platforms
NIV	New improved varieties
NLDB	National Livestock Development Board

NPPOS	National Plant Protection Organization
NSC	National Seed Council
NSP	National Seed Policy
OFC	Other field crops
OIV	Old improved varieties
PAGE	Pilot Analysis of Global Ecosystems
PCR	Polymerase Chain Reaction
PeCroDeP	Perennial Crops Development Project
PGRC	Plant Genetic Resources Centre
QTL	Quantitative Trait Loci
RARC	Regional Agricultural Research Centre
RDD	Rubber Development Department
RRDI	Rice Research and Development Institute
RRI	Rubber Research Institute
SAARC	South Asian Association for Regional Cooperation
SAEP	Second Agricultural Extension Project
SAVERNET	South Asian Vegetable Research Network
SCPPC	Seed Certification and Plant Protection Centre
SCS	National Seed Certification Service
SDG	Sustainable Development Goals
SHB	Shot hole borer
SIDA	Swedish International Development Cooperation Agency
SILEP	Smallholder Integrated Livestock Extension Project
SL-ADB	Sri Lanka-Asian Development Bank
SLAEA	Sri Lanka Agricultural Extension Association
SLCAREP	Sri Lanka Council for Agricultural Research and Extension Policy
SLCARP	Sri Lanka Council for Agricultural Research Policy
SLR	Sri Lankan Rupees
SMO	Subject Matter Officer
SNP	Single Nucleotide Polymorphism
SPMDC	Seed and Planting Material Development Centre
SRI	Sugarcane Research Institute
SRICANSOL	Sri Lanka–Canada Soil Resource Project
SSR	Simple Sequence Repeats
STF	Seed Task Force
T&V	Training and visit
TA	Technical Assistant
TMR	Total Mixed Rations
TRI	Tea Research Institute
TSHDA	Tea Small Holdings Development Authority
TSP	Triple Superphosphate
TYLC	Tomato Yellow Leaf Curl Virus
UETS	Unified Extension and Training System
UN	United Nations
UP	Uva Province

USA	United States of America
VAT	Varietal Adaptability Trial
VMS	Vessel Monitoring System
VRI	Veterinary Research Institute
WCLWD	Weligama Coconut Leaf Wilt Disease
WHO	World Health Organization
WIBI	Weather Index-Based Insurance
WRI	World Resources Institute
WZ	Wet Zone

Food Systems in Sri Lanka: Components, Evolution, Challenges and Opportunities

1

Jeevika Weerahewa, Warshi S. Dandeniya, and Buddhi Marambe

Abstract

Many countries are in the process of transforming their food systems to be more environmentally sustainable and resilient and able to deliver healthy and nutritious diets to all. Sri Lanka is no exception to this process. This chapter describes the key features of different subcomponents of food systems of Sri Lanka and presents the opportunities and challenges faced by Sri Lanka in order to make food systems more sustainable. Authors argue that interrelationships among various elements of the food system should be given due consideration in designing and redesigning policies, institutions and technologies to address complex issues governing the food systems of the country.

Keywords

Food systems · Sustainability · Agriculture · Nutrition · Sri Lanka

J. Weerahewa (✉)
Department of Agricultural Economics and Business Management, Faculty of Agriculture, University of Peradeniya, Peradeniya, Sri Lanka
e-mail: jeevikaw@pdn.ac.lk

W. S. Dandeniya
Department of Soil Science, Faculty of Agriculture, University of Peradeniya, Peradeniya, Sri Lanka

B. Marambe
Department of Crop Science, Faculty of Agriculture, University of Peradeniya, Peradeniya, Sri Lanka

B. Marambe et al. (eds.), *Agricultural Research for Sustainable Food Systems in Sri Lanka*, https://doi.org/10.1007/978-981-15-2152-2_1

1.1 Food System: The Concept

1.1.1 What Is a Food System?

A food system goes beyond a farming system, the predecessor of the concept. A farming system is predominantly an agricultural production system which generally consists of crop, livestock, aquaculture, agroforestry and fruit crops to which farm family allocates its scarce resources to reach a family goal. A food system comprises farming systems, their input supply systems and waste management systems as well as the trade system and health system. The Food and Agriculture Organization (FAO) of the United Nations defines a food system as "the entire range of actors and their interlinked value-adding activities involved in the production, aggregation, processing, distribution, consumption and disposal of food products that originate from agriculture, forestry or fisheries, and parts of the broader economic, societal and natural environments in which they are embedded". Figure 1.1 illustrates the key subcomponents of a typical food system. The characteristic feature in a food system, as in any other system, is the interconnectedness among its subsystems. Due to this characteristic, a change that originates from one subsystem will impact on all other subsystems.

1.1.2 Why Sustainable Food Systems?

A sustainable food system is a food system that produces and accesses food in a sustainable manner and provides food and nutritional needs of the present as well as future generations. By definition, sustainable food systems generate economic

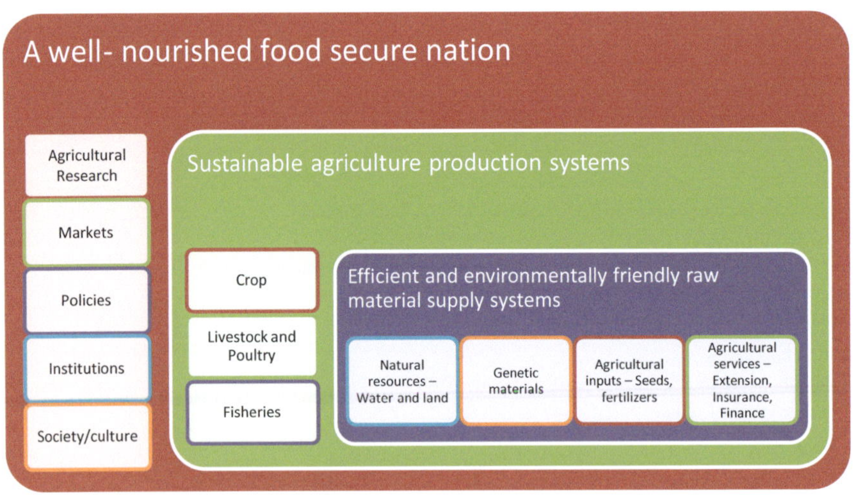

Fig. 1.1 Key components of a sustainable food system (Source: Authors' compilation)

returns to all the actors involved, provide benefits to the society and do less or no harm to the natural environment. Sustainable food systems are needed to achieve most of the targets of the United Nations' Sustainable Development Goals (SDGs). Reshaping of unsustainable or not-so-sustainable food systems into sustainable food systems is one of the major challenges faced by the globe in the present era. The challenges are multiple and multifaceted. For some countries, the critical challenge is to make the food system more productive, whereas for others, it is to make them more inclusive of poor and marginalized populations (see Chap. 2 for issues in Sri Lanka). Many countries are in the process of transforming their food systems to be more environmentally sustainable and resilient and able to deliver healthy and nutritious diets to all. All these require combinations of interconnected actions at the local, national, regional and global levels.

1.1.3 Why Food System Approach Is Needed to Achieve Development Targets?

A system approach is required to address problems in a system as the given solution would have implications on all the components in the system. Food systems are not exceptions to this rule. Better solutions for many food and agricultural issues, i.e. food and nutrition insecurity, climate change, natural disasters, etc., can be provided when the interrelationships among various elements of the food system are given the due consideration. This enables to examine all social, environmental and economic impacts of a given solution. This also facilitates multi-stakeholder collaboration and policy coordination at different levels which is the order of the present era.

1.2 Measuring Performance and Drivers of Food Systems

Performance of a food system is generally measured using agricultural productivity, quality and safety of the produce, nutrition and health, farm income and employment, women empowerment, degree of agro-industrialization, food and nutrition security of the nation it serves and the degree of depletion of natural resources, particularly land and water. There is an increased concern over certain other aspects of the food system, and measures are being developed to assess the nature of food items produced (whether they are highly processed, high-calorie and low nutritional valued; see Chap. 2), provision of market access to small-scale producers and agri-enterprises, food loss and waste, incidences of food safety (see Chaps. 7–9), animal and human health issues (see Chap. 11), and energy-intensity and ecological footprint associated with the lengthening and industrialization of food supply chains.

Identification of drivers of food system is of paramount importance as it paves the way to correct the weaknesses in the food systems and make them sustainable. Rapid population growth, urbanization, growing wealth, changing consumption patterns, globalization and climate change are the key natural drivers of food systems. The man-made drivers include technologies, policies and institutions.

1.3 Evolution of Food Systems of Sri Lanka

1.3.1 The Key Components of the Food System

Foods systems in Sri Lanka are diverse and can be broadly classified as traditional food systems, mixed food systems and modern food systems. Chena cultivations and home-garden systems are examples for traditional food systems (see Chap. 3), which are characterized by subsistent production and/or exchanges of primary produce in informal markets with short and local supply chains. Mixed food systems are characterized by semi-subsistence or commercial production and food manufacturing where the processed foods are packaged, labelled and sold in both formal and informal markets with frequent branding and advertising. Modern food systems comprise more diverse food production options all year long, with significant processing and packaging to extend food's shelf life, food safety being monitored and enforced, and storage and transport infrastructures such as cold chains are generally prevalent and reliable. The systems involving food production by large farmers, cultivations in protected agriculture (see Chap. 9), food processing in factory industries, retailing in supermarket chains, connections with global value chains, etc. can be treated as modern food systems of Sri Lanka.

1.3.2 Agricultural Production System in the Economy of Sri Lanka

Sri Lanka had been a self-sufficient economy comprising a peasant agricultural sector with little international trade, till the country fell totally under British colonial rule in 1815 (see Chap. 7). The interest of the British rulers was to establish an export agricultural sector consisting of large plantations. For this purpose, an overt land grabbing campaign sanctioned by the government through the Crown Land Ordinance of 1840 was launched. This restricted traditional land use and ownership patterns making a large proportion of hitherto used and unused land available for European investors (IPS 2004). Foreign capital was invested on establishing large plantations in such acquired lands resulting in a steep rise in the acreage under the "Export Agricultural Sector", while the neglected "Peasant Agricultural Sector" stagnated producing mainly for household consumption with a little surplus of food crops aimed at local markets (Samaratunga 2012). In fact, the role of the domestic food production was further marginalized by the depressed prices in these markets emerged following the widespread promotion of cheap imported foods. Following these changes, the country's gross domestic product (GDP) grew fast with the export earnings of the plantation agriculture, while the contribution of the peasant food crops sector gradually shrank.

The first change in this pattern appeared as a result of "nationalistic thinking" emerged within the local political movement. In order to promote domestic food production and to distribute land among the landless in the wet zone of the country, the Land Development Ordinance of 1933 was enacted under which large irrigation projects were undertaken in the dry zone. Further, the developed lands were alienated to the landless from other areas, chiefly for production of rice, the country's main staple food. This started a marginal upward trend in the domestic food production sector. This trend was further reinforced by the scarcity and high prices of imported food, mainly rice, during the World War II from the early 1930s to the mid-1940s. It was in this backdrop that Sri Lanka gained her political independence in 1948 which had serious implications on food production and consumption in the country.

From this point onwards, the drive for domestic food production was given prominence with self-sufficiency in rice as the prime objective. With continued land development and support prices for rice, the domestic food production sector growth picked up some momentum, and this was further strengthened after 1960 by the increasing prices of imported food items. The start of the present epoch of growing domestic food production was therefore marked around this period. However, in spite of the absolute growth of agriculture (comprising both food and plantation crops), its share in the total GDP has been steadily declining since 1950 onwards. The economic policy paradigm adopted by the Sri Lankan government during these years was clearly of import substitution orientation, and restrictions on imports and sporadic price supports imposed on domestic food crops supported further the growth of domestic food production sector. Nevertheless, this change did not help the country in alleviating her food importation burden. Figure 1.2 depicts that the food imports bill has been, on a long run trend, rising in both absolute terms and as a percentage of the total exports. However, the food imports as a percentage of total imports have remained constant, due to the faster increasing non-food imports. All these point to the unabated import dependency problem in the food sector of Sri Lanka.

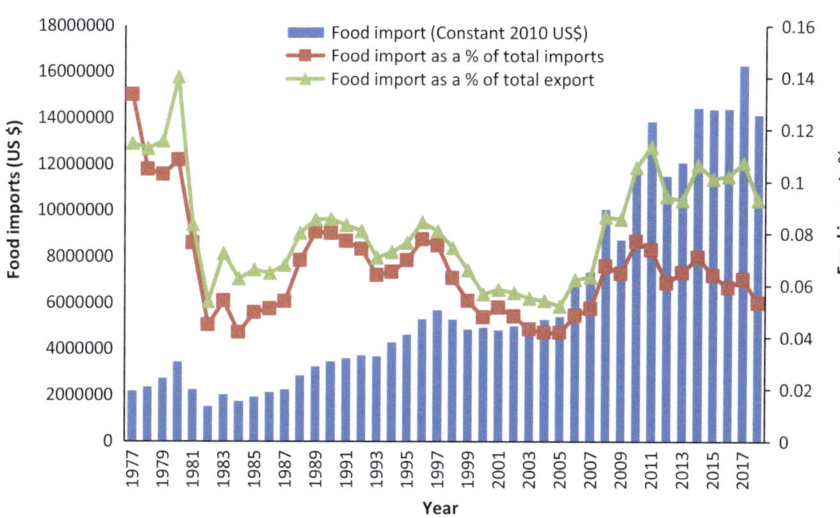

Fig. 1.2 Values of food imports in absolute and proportional terms (Source: Central Bank Annual Report (various years) (n.d.))

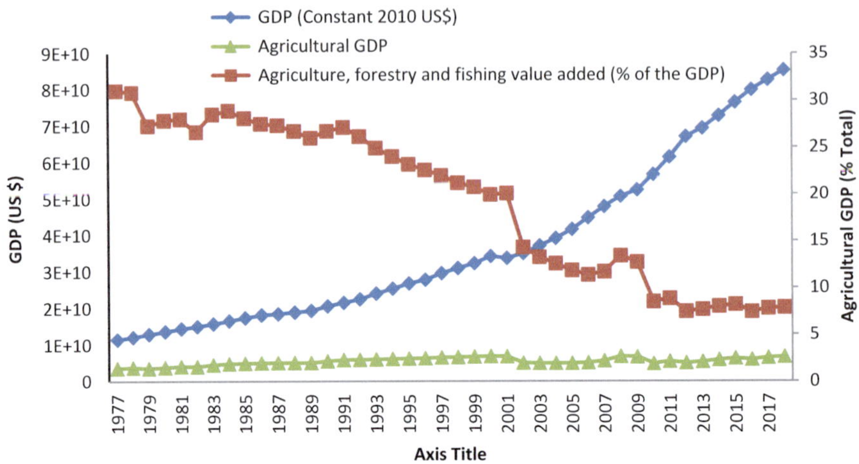

Fig. 1.3 Trends in GDP and agricultural GDP (Source: World Bank (various years) (n.d.))

The strategy of import substitution came to an end following the radical political change of 1977 (see Fig. 1.3 for changes in GDP and agriculture GDP during 1977–2018). The new regime immediately liberalized trade and adopted an export promotion development strategy in the years that followed. This changed the relative price structure facing domestic agriculture, bringing about drastic changes in food production and consumption. In fact, it is the paradigm following this change that initiated and still governs the trends in the food production and consumption sector of Sri Lanka.

1.4 Performance of Food Systems in Sri Lanka

1.4.1 Agricultural Production System

Rice is a success story in Sri Lanka. According to the Department of Census and Statistics of Sri Lanka (2019), paddy yields have increased steadily from 1588 to 4446 kg/ha during 1952–2015, and this was associated with an increase in the total production of paddy from 604 to 4819 million kg during the same period (see Chaps. 5 and 7).

Lower land productivity of many crops and livestock, except for paddy, compared to those of neighbouring countries has been highlighted by many experts. The growth in agricultural output has been found to be driven by input growth more than productivity growth. Though the total factor productivity has grown over 1961–2013 period in Sri Lanka, its performance compared to those of other developing countries is found to be not satisfactory. It has been argued that agricultural value chains in Sri Lanka are characterized by a large number of scatted small-scale producers, concentrated food processing sector, inadequate linkages with the global value chains (despite the unique characteristics of certain agricultural products) and inadequate diversification at the farm level and export level (despite the diversification of diet). Concerns have been expressed regarding malpractices in agrochemical application (see Chap. 7) in certain production systems claiming the association between the same and human health hazards such as chronic kidney diseases.

Limited land and labour availability for agricultural production and the spread of agricultural production among a large number of small semi-subsistence farmers are two key challenges to enhance the efficiency of the agricultural production system.

1.4.2 Agricultural Marketing System

Traditionally, agricultural marketing system is a complex process, involving a large number of intermediaries handling a number of agricultural commodities, which are seasonal and bulky and some are highly perishable. This process involves assembling, grading, processing, storage, transportation, wholesaling and retailing apart from pre- and postharvest operations. Supporting services such as financing, market research is also important. Since the early 1970s, successive governments in Sri Lanka have intervened in agricultural marketing by various ways that include offering guaranteed prices to farmers for selected food crops. The two main objectives of government intervention in agricultural marketing were to stabilize food prices and to ensure that farmers receive remunerative prices for their produce (Weliwita and Epaarachchi 2003). Since 1977, the role of the state intervention gradually diminished, and the private sector participation has spread in every facet of the agricultural marketing partly due to the obvious failure in state intervention and partly due to renowned interest on market-based alternatives and innovation mechanisms. The key advantages of market-based instruments were assurance of future revenues, comparatively low cost of implementation and shifting the risks to traders who are willing to take the price risk (Varangis and Larson 1996).

Currently, the agricultural marketing is a private sector operation except for necessary government intervention during market failures especially during the best performing years and the bad weathered years. Traditional agricultural marketing system is plagued with the issues like high price fluctuation; high marketing margins; lack of integration between input supply, production and marketing; and heavy postharvest losses (Karunagoda 2010). However, rapid and ongoing changes are occurring in domestic and export procurement systems, with the growing dominance of supermarkets, high-quality retail, bulk procurement by domestic manufacturing firms and increased quality awareness in the global trading system. This phenomenon has been further influenced by urbanization, income increases, increased migration of Sri Lankan citizens, increased female participation in the labour force, etc. (Samaratunga 2007). Consequently, new demand patterns have emerged, shifting the focus towards continuous supply of high-quality, value-added products with improved processing, packaging and labelling. The most intractable issue arising from the global market is public concern regarding food safety in industrialized countries that has resulted in increased and tightened public standards over the past two decades (IPS 2015). As in other developing countries, the food retail market in Sri Lanka has gradually evolved from fragmented local markets to centralized wholesale markets and eventually to the emergence of supermarkets (Samaratunga and Marawila 2006). This might have simplified the traditional complex value chains of domestically produced food commodities and developed modern value chains that integrate small producers with the dynamic markets.

1.4.3 Food and Nutrition Security

The average dietary energy supply adequacy increased from 103 to 115, and prevalence of undernutrition decreased from 3.5 million to 1.9 million from 1999/2001 to 2016/2018 (FAO 2019). Poverty is a rural phenomenon in Sri Lanka, and the head count index of poverty decreased from 26.1% to 4.1% from 1990/1991 to 2016 (Department of Census and Statistics 2017), which is highly attributable to increase in agricultural wages that grew annually by an average of 5.7% during 2006–2013. However, child malnutrition as measured by underweight, stunting and wasting has not significantly decreased during the period 2006–2016 (Department of Census and Statistics 2019) (see Chap. 2).

For a country that suffers no significant food shortages and demonstrates high scores for education and health services, it is rather paradoxical to note the status with respect to some of the indicators of undernutrition. Though there is a clear decline in numbers, 1.9 million of the Sri Lankan population does not have access to their energy requirement. Sixteen per cent children are born with low birthweight—and around same per cent of under-fives are reported to be stunted, rising as high as 32.4%, in some deprived districts (Department of Census and Statistics 2019). According to the Food Security Survey conducted by the Department of Census and Statistics (2014a), among the 10.3% who are food insecure, 30.1% are severely food insecure, 39.3% are moderately insecure, 21.7% are mildly food insecure and 8.9% are marginally food insecure.

The second burden of malnutrition, i.e. micronutrient deficiencies is also widespread in Sri Lanka. In a recent survey, results show that of all anaemic children 52.3% are iron deficient and of the children population 5.1% are zinc deficient (Ministry of Health and UNICEF 2012). The study also shows the relatively high prevalence (47.6%) of calcium deficiency with a wide interdistrict variation ranging from 23.1% in Badulla District and 70.3% in Matale District.

Overweight and obesity, which is the cause of many non-communicable diseases, is the third burden of malnutrition. Overweight was observed even among schoolchildren of Sri Lanka (Mohamed 2015). According to the survey on self-reported health, the highest number of cases reported is on high blood pressure (39.0%), and diabetes cases are the second highest (30.4%). The prevalence of diabetes in the population age 15 years and above is higher in districts of Colombo, Kalutara and Puttalam. The prevalence of high blood pressure for the same population is higher than 10% in several districts. The prevalence for all selected chronic illnesses is clearly increasing with age (Department of Census and Statistics 2014b).

1.4.4 Policies, Institutions and Regulations

As stated earlier, Sri Lanka embarked on an extensive economic liberalization process in 1977, and it had dramatic effects on all components of food system. The first round of reform measures covered most aspects of economic policy, including trade policy. For the agricultural commodities however Sri Lanka has traditionally pursued a two-pronged agricultural policy. One was applied to the plantation sector, which grows export crops such as tea, rubber and coconut. The sector benefited from some incentives provided for export expansion and foreign exchange. The other was applied to the non-plantation sector, based on smallholders' production of mainly basic foods. The sector, sometimes referred to as "subsistence agricultural sector", was also provided with a fair degree of protection from imports. It was also assisted with subsidized inputs, particularly of fertilizer, seeds and planting material, and through other support measures, such as low interest credit, guaranteed marketing through virtually free irrigation water. With the openness to trade, agricultural exports remain one of the most important sectors in the total exports and contribute around a quarter of the total exports.

1.5 The Way Forward for Sri Lanka: Opportunities and Challenges

Scattered nature of production which made the quality assurance across agricultural producers difficult and a poor adherence to quality standards are observed in many agricultural value chains. Collective behaviour among farmers is lacking in Sri Lanka and the bargaining power of farmer organizations is poor. The conflicting political ideologies of the ruling governments also block design and implementation of rational and pragmatic development policies for the country to a greater degree. Poor governance with respect to public provisions has also been observed. Being competitive in the export market is a challenge due to rising labour costs. Value addition is largely done by the small and medium enterprises which do not have sufficient access to finance. Most large enterprises do not wish to invest in Sri Lanka due to policy uncertainty. Outmigration of young and males from the agricultural sector, non-communicable disease burden and health risks associated with chronic kidney disease of uncertain aetiology (CKDu), land and soil degradation and climate change are the other challenges to develop its agricultural sector.

The growth in the agricultural sector could help to a greater extent in alleviating poverty in rural areas. The incidences of poverty have reduced significantly over the past decade in Sri Lanka despite its stagnant economic growth and productivity growth. Interestingly, an overall widening of inequality particularly across geographical areas can be observed and agricultural provinces being the least developed regions in the country. On one hand, it is claimed that the country relies heavily on natural resources to support economic growth, and hence there is an urgent need to evaluate environmental impacts of fast economic growth rate. On the other hand, some experts argue that stringent environmental regulations prevented the establishment of large-scale agricultural production units of the country.

Despite the above challenges, Sri Lanka is blessed with myriad of opportunities to strengthen its food systems. Sri Lanka's strategic location provides ample opportunities to trade with the growing Asia, which would benefit its local agribusiness and consumers alike. The growing middle class of the country and increased tourist arrivals enhance domestic demand for high-value agricultural products and value-added products as well. The country is situated in a strategic location, and it can be the maritime hub in the region. There is a large room for agricultural value addition through establishment export processing zones. The country has already an established name for quality products (intrinsic qualities of tea, cinnamon, pepper, etc.), and it can build on this reputation in making its exports more competitive. Furthermore, the country possesses a group of highly qualified and renowned agricultural professionals. Most of all, the successive governments have been treating its agricultural sector in a very special manner, and the needed political commitment is there.

References

Central Bank of Sri Lanka (various years) (n.d.) Central Bank Annual Report

Department of Census and Statistics (2014a) Household food security survey report. http://www.statistics.gov.lk/agriculture/Publications/HFS_Survey2014.pdf. Accessed 26 Aug 2019

Department of Census and Statistics (2014b) National survey on self-reported health. http://www.statistics.gov.lk/social/National%20Survey%20on%20Self-reported%20Health-2014.pdf. Accessed 26 Aug 2019

Department of Census and Statistics (2017) Poverty indicators. ISSN 1391 – 4693. http://www.statistics.gov.lk/poverty/Poverty%20Indicators_2016.pdf. Accessed 26 Aug 2019

Department of Census and Statistics (2019) Demographic and Health Surveys. http://www.statistics.gov.lk/page.asp?page=Health. Accessed 26 Aug 26 2019

Department of Census and Statistics of Sri Lanka (2019) Paddy statistics. http://www.statistics.gov.lk/agriculture/Paddy%20Statistics/PaddyStatsPages/PaddyAnnualSHYP.html. Accessed 26 Aug 2019

Food and Agriculture Organization (2019) FAOSTAT—Sri Lanka. http://www.fao.org/faostat/en/#country/38. Accessed 26 Aug 2019

IPS (2004) Land policy in Sri Lanka. In: State of the economy 2004. The Institute of Policy Studies, Colombo

IPS (2015) State of the economy report 2015. The Institute of Policy Studies, Colombo

Karunagoda K (2010) Sri Lanka—value chain of rice, coconut, fruits and vegetables and milk. Background paper prepared for FAO project on articulating and mainstreaming trade policies, Colombo, Sri Lanka

Ministry of Health and UNICEF (2012) National Nutrition and Micronutrient Survey. https://geonode.wfp.org/documents/2454. Accessed 26 Aug 2019

Mohamed SM (2015) Childhood obesity: epidemiology, determinants, and prevention. J Nutr Disord Ther 5:156. https://doi.org/10.4172/2161-0509.100015

Samaratunga P (2007) Sri Lanka: innovative practice in integrating small farmers into dynamic supply chains: a case study of Ma's Tropical Food Company. Regoverning markets innovative practice series. International Institute for Environment and Development, London

Samaratunga P (2012) Achieving economic efficiency and equitable distribution of land use through reforms and/or market discipline. In: Kelegama S, Gunewardena D (eds) Economic and social development under a market economy regime in Sri Lanka—Buddhadasa Hevavitharana Felicitation, vol 2. Vijitha Yapa Publications, Colombo

Samaratunga P, Marawila D (2006) Major characteristics and their determinants of the rural land sector in Sri Lanka and implications for policy, research studies. Agricultural Policy Series No. 7. The Institute of Policy Studies, Colombo

Varangis P, Larson DF (1996) Dealing with commodity price uncertainty. Policy Research Working Paper 1667. The World Bank, Washington, DC

Weliwita A, Epaarachchi R (2003) Forward contracts: a market based alternative to government intervention in agriculture marketing in Sri Lanka. Agricultural policy series. Institute of Policy Studies of Sri Lanka, Colombo

World Bank (various years) (n.d.) World development indicators. http://datatopics.worldbank.org/world-development-indicators/. Accessed 24 Aug 2019

Nutrition Transition in Sri Lanka: A Meta-Analysis of the Nutrition Profile

2

Jeevika Weerahewa, Pradeepa Korale-Gedara, and Chatura Sewwandi Wijetunga

Abstract

The aim of this review is to diagnose the nutrition transition in Sri Lanka with special emphasis on the nutrition profile of the country. The estimates reported in peer-reviewed journal articles and relevant national surveys conducted by the government agencies from 1977 to 2017 were synthesized. The estimates were summarized using tabular analysis, and trends over time were estimated using meta-regression models. The synthesis revealed that a considerably higher proportion of Sri Lankan population suffers from at least one form of malnutrition and the problem of undernutrition and hidden hunger is overlapping with the increasing threat of overnutrition despite the better performance shown by the economy as per various social and health indicators. The results of the meta-regression analysis revealed declining trends in many forms of undernutrition, particularly among females, over time. While the estate sector is more vulnerable to undernutrition among preschool children, the urban sector is more vulnerable to rising level of overnutrition across all age cohorts. The society is on the verge of entering into the fourth stage of nutrition transition, which is characterized by the consumption of energy-dense food, overweight and obesity, and rising non-communicable diseases (NCDs). It is recommended to make food systems more nutritionally sensitive through appropriate research, pricing, trade, food and nutrition policies to promote production and consumption of nutritious and healthy foods.

J. Weerahewa (✉) · P. Korale-Gedara
Department of Agricultural Economics and Business Management, Faculty of Agriculture, University of Peradeniya, Peradeniya, Sri Lanka
e-mail: jeevikaw@pdn.ac.lk

C. S. Wijetunga
Faculty of Animal Science and Export Agriculture, Uva Wellassa University of Sri Lanka, Badulla, Sri Lanka

B. Marambe et al. (eds.), *Agricultural Research for Sustainable Food Systems in Sri Lanka*, https://doi.org/10.1007/978-981-15-2152-2_2

Keywords
Undernutrition · Overnutrition · Meta-regression analysis · Government policy

2.1 Introduction

The existing body of literature documents two-way interrelationship between nutrition and economic development (Ranis et al. 2000; Suri et al. 2011). Better nutritional status of the population is considered as an end as well as a critical input to an economic development (Strauss and Thomas 1998; Ranis et al. 2000; Bhargava et al. 2001; Smith and Haddad 2002; Bloom et al. 2004; Well 2007; Narayan et al. 2010). Evidence from Sri Lanka indicates a strong causal relationship between the nutritional status and school achievement (Wisniewski 2010) and labor productivity (Sahn and Alderman 1988) reinforcing the importance of nutrition as an input to formation of human capital and in turn economic growth. Hartwig (2010) reported that the importance of nutrition to economic development is more important to middle-income countries as their marginal productivity of health is higher than that in the high-income countries.

Coexisting to this literature, Popkin (1994, 1998, 1999, 2001, 2002, 2003, 2004) illustrated a dynamic nature of the relationship between income growth and nutritional status of the people. It has been argued that as the income of a country rises and the rate of urbanization increases, its people enter into different stages of consumption and physical activity levels (Fig. 2.1). The term used to classify this phenomenon is "nutrition transition." When the income of a country rises, people will replace their traditional diet, which is low in fat and high in cereal with more fruit and vegetable and products of animal origin. However, when income further rises, diet of the people shifts more toward high-fat, more sugar, and processed foods which are low in fiber. The outcome of this shift is an epidemiological transition where people move away from nutrient deficiency diseases (undernutrition) toward higher rates of non-communicable diseases (NCDs) and overnutrition. With further rise in income and changes in life style, people will realize the adverse health effects of the high-fat, high-sugar diet and consequently will move toward a simple diet based on grains, vegetables, and fruits. It is further argued that at the presence of inequality in socioeconomic status across different geographical and socioeconomic sub-groups in a country, it is possible that different sub-communities will be at different stages of the transition. The outcome of this unequal rate of transformation is the coexistence of undernutrition and overnutrition at a cross section of time. Since inequality in economic development is a norm rather than an exception, this overlapping of phases is a common phenomenon in both developing and transitional economies.

It is evident that many developing and transitional economies have undergone nutrition transition in varying degrees. For example, in China, consumption has shifted more toward a diet high in fat resulting in increased overweight and obesity (Bell et al. 2001; Popkin 2001). In South Korea, consumption of animal-based food

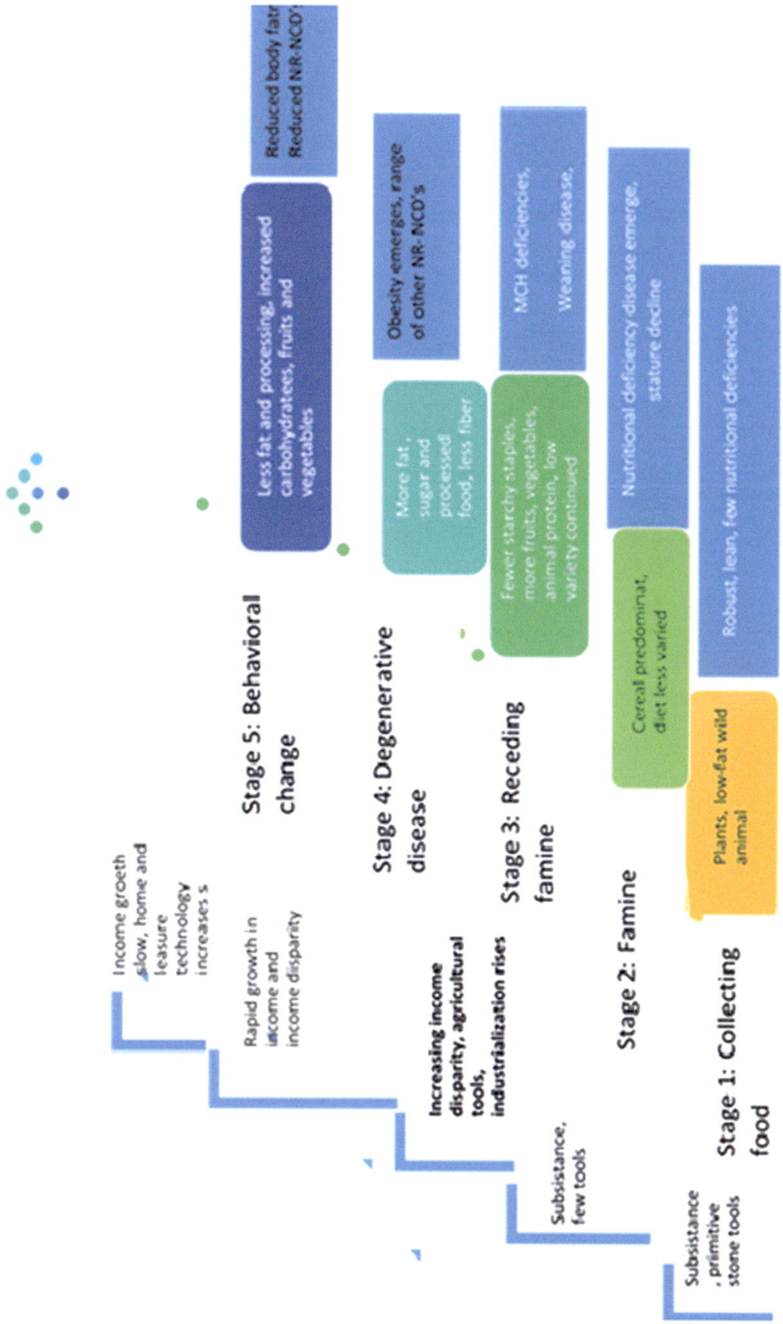

Fig. 2.1 Nutrition transition (Source: Adopted from Popkin (various years))

and vegetable and fruits has increased over the years, but the total fat consumption was maintained at a steady level. Thus the prevalence of overweight and obesity has not changed (Lee et al. 2002). As reported by Baker and Friel (2014), the consumption of fat, sugar, and salt has increased in the recent past in many Asian countries due to the tendency toward consuming more processed foods. Accounting for the demographic and epidemiological changes in Sri Lanka, it is apparent that the country, too, has a conducive environment for a nutrition transition and its consumption pattern is changing (Wijesekere 2015; Weerahewa et al. 2018a, b).

Against this background, the aim of this chapter is to examine the nature of nutrition transition of Sri Lanka. More specifically, this review (i) synthesizes the published literature on undernutrition, overnutrition, and dietary pattern of Sri Lanka for the time period from 1977 to 2017 with a view to examine the nature of nutrition transition of Sri Lanka, and (ii) documents the empirical evidence on the underlying causes of the undernutrition and overnutrition problems in Sri Lanka. The review attempts to inform policy makers on the areas where interventions are required and to inform researchers on the gaps in literature. Furthermore, the analysis will enable to track the process of achieving the second and third Sustainable Development Goals (SDGs) on Zero hunger, and Global Health and Wellbeing, respectively.

The chapter is organized as follows. The next section sets the background of the study summarizing the economic situation, food policies, safety net programs, and diet transformation in the country. Section 3 introduces the methodology of the review paper, while Sect. 4 presents the results of review. Section 5 describes the nutrition transition in Sri Lanka and the final section concludes the paper with a way forward.

2.2 Contextual Background

2.2.1 Key Characteristics of the Sri Lankan Economy

Sri Lanka is a lower middle-income country with a per capita GDP of USD 4065 in 2017 (Central Bank of Sri Lanka 2017). It is evident that the country's economy has been shifted toward industry- and service-oriented economy over the years, from a predominant agriculture base. Despite the declining role of agriculture sector to the economy, the sector provided direct employment to 26.1% of the labor force in 2017. However, due to low labor productivity in the agriculture sector, the share of the agriculture sector in total GDP is three times less than its share in the labor force.

Sri Lanka has experienced a remarkable increase in per capita GDP over the last few decades. In line with this economic growth, the rate of poverty has declined to a level of 4.1% in 2016 from being 26.1 in 1990/1991 (DCS 2017). Over the same period, the quality of the human capital, in terms of education and health, has increased. However, despite these achievements in the social sphere, Sri Lanka has continued to struggle to combat food insecurity and malnutrition. A considerable proportion of women and children in Sri Lanka still suffers from malnutrition (DCS 2017).

The percentage of the population living in urban areas has been stagnant in Sri Lanka, and approximately 18% of the population live in urban areas (World Bank 2018). However, there is a rapid increase in the expansion of urban areas in Sri Lanka that has led to an increase in population surrounding major urban areas (Ellis and Roberts 2015). Unlike in many South Asian countries, Sri Lanka has been able to alleviate urban poverty in a significant manner. The poverty incidences in urban areas reduced from 16.3% in 1990/1991 to 1.9% in 2016 (DCS 2016).

As in many developing countries, government intervenes in food systems, food pricing, and nutritional programs in Sri Lanka. Thus, for a better understanding of the changes in diet and malnutrition, one needs to have a better idea about the food and nutrition policies, food assistance programs, and nutritional interventions carried out by the government in the country. A brief description of those interventions is presented below.

2.2.2 Food and Nutrition Policy Framework During 1977–2017

With the liberalization of the economy in 1977, Sri Lanka shifted its policy orientation from protectionist to outward. Subsidies on many food commodities were removed and taxes on imports were cut down. However, country continued to protect the rice producers. To boost the local paddy production, huge public investments were made on irrigation projects. This open economic policy continued for two decades and in 2005, a major shift in outward-oriented policy took place with the change in political regime in the country. Several trade-related taxes were imposed on imports, though essential goods were levied from import tax. Several national-level programs were implemented during this period to promote home gardening and local food production. "*Api wawamu rata nagamu*" ("Let us grow and uplift our nation") is one such programs (Weerahewa et al. 2017).

Extensive bodies of empirical research have been focused on the evolution of food and nutrition policy framework in Sri Lanka during the past several years. Among them, Edirisinghe (1987), Bhalla (1991), Ellis et al. (1997), Shekar et al. (2007), and Weerahewa et al. (2017, 2018a, b) provided detailed reviews, and the following subsections provide some salient features in those literature.

2.2.2.1 Food Assistance and Safety Net Programs

Sri Lanka has a long history of food assistance programs. At its inception was "Universal rice ration scheme," which covered everyone in the country. However, most of the food assistance programs implemented after 1970 were targeted interventions. One of the earliest of this sort was the food stamp program. The next salient change to food assistance program was the replacement of direct food assistance program with social safety net program, which included food assistance as a main component. "*Janasaviya*," which was in operation from 1989 to 1994 was the first of such kind. Later in 1995, this program was replaced by another income transfer program called "*Samurdhi*."

2.2.2.2 Nutritional Interventions

In parallel to food assistance and social safety net programs, nutritional intervention programs were implemented even before 1977 in the country to combat both macro and micronutrient deficiencies among preschool children, primary school children, and pregnant mothers. Among them, *Thriposha* program and school midday meal program were the oldest. Salt iodization program to control Iodine Deficiency Disorders (IDD), provision of iron supplements to pregnant mothers and preschool children, provision of *Poshana Malla* ("nutrition bag"), a package containing nutritious food to pregnant women, are more recent interventions that the government has launched. Moreover, a program to provide a nutrition allowance of Rs. 20,000 to pregnant and lactating mothers was introduced in 2015 with the aim of minimizing the number of low birthweight babies in Sri Lanka. Its mode of implementation changed in May 2016 where a new voucher was introduced to replace the coupons.

With all these changes in the economy, social sphere and in policy arena, diet of the people has transformed, and the next section will outline these diet transformations in a nut shell.

2.2.3 Diet Transformation

This section depicts the dietary shift in terms of the intake of calorie, protein, and fat and the sources of these macro-nutrients focusing on the change in food basket of consumers. It examines the changes in the share of dietary energy, protein and fat coming from animal-origin foods. Change in vegetable and fruit consumption and sugar consumption are the indictors used in depicting the shift in diet.

2.2.3.1 Dietary Nutrient Intake

In the absence of continuous data series on dietary energy, protein, and fat intake, The FAO food balance sheet data, which provides time series data on per capita availability of calorie, protein, and fat till 2013, were extracted to present how availability of major nutrients has changed during 1977–2013. The per capita availability of dietary energy in Sri Lanka has increased from 2199 to 2539 kcal per capita per day from 1977 to 2013. This is higher than the recommended intake of 2030 kcal per day in 2002 as given by the DCS (recommended figure for recent years is unavailable). The FAO indicated that Sri Lankans still derive more than 90% of their dietary energy from plant origin foods, though the share has slightly decreased from 96% to 93% during this period. During the period 1977–2013, the per capita supply of dietary energy from animal-origin product has increased from 95 to 167 kcal per capita per day.

Per capita availability of protein for a Sri Lankan has increased from 46.05 g to 59.71 g from 1977 to 2013. Even though 72% of the total protein is still derived from food of plant origin, over the years, protein derived from food of animal origin has more than doubled (7.44 g to 16.31 g per capita per day) from 1977 to 2013. The share of food of animal origin to total protein intake has increased from 16.16% in 1977 to 27.32% in 2013. According to household Income and expenditure Surveys

(HIES), this increased share of food of animal origin to total protein has come from the increased consumption of egg, fish, and meat (DCS 2018).

There was a marginal increase in the daily per capita availability of fat in Sri Lanka. It has increased from 42.18 g in 1977 to 49.71 g in 2013 (Fig. 2.2). Although the availability of fat from foods of plant and animal origin has increased over the study period, the major contribution has come from food of plant origin. The share of food of animal origin in total daily fat has increased from 12.92% to 17.24% from 1977 to 2013.

In summary, it can be concluded that the Sri Lankan diet has transformed to accommodate more calories, more protein, and more fat during the past few decades (Fig. 2.2). In terms of the changes in the contribution of food commodity to total energy during late 1970s, little over 60% of the calorie comes from starch, cereal, and tuber crops. According to Bogahawatte and Kailasapathy (1986), by 1970, the starchy staples had provided 75% of the total calorie requirement of the population. However, over the years, their contribution has declined to 56% (FAOSTAT). In 1980, the most important food commodities in Sri Lankan diet were rice and coconut (Sahn 1988). However, analysis of data from HIES suggests that, with time, consumption of rice, coconut as well as wheat has decreased and that of protein-rich commodities such as pulses, meat, fish, and egg has increased. The consumption of fresh milk was an exception, which has declined over the years, due to increased consumption of powdered milk.

In some geographical areas, e.g., Colombo (which houses one-tenth of the population), people are moving away from cereal-based diet (rice- and wheat-based) to more vegetable-, meat-, and fish-based diets (FAO 2016). Less than 30% of the weight of their diet is comprised of rice and wheat. In contrast, closer to 60% of the diet of Nuwara Eliya consumers is from rice and wheat.

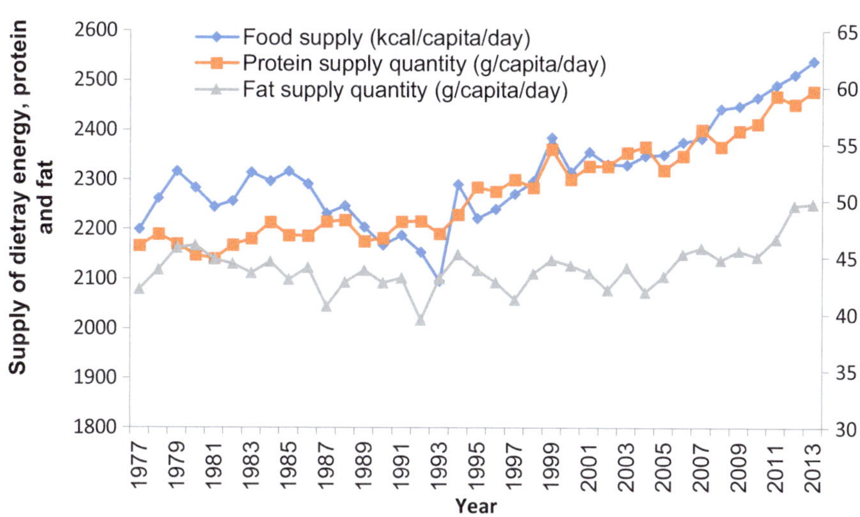

Fig. 2.2 Changing availability of major nutrients in Sri Lanka (Source: FAOSTAT)

2.2.3.2 Vegetable and Fruit Consumption

The intake of vegetable by an average Sri Lankan has remarkably increased over the years as 16.89 kg per capita per year in 1977 to 46.24 kg per capita per in 2013. However, the consumption of fruit has decreased over the study period from 64 kg per capita per annum (1970) to 36 kg per capita per annum (2013). With these opposite trends, the intake of vegetables and fruits as one category has only shown a marginal increase over the years (Fig. 2.3).

2.2.3.3 Meat and Fish Consumption

There was a considerable increase in meat intake by a Sri Lankan though the consumption of certain meat types such as bovine meat, mutton, goat meat, and pork has decreased. Increase in per capita availability of poultry meat has increased the per capita availability of meat from 3.62 in 1977 to 6.48 in 2013 kg per capita per year. Fish consumption also has more than doubled during the same period where the consumption has increased from 10.09 to 25.65 kg per capita per year during 1977 to 2013. Similarly, the marine fish consumption and fresh water fish consumption also have increased from 0.9 to 3.2 kg per capita per year.

2.2.3.4 Sugar and Salt Intake

Sugar and sweetener intake has increased over the study period (1977–2013), from 8.67 to 26.34 kg per capita per year. As a percentage, the energy intake from sugar was 0.39% of the total energy intake in 1977 and 10.47% in 2013, surpassing the WHO recommended level, which is below 10%.

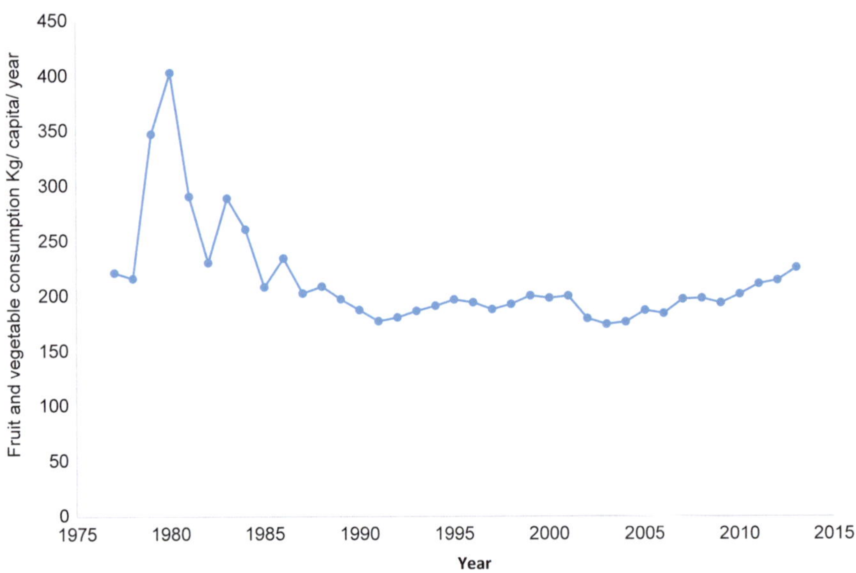

Fig. 2.3 Fruit and vegetable consumption in Sri Lanka from 1977 to 2014 (Source: FAOSTAT)

There was no national figure available for salt intake. The WHO (2012) recorded the level of salt intake by Sri Lankan as 9–11 g per day, while a national survey conducted by Jayawardena (2014) reported sodium intake, which is the main component of salt, as 3.26 g for male and 2.51 g for female. Salt intake in 1970 was recorded as 7 g per day per capita, suggesting that even five decades ago, salt intake by a Sri Lankan was higher than the recommended level of 5 g per capita per day. With the time salt intake has further diverged from the recommended intake. As the *WHO step survey* indicates (WHO 2015) this may be due to the increased consumption of processed foods, which are high in sugar and salt.

With this general background of the country, the methodology adopted in this review to diagnose the nutrition transition in Sri Lanka is presented next, with special emphasis on nutritional status.

2.3 Methodology

2.3.1 Approach

In achieving the objectives of the study, the review synthesized the published literature focusing on the key features of each stage of nutrition transition (Fig. 2.1). In particular, the study utilized indicators of undernutrition (stunting, wasting, and underweight), overnutrition (overweight and obesity), and micronutrient deficiencies (iron deficiency anemia and vitamin A deficiency) in exploring the trend of the prevalence rate. Table 2.1 provides the definitions of the parameters. In determining the causal linkage between socioeconomic condition and malnutrition, the study hypothesized that poverty, sector of residence, gender of the child, age of the child, and education level of the mother are strongly associated with the incidence of malnutrition, and it was tested using a meta-regression. The stage of nutrition transition was determined using the characteristics associated with each stage as mentioned in Fig. 2.1.

2.3.2 Meta-Database and Eligibility Criteria

A general Google search was run to obtain key relevant research in the areas of nutrition and food consumption. Key terms and phrases included in the search were malnutrition, child malnutrition, overweight and obesity, nutrition transition, food consumption behavior, anthropometric measures of nutrition, nutritional status of preschool children, nutritional status of adolescents, nutritional status of adults, dietary diversity, micronutrient deficiency, and anemia. In addition to the search in the Google Scholar, the study scanned the reference lists of newly included studies and previous review studies (Rajapaksa et al. 2011; Bandara and Weerasinghe 2015). The database included Well-Being and Social Policy Journal, Obesity Reviews, Asia Pacific Journal of Public Health, Tropical Agricultural Research, Food and Nutrition Bulletin, Poverty and Economic Policy Network, BMC Public

Table 2.1 Nutritional indicators used in the analysis

Form of malnutrition	Indicator	Definition
Undernutrition	Low birthweight incidence (%)	Percentage of live births that weighed less than 2500 g at birth
	Under-5 stunting (%)	Percentage of children 0–59 months who are below −2 (moderate and severe) standard deviations from median height-for-age of the WHO Child Growth Standards
	Under-5 wasting (%)	Percentage of children 0–59 months who are below −2 (moderate and severe) standard deviations from median weight-for-height of the WHO Child Growth Standards
	Under-5-underweight	Percentage of children 0–59 months who are below −2 (moderate and severe) standard deviations from median weight-for-age of the WHO Child Growth Standards
	Women of reproductive age short stature (%)	Maternal short stature is defined as height less than 145 cm. Maternal is defined as women who had a birth in the 3 (5) years preceding the survey
Micronutrient deficiencies	Women of reproductive age anemia (%)	Women of reproductive age (15–49 years), both pregnant and nonpregnant, with hemoglobin levels below 12 g/dL for women of reproductive age and below 11 g/dL for pregnant women
	Vitamin A deficiency in preschool-age children (%)	Preschool-aged children is defined by the majority of the countries as children >6 months and less than 5 years of age; however, 27 surveys used the age limits ranging from 5 to 6 years, and China used an upper age limit of 12 years. Vitamin A deficiency is defined as serum retinol below 0.70 μmol/L
Overweight and obesity	Under-5 overweight (%)	Percentage of children 0–59 months who are above two standard deviations from median weight-for-height of the WHO Child Growth Standards
	Adolescent overweight (%)	Percentage of adolescents aged 13–15 years who are above one standard deviation (+1 SD) from the median BMI-for-age of the WHO Growth Reference for School-Aged Children and Adolescents
	Adolescent obesity (%)	Percentage of adolescents aged 13–15 years who are above two standard deviations (+2 SD) from the median BMI-for-age of the WHO Growth Reference for School-Aged Children and Adolescents
	Adult overweight (%)	Percentage of defined population (adults 20+) with a body mass index (BMI) of 25 kg/m^2 or higher
	Adult obesity (%)	Percentage of defined population (adults 20+) with a body mass index (BMI) of 30 kg/m^2 or higher

Health, European Journal of Preventive Medicine, Sri Lanka Journal of Diabetes Endocrinology and Metabolism, International Journal of Environmental Research and Public Health, Ceylon Journal of Medical Sciences, Ceylon Medical Journal, Preventive Medicine, Galle Medical Journal, Public Health Nutrition, Gerodontology,

BMC Pediatrics, Asia Pacific Journal of Clinical Nutrition, and BMC Obesity. Apart from the original research articles, a number of national study reports were added to the meta-sample. These included reports of Demographic and Health Survey (DHS) for various years, annual reports of the Family Health Bureau, and various health reports published by the DCS, Sri Lanka.

Once the relevant peer-reviewed articles and national reports were complied, authors screened the abstracts of all relevant studies to decide whether to include them into the database. In constructing the meta-database, restriction was set to obtain original research studies carried out in Sri Lanka to assess the nutritional status and food consumption behaviors during the period from 1979 to 2018. The study included only the published peer-reviewed articles in economics, medicine, and nutritional disciplines. It excluded gray literature (working papers, conference proceedings, and technical reports) and studies that focused on nutritional status of patients. Only research articles written in English were included to the meta-database. All the primary studies included had either estimated malnutrition among preschool children, primary school children, adolescents, and adults or determinants of malnutrition in Sri Lanka. All entries have also assessed at least one of the following nutritional outcomes: stunting, wasting, underweight, micronutrient deficiencies, overweight, and obesity. The Prisma flow diagram is depicted in Fig. 2.4.

2.3.3 Tabular Analysis

Once the meta-database was constructed, authors extracted information from each research study, namely author, year of the study, population of interest, sample size, measure of malnutrition, micronutrient deficiencies, determinants of nutritional status, title of the article, and the journal. Then, the data were systematically tabulated for synthesis.

The stage of nutrition transition was diagnosed with a change in the prevalence rate of undernutrition (stunting, wasting, underweight, thinness), overnutrition (overweight and obesity), and the prevalence rate of anemia among preschool children, adolescents, and adults.

2.3.4 Regression Analysis

The regression analysis was performed to test the study hypothesis related to the causal factors of child malnutrition in Sri Lanka. The sector of the residence (x_1), gender of the child (x_2), age of the child (x_3), income level of the household (x_4), year of the survey (x_5), and the education level of the mother (x_6) were treated as explanatory variables in the regression model (Eq. (2.1). All the explanatory variables are dummy variables. Here, the *i*th subscript denotes the year of the survey.

$$Y_i = \beta_0 + \beta_1 x_{1i} + \beta_2 x_{2i} + \beta_3 x_{3i} + \beta_4 x_{4i} + \beta_5 x_{5i} + \beta_6 x_{6i} + \varepsilon_i \tag{2.1}$$

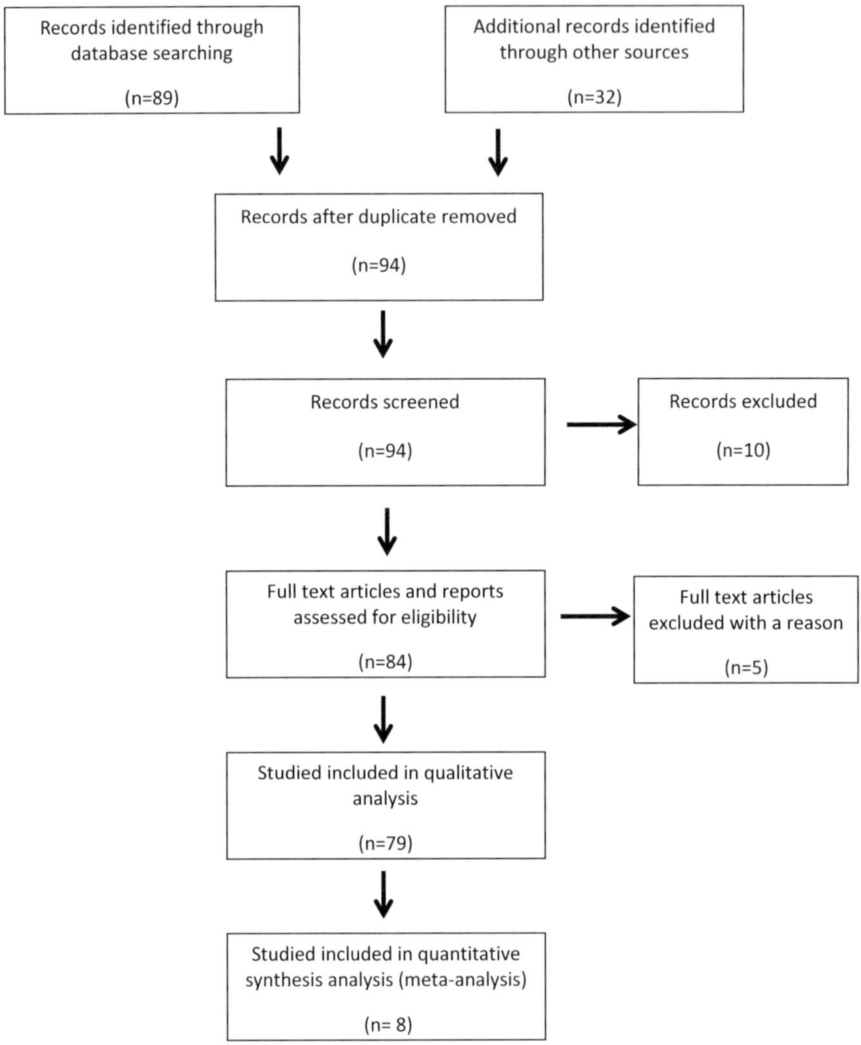

Fig. 2.4 Prisma flow diagram

where, the dependent variable of the model (Y_i) is the rate of prevalence of malnutrition in the ith year, β_0 is the constant term and β_j are the estimated coefficients, and ε_i is the error term. The Eq. ((2.1) was tested for the pool sample as well as for each sector of residence and gender of the child. In total, ten econometric models were estimated.

In obtaining the published data on malnutrition to the regression analysis, the study only utilized the national-level findings published by DCS, Sri Lanka. Data available on Demographic and Health Survey (DCS 2007, 2016) report on "Nutritional status of preschool children in Sri Lanka by DCS, undated" and report on "Social Conditions of Sri Lanka" were extracted.

2.4 Results

2.4.1 Trends in Nutritional Status and Determinants

This section presents the findings of the review using tabular analysis. In this exercise, the result section is divided into three main sub themes: Undernutrition, prevalence of anemia, and overweight and obesity. In each subsection, the paper will first summarize the studies reviewed and then presents and discusses the findings. Each section will conclude with a review of studies carried out to determine specific nutritional problem.

2.4.1.1 Status of Undernutrition among Preschool Children

The review consisted of 29 published reviewed articles or national reports on child malnutrition in Sri Lanka. Out of them, nine studies are national representative studies (Gavan and Chandrasekera 1979; DCS 1988, 1995, 2002, 2009, 2017; FHB 2009, 2010, 2011). Six studies were exclusively on malnutrition among children between 0- and 5-year old (DCS 2009, 2017; FHB 2011; Perris and Wijesinghe 2011; Jayatissa et al. 2012a, b, c). Some studies have limited their focus to study malnutrition in children between the age of 3 and 60 months (DCS 2002; Jayasekera 2006). Some other studies have excluded the babies aged less than 6 months, 1 year, and 2 years old in studying child malnutrition in Sri Lanka (Thamilini et al. 2015; Kandeepan et al. 2016; Jayatissa and Wickramage 2016). This variation in the age group presented a difficulty in making comparison between studies. Though the DHS reports are available from 1988 to 2017, the difference in their target population makes it harder for a one-to-one comparison across figures. Except few (Malkanthi et al. 2007, Gavan and Chandrasekera 1979; Pathmeswaran et al. 2005), all the other national and sample surveys have estimated the prevalence of malnutrition through three widely used anthropometric measures: stunting, wasting, and underweight (Tables 2.2, 2.3 and 2.4).

All national studies have witnessed a snail pace in progress in tracking all forms of malnutrition in Sri Lanka. In fact, the prevalence of wasting has slightly increased in Sri Lanka over the study period (DCS 1988, 1995, 2002, 2017). However, there was a geographical inequality in the rate of change of malnutrition. In comparison to other two sectors, estate sector has shown a remarkable progress in curtailing the proportion of children who were stunted (DCS 1988, 2017; Gavan and Chandrasekera 1979). By 1988, two in every three children were stunted in the estate sector and by 2016, only one in three children were stunted.

In 2016, the prevalence of underweight among preschool children in Sri Lanka was 20.5% (DCS 2017), which was relatively higher as compared to stunting (chronic malnutrition) and wasting. The prevalence of stunting in Sri Lanka (17.3%) is less than the corresponding figure for South Asia and in the world (Tables 2.5 and 2.12). However, wasting (15.1%) is relatively high in Sri Lanka as compared to the World and South Asia (Table 2.5). Considering the regional disparities, prevalence of wasting among urban children is lower compared to estate and rural children (DCS 2017).

Table 2.2 A summary of studies reporting the prevalence of stunting among preschool children (%)

Author and year	Study population	Sample size	Year of survey	Age cohort	Rural	Urban	Estate	Sri Lanka
Gavan and Chandrasekera (1979)	National survey		1975		31.4	27.8	56.3	
DCS (1988)	National survey	1995	1987	3–36 mon.	26.2	Colombo 21.8 (other urban 16.3)	60.0	27.5
DCS (1995)	National survey		1993	<5 ys.	22.9	Colombo 19.7 (other urban 16.8)	53.7	23.8
DCS (2002)	National survey		2000	3–59 mon.	12.8	Colombo 7.4 (other urban 8.6)	33.8	13.5
Jayasekera (2006)	State-operated foster care institutions in Sri Lanka	77	2004	3–60 mon.				51.9
Malkanthi et al. (2007)	Farming community	80	2005	<5 ys.	19.0			
DCS (2009)	National survey	6567	2006–2007	0–59 mon.	16.2	13.8	40.2	17.3
Jayatissa (2009)	National survey	2865	2009	<5 ys.	17.4	14.3	46.7	19.2
Jayatissa and Hossaine (2010)	Kurunegala	224	2009	6–59 mon.	12.4	18.8	–	12.9
				Male (0–59 mon.)				14.8
				Female (0–59 mon.)				11.0
FHB (2011)	National survey		2010	<5 ys				15.9
Nanayakkara and Silva (2011)	Pannala	200	n.a.	4 mon.	15.5			
				6 mon.	13.8			
				9 mon.	16.1			
				12 mon.	22.4			
Peiris and Wijesinghe (2011)	Weeraketiya	1219	n.a.	0–5 ys.	11.8			

(continued)

Table 2.2 (continued)

Author and year	Study population	Sample size	Year of survey	Age cohort	Rural	Urban	Estate	Sri Lanka
FHB (2012)	National survey		2011	<5 ys.				12.8
			2011	Infants ≤1 yr.				7.3
			2011	2–5 ys.				15.1
Jayatissa et al. (2012a, b, c)	National survey	7306	2012	6–59 mon.				13.1
Jayatissa et al. (2012a, b, c)	Jaffna	282	2007	0–5 ys.	15.2			
FHB (2013)	National survey		2012	<5 ys.				11.3
			2012	Infants				6.6
			2012	2–5 ys.				13.2
Prasanna (2015)	Fishing community in Ambalangoda	189	n.a.	1–5 ys.	23.0			
FHB (2015)	National survey		2014	<5 ys.				10.5
			2014	Infants				6.3
			2014	2–5 ys.				11.6
Pathmeswaran et al. (2005)	National survey	2528	2003	9–10 ys.				15.5
Thamilini et al. (2015)	Puttalam, Kurunegala and Matale	380	n.a.	2–5 ys.	15			
Ubeysekara et al. (2015)	Akurassa	428	n.a.	6–24 mon.	17.0			
FHB (2016)	National survey		2015	<5 ys.				9.6
				Infants				5.5
				2–5 ys.				10.5
Jayatissa and Wickramage (2016)	National (migrant worker fathers)	321	2012	6–59 mon.				11.6
Kandeepan et al. (2016)	Jaffna	846	n.a.	1–5 ys.				26.4
DCS (2017)	National survey	7870	2016	0–59 mon.	17	14.7	31.7	17.3

There were ten studies focusing on child malnutrition in sub-populations. One study has focused on the child malnutrition in farming communities (Malkanthi et al. 2007) and their reported value of stunting (19% in 2005) was higher than the closest national survey value of 13.5% reported by DCS in 2002 (DCS 2002). Prasanna (2015) also reported a higher stunting value of 23% for the fishing community. Those two studies reflect higher vulnerability of farming and fishing communities to chronic malnutrition. However, the prevalence of wasting (11.2%) in the fishing community was lower than the national average (15.1%) reported in DCS (2017). In contrast, Peiris and Wijesinghe (2011) reported that the prevalence of

Table 2.3 A summary of studies reporting the prevalence of wasting among preschool children (%)

Author and year	Study population	Sample size	Year of survey	Age cohort	Rural	Urban	Estate	Sri Lanka
DCS (1988)	National survey	1995	1987	3–36 mon.	13.6	Colombo 13.4 (other urban 10.2)	7.1	12.9
DCS (1995)	National survey		1993		16.4	Colombo 12.2 (other urban 16.8)	9.5	15.5
DCS (2002)	National survey		2000	3–59 mon.	15.9	Colombo 10.1 (other urban 6.3)	11.8	14.0
Jayasekera (2006)	State-operated foster care institutions in Sri Lanka	77	2004	3–60 mon.				25.0
DCS (2009)	National survey	6567	2006–2007	0–59 mon.	14.8	14.7	13.5	14.7
Jayatissa (2009)	National survey	2865	2010	<5 ys.	12.0	11.0	12.3	11.8
Jayatissa and Hossaine (2010)	Kurunegala	224	2009	6–59 mon.	13.1	25.0		14.0
				6–59 mon. male				15.9
				6–59 mon. female				12.0
Peiris and Wijesinghe (2011)	Weeraketiya	1219	n.a.	<5 ys.	42.7			
Nanayakkara and Silva (2011)	Pannala	200	n.a.	4 mon.	10.3			
				6 mon.	8.0			
				9 mon.	8.0			
				12 mon.	11.5			
FHB (2012)	National survey		2011	<5 ys.				15.3
				2–5 ys.				9.8
				Infants				7.3
				<5 ys.				14
Jayatissa et al. (2012a, b, c)	Jaffna	282	n.a.	0–5 ys.	14.3			

(continued)

Table 2.3 (continued)

Author and year	Study population	Sample size	Year of survey	Age cohort	Rural	Urban	Estate	Sri Lanka
Jayatissa et al. (2012a, b, c)	National survey	7306	2012	6–59 mon.				19.6
FHB (2013)	National survey		2012	<5 ys.				13.6
				Infants				8.6
				2–5 ys.				15.6
FHB (2014)	National survey		2013	<5 ys.				13.0
				Infants				8.0
				2–5 ys.				14.9
Prasanna (2015)	Fishing community in Ambalangoda	189	n.a.	1–5 ys.	11.2			
FHB (2015)	National survey		2014	<5 ys.				12.2
				Infants				7.9
				2–5 ys.				13.7
Thamilini et al. (2015)	Puttalam, Kurunegala and Mathale	380	n.a.	2–5 ys.	21.8			
Ubeysekara et al. (2015)	Akurassa	428	n.a.	6–24 mon.	17.1			
FHB (2016)	National survey		2015	<5 ys.				12.2
				Infants				7.2
				2–5 ys.				13.5
Jayatissa and Wickramage (2016)	National survey	321	2012	6–59 mon.				18.2
Kandeepan et al. (2016)	Jaffna	846	n.a.	1–5 ys.				21.6
DCS (2017)	National survey	7870	2016	0–59 mon.	15.6	12.9	13.4	15.1

stunting was lower and that of wasting was far higher in the *Weeraketiya* area compared to the national average figure for the recent year. As wasting is an outcome of the recent lack of food or illness that prevents children from eating or absorbing nutrient and stunting is an indicator of chronic undernutrition, the latter is less surprising.

Another study reported that in state-operated foster care institutions, half of the children between the ages of 3 and 60 months were stunted, one-fourth was wasted and two-thirds were underweight (Jayasekera 2006). As expected, children in the war-affected areas were more malnourished than others with respect to all the three forms of malnutrition (Kandeepan et al. 2016). When the age range of children and the risk of malnutrition are considered, children less than 6 months of age seem to be at a minimum risk (Nanayakkara and Silva 2011). Further, the children in households having a migrant worker father were found to be less stunted (Jayatissa and Wickramage 2016).

Table 2.4 A summary of studies reporting the prevalence of underweight among preschool children (%)

Author and year	Study population	Sample size	Year of survey	Age cohort	Rural	Urban	Estate	Sri Lanka
DCS (1988)	National survey	1995	1987	3–36 mon.	38.7	Colombo 27.6 (other urban 26.5)	52.9	38.1
DCS (1995)	National survey		1993		38.3	Colombo 31.2 (other urban 29.9)	52.1	37.7
DCS (2002)	National survey		2000	3–59 mon.	30.8	Colombo 18.2 (other urban 21.3)	44.1	29.4
Jayasekera (2006)	State-operated foster care institutions in Sri Lanka	77	2004	3–60 mon.				63.5
DCS (2009)	National survey	6567	2006–2007	0–59 mon.	21.2	16.5	30.1	21.1
FHB (2010)	National survey		2009	<5 ys.				27.4
Jayatissa and Hossaine (2010)	Kurunegala	224	2009	6–59 mon.	19.4	17.6	–	19.3
				6–59 mon. male				20.4
				6–59 mon. female				18.2
Jayatissa (2009)	National survey	2865	2010	<5 ys.	20.8	17.7	37.9	21.6
FHB (2011)	National survey		2010	<5 ys.				23.8
Nanayakkara and Silva (2011)	Pannala	200	n.a.	4 mon.	11.5			
				6 mon.	13.8			
				9 mon.	14.9			
				12 mon.	17.2			
Peiris and Wijesinghe (2011)	Weeraketiya	1219	2011	<5 ys.	41.2			

(continued)

Table 2.4 (continued)

Author and year	Study population	Sample size	Year of survey	Age cohort	Rural	Urban	Estate	Sri Lanka
FHB (2012)	National survey		2011	<5 ys.				19.8
				Infants				9.5
				2–5 ys.				24.2
Jayatissa et al. (2012a, b, c)	National survey	7306		6–59 mon.				23.5
Jayatissa et al. (2012a, b, c)	Jaffna	282	2012	0–5 ys.	9.5			
FHB (2013)	National survey		2012	<5 ys.				17.3
				Infants				8.5
				2–5 ys.				20.8
FHB (2014)	National survey		2013	<5 ys.				17.0
				Infants				8.7
				2–5 ys.				20.4
Prasanna (2015)	Fishing community in Ambalangoda	189	n.a.	1–5 ys.	31.0			
FHB (2015)	National survey		2014	<5 ys.				16.4
				Infants				8.8
				2–5 ys.				19.3
Thamilini et al. (2015)	Puttalam, Kurunegala and Mathale	380	n.a.	2–5 ys.	21			
Ubeysekara et al. (2015)	Akurassa	428	n.a.	6–24 mon.	21.3			
FHB (2016)	National survey		2015	<5 ys.				15.6
				Infants				8.4
				2–5 ys.				19.0
Jayatissa and Wickramage (2016)	National survey	321	2012	6–59 mon.				24.0
Kandeepan et al. (2016)	Jaffna	846	n.a.	1–5 ys.				33.1
DCS (2017)	National survey	7870	2016	0–59 mon.	20.8	16.4	29.7	20.5

2.4.1.2 Status of Undernutrition Among Adolescents and Adults

There is a dearth of research on malnutrition among school-age children and adults in Sri Lanka. However, as revealed by Jayatissa et al. (2002), the rate of malnutrition among primary school children was as same as preschool children in 2002 (Table 2.6). Malnutrition continued to prevail at the same rate among preschool and primary school children. However, a recent study signaled a rising rates of both underweight and stunting among primary school children (Naotunna et al. 2017). There was a marginally lower prevalence rate of stunting. The same study suggested that one in

Table 2.5 Nutrition status of Sri Lanka compared with South Asia and World

Category	Indicator		Sri Lanka (national average)	South Asia (regional average)	World (global average)
Undernutrition	At birth	Low birth weight	15.7[a]	27	15
	Preschool	Stunting	17.3[a]	34	23
		Wasting	15.1[a]	15	8
		Underweight	20.5[a]		
	Women in reproductive age	Thinness	14.9[b]	31	21
		Short stature	7.2[a]	11	7
Micronutrient deficiencies	Women in reproductive age	Anemia	34.1[c]	49	33
	6–59 months	Vit A deficiency	6	42	25
Overweight and obesity	Preschool	Overweight	1.0[a]	4	6
	Adolescent	Overweight	4.6[b]	10	15
		Obesity	1.4[b]	2	4
	Adult-men	Overweight	22[b]	18	36
		Obesity	4.2[b]	3	10
	Adult-women	Overweight	31.9[a]	24	38
		Obesity	13.3[a]	7	14
Food supply	Undernutrition	1990	31[d]	24	25
		1999	30	18	20
		2009	26	16	16
		2014	22	16	14
	Availability of fruits and vegetables	1990	188	221	342
		1999	198	266	462
		2009	202	338	551
		2014	226	373	588
	% calories from staples	1990	40	33	41
		1999	44	36	44
		2009	42	39	48
		2014	43	40	44

[a]DCS (2017)
[b]Jayatissa et al. (2012b)
[c]DCS (2007) (refers to mild anemia), all others: Hawkes and Fanzo (2017)
[d]Shekar et al. (2007) Malnutrition in Sri lanka: scale, scope, causes, and potential response

four of the primary school children are thin and that the proportion of boys who are thin was statistically higher than that of girls. Study of Pathmeswaran et al. (2005) revealed that none of the students in a selected private school were stunted. In another study, it was revealed that primary school children in Colombo had a relatively lower rate of malnutrition prevalence (Thilakarathne and Wijesinghe 2011).

Study on malnutrition among adolescents have given arise to a large range of estimates, starting from 16.4% (wasting) to 28.5% stunting and 47.2% underweight (Jayatissa and Ranbanda 2006; FHB 2015). The FHB (2012, 2015) reported stunting and wasting in children among different school grades, suggesting that over a

Table 2.6 A summary of studies reporting the prevalence of undernutrition among school children, adolescents, and adults (%)

Author and year	Study population	Sample size	Year of survey	Age cohort	Category	Stunting	Wasting	Underweight	Thinness
Pathmeswaran et al. (2005)	National survey	2528	2002	9–10 ys.				15.5	
Pathmeswaran et al. (2005)	National survey	2528	2003	9–10 ys.	State school				15.5
Jayatissa and Ranbanda (2006)	National survey	6264	n.a.	10–15 ys.		28.5	16.4	47.2	
FHB (2008)	National survey		2007		Pregnant mothers				26.1
FHB (2009)	National survey		2008		Pregnant mothers				26.3
FHB (2010)	National survey		2009		Pregnant mothers				25.4
Fernando and Wijesinghe (2010)	Kandy	105	2009		Institutionalized elders	59.1			
Jayatissa and Hossaine (2010)	Kurunegala	185	2009	15–49 years nonpregnant women	Urban			13.3	
					Rural			19.8	
					All			19.2	
					Girls	4.6	33.0	20.0	
	National survey			Women	Urban				11.2
					Rural				18.7
					Estate				42.1
					All				18.2
FHB (2011)	National survey		2010		Pregnant mothers				25.4

(continued)

Table 2.6 (continued)

Author and year	Study population	Sample size	Year of survey	Age cohort	Category	Stunting	Wasting	Underweight	Thinness
FHB (2012)	National survey		2011		Pregnant mothers				24.6
FHB (2012)	National survey	3,898,259	2011	Grade 1		8.1	16.3		
				Grade 4		6.9	16.7		
				Grade 7			20.9		
				Grade 10			15.0		
Jayatissa et al. 2012a, b, c	National survey	7606	2012	15–49 ys.	Nonpregnant, non-lactating women				18.2
		1135		5–9.9 ys.		8.9		28.3	
		405		10–18 ys.		17.3			25.1
		3019		18–59 ys.	Women				14.9
		3019		18–59 ys.	Men				16.2
Perera and Ekanayake (2012)	Colombo	437	n.a.		Institutionalized elders			19.9[a]	
FHB (2013)	National survey		2012		Pregnant mothers				23.8
FHB (2014)	National survey		2013		Pregnant mothers				23
FHB (2015)	National survey		2014	Grade 1		7.9	17.9		
				Grade 4		6.5	17.4		
				Grade 7			19.9		
				Grade 10			14.9		
FHB (2015)	National survey		2014		Pregnant mothers				23.5
Rathnayake et al. (2015)	Six provinces	311	n.a.		Institutionalized elders				30

FHB (2016)	National survey	1,729,268	2015	Grade 1			8.7	20.3		
				Grade 4			6.9	20.3		
				Grade 7				20.8		
				Grade 10				16.4		
DCS (2017)	National survey	16,806	2016	15–49 ys. married women	Urban					5.6
					Rural					9.1
					Estate					22.0
					All					9.1
De Silva et al. (2017)	Galle	396		>60 ys.				31[b]		
Naotunna et al. (2017)	National survey	4469	n.a.	5–10 ys.	Rural	43.9	n.a.	25.9		

[a]Undernourished
[b]Risk of malnourished

period of 3 years, prevalence of both stunting and wasting has marginally increased, except among Grade 7 students. Only very few studies are available on the adult anthropometry and among them one indicates that 14.9% of women and 16.2% of the men between the age category of 18 and 59 are thin (Jayatissa et al. 2012a, b, c). Considering the regional disparities, 22% of the estate women were thin in 2016 (DCS 2017).

2.4.1.3 Determinants of Undernutrition

Fifteen studies examined the association or causal relationship between household income, mother's nutritional status, socioeconomic status, feeding behavior, and child malnutrition in Sri Lanka. Of the 15, 5 reported a significant negative association between (causality) household income and/or wealth and child malnutrition (Ekanayake et al. 2003; Rathnayake and Weerahewa 2005a, b; Rannan-Eliya et al. 2013; Jayatissa and Wickramage 2016). Of these five studies, one study (Rathnayake and Weerahewa 2005b) has researched the association between mother's income and child nutritional status (Table 2.7). The findings of this study indicate that when maternal employment generates extra income, the calorie intake of all individuals of the household increases, yet the allocation of calories to the children of the household is reduced.

Six of the 15 studies listed in Table 2.7 have investigated the association between mother's knowledge and child malnutrition (Ekanayake et al. 2003; Rathnayake and Weerahewa 2005a, b; Jayawardena 2012, 2014; Prasanna 2015). Some studies have conceptualized the maternal knowledge about nutrition through their literacy rate (Prasanna 2015) and number of years of schooling (Rathnayake and Weerahewa 2005a, b; Jayawardena 2012), while Ekanayake et al. (2003) has specifically looked at awareness of the mother on nutritional aspects. Irrespective of the different conceptualization, except one almost all of the studies reported a significant negative causality between mother's knowledge/education and child malnutrition. The association between mother's nutritional status and age, and child malnutrition has been assessed in three studies using maternal height and maternal BMI as proxy for maternal nutritional status (Ekanayake et al. 2003; Rannan-Eliya et al. 2013; Ubeysekara et al. 2015). All three studies reported a positive association between age of the mother and the child malnutrition. The association between the maternal height and child malnutrition is negative (Rannan-Eliya et al. 2013).

The significant association between the feeding practices and child malnutrition was reported in four studies (Rannan-Eliya et al. 2013; Jayawardena 2014; Ubeysekara et al. 2015; Perkins et al. 2018). Examples of feeding practices included to the econometric models were, late introduction of fat and oil to child food, continuation of breastfeeding at the age of one, early initiation of breast feeding, and feeding with a low protein diet.

Apart from the above determinants, some reported studies have looked at the association between child malnutrition and alcoholism of the farther or smoking habit of the farther (Jayawardena 2014; Prasanna 2015), low birthweight, altitude, ethnicity, number of children in the family, and the household size. The effect of low birthweight will continue to hamper the growth of child in later state as well (Ekanayake et al. 2003; Jayawardena 2012; Rannan-Eliya et al. 2013; Jayatissa and Wickramage 2016).

Table 2.7 A summary of studies reporting determinants of undernutrition

Author and year	Association or causality	Outcome variable	Food-related factors	Income	Education	Mother's characteristics	Other factors
Ekanayake et al. (2003)	Causal	Pre-school children Malnutrition		Household income	Nutrition awareness of the mother	Age	Mother's interest in media, low birthweight
Rathnayake and Weerahewa (2005a)	Causal	Child malnutrition		Household income	Mother's education		Area of residence, household size, age of the child, birthweight of child
Rathnayake and Weerahewa (2005b)	Causal	Calorie adequacy		Mother's income	Mother's education		
Nanayakkara and Silva (2011)	Correlation	Stunting					Low birthweight
Jayawardena (2012)	Causal	Child malnutrition			Mother's knowledge and mother's education		Low birthweight, gender (male)
Perera and Ekanayake (2012)		Underweight					Tooth loss
Rannan-Eliya et al. (2013)	Causal	Stunting (9–23 mon.)				Height, BMI, age	Female, birthweight, altitude, number of children in the household
Rannan-Eliya et al. (2013)	Causal	Stunting (24–59 mon.)	Early initiation of breast feeding, continue breast feeding at the age of 1 year	Wealth		Age, height, BMI	First child, birthweight, ethnicity, number of children in the household, altitude

(continued)

Table 2.7 (continued)

Author and year	Association or causality	Outcome variable	Food-related factors	Income	Education	Mother's characteristics	Other factors
Jayawardena (2014)	Causal	Child malnutrition	Protein-low diets		Mother's knowledge		Alcoholism
Prasanna (2015)	Causal	Age 1–5 ys. Underweight, wasting, and stunting			Maternal literacy		Father's smoking habit
Rathnayake et al. (2015)		Undernourishment	No appetite, skipping meals, having food allergies				Symptoms of depression, lack of leisure activities, and use of dentures
Thamilini et al. (2015)	Association	Stunting/wasting and underweight	Food insecurity				
Ubeysekara et al. (2015)		Wasting and stunting	Late introduction of oil and fat			Age	Gender: Male babies high incidence of malnutrition
Jayatissa and Wickramage (2016)	Causal	Wasting and stunting		Household wealth			Farther being a migrant worker Stunting: Low birthweight wasting: Diarrhea Caring
Perkins et al. (2018)		Stunting	Dietary diversity negatively				

2.4.2 Micronutrient Deficiencies

Nineteen studies included in this review have estimated the prevalence of anemia among preschool children, adolescents, and adults (Table 2.8). Among them, 12 were national-level surveys (Jayatissa (2009); Mudalige and Nestel 1996; Piyasena and Mahamithawa 2003; Pathmeswaran et al. 2005; Jayatissa and Ranbanda 2006; Hettiarachchi et al. 2006; DCS 2007; Jayatissa and Hossaine 2010; Jayatissa et al. 2012a, b, c; Naotunna et al. 2017; Allen et al. 2017). As per the findings of the national surveys conducted by the MRI in 1996, 45% of the preschool children in Sri Lankan were anemic (45%) (Jayatissa et al. 2012a, b, c). However, with the time, deficiency has declined to a level of 15.1% (Jayatissa et al. 2012a, b, c). By 1994, the prevalence rate of anemia among primary school children and adolescents was closer to 50% (Mudalige and Nestel 1996). According to DCS (2007), there was a marginal decrease in the prevalence rate in 2006/2007 for all age groups. However, even by mid of the 20th century closer to one-third of the preschool children and more than one-third of pregnant and nonpregnant women were anemic. An estimate coming from a geographic subgroup (Galle in the Southern Province of Sri Lanka) indicated that in 2003, the prevalence rate of anemia is as high as 54% among adolescents (Hettiarachchi et al. 2006). The prevalence rate among primary school children in 2014 in North Central province was only 17.3% (Allen et al. 2017; Naotunna et al. 2017). Many sample surveys carried out after 2006 indicated that the rate of anemia prevalence was less than 20% for several communities (Jayatissa et al. 2012a, b, c; Senadheera et al. 2017) except for Jaffna, which was a war-affected area (Kandeepan et al. 2016).

The prevalence of anemia among boys has been slightly higher than that among girls in both preschool children and primary school children (Pathmeswaran et al. 2005; DCS 2007; Jayatissa and Hossaine 2010; Jayatissa et al. 2012a, b, c). However, adolescent females had shown a higher prevalence compared to males (Hettiarachchi et al. 2006; Allen et al. 2017).

2.4.3 Overweight and Obesity

2.4.3.1 Overweight and Obesity Among Preschool Children

Twenty of the included studies examined overweight and obesity among Sri Lankan population (Table 2.9). Out of them, 11 had explored the prevalence of overweight among preschool children and adolescents (Pathmeswaran et al. 2005; Jayatissa and Ranbanda 2006; Thilakarathne and Wijesinghe 2011; Jayawardena et al. 2012; Jayatissa et al. 2012a, b, c; Rathnayake et al. 2014a, b; FHB 2015; DCS 2017), while the rest studied the prevalence of overweight and obesity among adults (Fernando et al. 1994; Katulanda et al. 2010; Fernando and Wijesinghe 2010; Jayatissa et al. 2012a, b, c; Perera and Ekanayake 2012; Rathnayake et al. 2014a, b; Jayawardena et al. 2012, 2013a, b, 2016, 2017; DCS 2017).

Table 2.8 A summary of studies reporting the prevalence of anemia

Author (year)	Study population	Sample size	Year of survey	Age cohort	Category	Rural	Urban	Estate	Sri Lanka
Jayatissa (2009)			1970	<5 ys.	Children				52
				5–10 ys.	Children				70
					Adult female				68
					Pregnant women				73
Jayatissa (2009)				<5 ys.	Children				45
Mudalige and Nestel (1996)	National		1994/1995	Pre- school children					44.8
				5–10.9 ys.					58.0
				Adolescent					36.0
				Nonpregnant					45.0
				Pregnant					39.0
Jayatissa (2009)				<5 ys.	Children				29.9
Piyasena and Mahamithawa (2003)	National		2001		All	22.9	33.4	24.4	29.9
					Male				29.0
					Female				30.6
Pathmeswaran et al. (2005)	National	1351	2003	9–10 ys.	All				12.1
					Male				12.5
					Female				11.6
Hettiarachchi et al. (2006)	Galle district		2003	12–16	Male				49.5
					Female				58.1
					All				54.8
Jayatissa and Ranbanda (2006)	National	6264	2002	10–15 ys.					11.1

DCS (2007)	National		2006/7	6–59 mon.	All	33.2	32.0	28.1	32.6
				9–11 mon.					61.0
					Male				33.4
					Female				31.9
				14–59	Nonpregnant women				39.0
					Pregnant				34.0
Jayatissa and Hossaine (2010)	Kurunegala	192	2010	6–59 mon.	Children	20.0	11.8	–	19.3
					Male				15.4
					Female				22.8
		174		15–49 ys.	Pregnant	7.1	0	–	6.7
					Lactating	19.5	0	–	18.5
					All nonpregnant	16.4	0	–	14.9
Jayatissa and Hossaine (2010)	National	2373	2009	6–59 mon.	All				25.2
					Male				27.3
					Female				23.2
Jayatissa (2009)	National	2865	2009	< 5 ys.	Children	24.7	26.7	25.2	25.2
		2139		15–49 ys.	Pregnant women	16.4	19.3	8.3	16.7
					Lactating	19.8	21.0	30.2	20.5
					All nonpregnant	21.5	21.3	33.6	22.2
Jayatissa et al. 2012a, b, c	National	7306	2012	6–59 mon.	All				15.1
					Male				17.1
					Female				13.1
FHB (2015)	National		2014		Mothers				9.4

(continued)

Table 2.8 (continued)

Author (year)	Study population	Sample size	Year of survey	Age cohort	Category	Rural	Urban	Estate	Sri Lanka
Kandeepan et al. (2016)	Jaffna	846	n.a.	1–5 ys.					36.4
Naotunna et al. (2017)	National	4469	n.a.	5–10 ys.					17.3
Allen et al. (2017)	National	7526	2009	15–16 ys.	Male				5.6
					Female				11.1
Amarasinghe et al. (2017)	North Central Province	4412	n.a.	60–131 mon.		147 anemia			
Senadheera et al. (2017)	Galle	350	2014–2015	Pregnant women	Pregnant women		16.6		

Table 2.9 A summary of studies reporting the prevalence of overweight and obesity among children and adolescents (%)

Author and year	Study population	Sample size	Year of survey	Age cohort	Category	Overweight	Obesity
Pathmeswaran et al. (2005)	National survey	2528	2003	9–10 ys.			3.1
Pathmeswaran et al. (2005)	Private school students 9–10 ys.	65	2003	Grade 5		21.5	
Jayatissa and Ranbanda (2006)	National survey	6264	2002	10–15 ys.	All	2.2	
Jayatissa and Hossaine (2010)	Kurunegala	224	2009	6–59 mon.	Children	0.9	
	Women	Nationally representative			Urban		15
					Rural		4.2
					Estate		7.4
					All		6.7
Thilakarathne and Wijesinghe (2011)	Colombo	1347	2008	9–10 ys.	All	8.9	5.1
					Boys	17.0	6.9
					Girls	10.1	7.8
Jayatissa et al. (2012a, b, c)	National survey	7306	2012	6–59 mon.	Infants	0.7	
		1135		5–9.9 ys.	Primary school children	3.1	1.6
		405		10–18 ys.	Adolescents	4.6	1.4
Jayatissa et al. (2012a, b, c)	Jaffna	282	2009	0–5 ys.	Children	34.0	
Jayawardena et al. (2012)	National survey	490	2006/2007	>5 ys.	All	17.1	28.8
					Men	14.2	21.0
FHB (2015)	School children in 300 MOH areas	3,976,852	2014	Grade 1	School children	1.5	
				Grade 4		2.6	
				Grade 7		4.4	
				Grade 10		3.7	
FHB (2016)	School children in 332 MOH areas	1,729,268	2015	Grade 1		1.7	
				Grade 4		2.9	
				Grade 7		4.8	
				Grade 10		4.3	
DCS (2017)	National survey	7870	2016	0–59 mon.	Urban	2.9	
					Rural	1.9	
					Estate	1.3	
					All	2.0	

As per the survey findings, the prevalence of overweight and obesity is less than 3% among the preschool children (Jayatissa et al. 2012a, b, c; DCS 2017). Considering the inequality, only 1.3% of the preschool children in the estate sector are overweight as compared to the urban sector (2.9%; DCS 2017). Male children have a higher tendency to be overweight than females (Jayatissa et al. 2012a, b, c). Primary school children have shown a higher tendency to be overweight and obese than preschool children (Pathmeswaran et al. 2005; Jayatissa et al. 2012a, b, c; FHB 2015). However, the difference is not that high except for the primary school children enrolled in private schools (Pathmeswaran et al. 2005) and primary school children in Colombo area in the Western Province (Thilakarathne and Wijesinghe 2011). As reported in Pathmeswaran et al. (2005), 20% of the private school students in the studied sample were overweight. In many studies the prevalence of overweight was found to be higher than that of obesity. Among the adolescents, nearly 5% are overweight (Jayatissa et al. 2012a, b, c; FHB 2015). As observed, the prevalence of overweight and obesity is relatively higher among boys than girls among primary school children and adolescents.

2.4.3.2 Overweight and Obesity Among Adult Men and Women

Prevalence of overweight and obesity among adults are far higher than that of preschool children and school-age children (Table 2.10). Despite the range of values reported by different studies, the prevalence of overweight among adults is at least four times higher than that of preschoolers and adolescents.

One study using a nationally representative sample reported that the overweight and obesity statistics related to women were 31.9% and 13.3%, respectively, for the year 2016 (DCS 2017). However, as suggested by DCS (2007), the prevalence of overweight and obesity among women was lesser a decade ago, i.e., 24% in 2016 and 7.2% in 2006 (DCS 2007). As both the studies have been carried out by the same institute and the age group was similar, these statistics are comparable and they reflect the rising burden of overnutrition among adults in Sri Lanka. Another national survey suggested that the prevalence was slightly lower among male in which 22.8% and 6.6% are overweight and obese, respectively (Jayatissa et al. 2012a, b, c).

In terms of the geographic distribution of the prevalence of overnutrition, three national studies (DCS 2000, 2007, 2017) have provided evidence to the positive association between urban living and the prevalence of overweight. In urban sector, one in three women are overweight and one in five women are obese. In particular, urban women are twice likely to be overweight as compared to estate women. However, with the rapid increase in the incidence of overweight in the estate sector women, the gap between two sectors is getting narrowed.

Apart from the nationally representative samples, several reports have presented overweight and obesity prevalence for a number of subgroups, e.g., adults in Colombo, women in Pannala and Negombo, institutionalized elders, men in Kandy (Fernando et al. 1994; Fernando and Wijesinghe 2010; Perera and Ekanayake 2012; Rathnayake et al. 2014a, b; Jayawardena et al. 2016; Jayawardana et al. 2017). The prevalence of obesity among cardiac patients, men in Kandy, and women in Pannala and Negombo is higher than the national-level statistics.

Table 2.10 A summary of studies reporting the prevalence of overweight and obesity among adults (%)

Author and year	Study population	Sample size	Age cohort	Year of survey	Category	Overweight (%)	Obesity (%)
Fernando et al. (1994)	Colombo suburb	633	30–64 ys.	n.a.	All		9.89
					Men		7.0
					Women		13.4
FHB (2008)	National survey			2007	Pregnant mothers	13.0	
FHB (2009)	National survey			2008	Pregnant mothers	13.1	
FHB (2010)	National survey			2009	Pregnant mothers	13.9	
Fernando and Wijesinghe (2010)	Kandy	105	>65 ys.	2009	Institutionalized elders	28.6	
Jayatissa and Hossaine (2010)	Kurunegala	185	15–49 ys.	Nonpregnant women	Urban	33.3	20.0
					Rural	24.6	4.8
					All	25.3	6.0
Katulanda et al. (2010)	National survey	4532	Above 18 ys.	2005–2006	All	25.2	9.2
					Urban: male	30.6	16.4
					Urban: female	34.8	20.7
					Urban: All	32.7	18.5
					Rural: male	20.4	4.7
					Rural: female	26.7	8.7
					Rural: all	23.3	6.7
FHB (2011)	National survey			2010	Pregnant mothers	14.1	
Chathurani et al. (2012)	Anuradhapura	990		Pregnant women		14.1	
FHB (2012)	National survey			2011	Pregnant mothers	15.2	
Jayatissa et al. (2012a, b, c)	National survey	7606	15–49 ys.	2012	Nonpregnant and non-lactating women	22.8	6.6
		3019	18–59 ys.		Women	24.6	6.9
		3019	18–59 ys.		Men	22	4.2
Jayawardena et al. (2012)	National survey	490	Adults	2011	All	17.1	28.8
					Men	14.2	21.0
					Women	18.5	32.7
Perera and Ekanayake (2012)	Colombo	437	≥60 ys.	n.a.	Elders	18.0	
FHB (2013)	National survey			2012	Pregnant mothers	16.2	

(continued)

Table 2.10 (continued)

Author and year	Study population	Sample size	Age cohort	Year of survey	Category	Overweight (%)	Obesity (%)
Jayawardena et al. (2013a, b)	National survey	600	>18 ys.	2005–2006		18.0	30.4
FHB (2014)	National survey			2013	Pregnant mothers	17.2	
Rathnayake et al. (2014a, b)	Pannala and Negambo	100	20–45 ys.	2012	Premenopausal women	38.0	34.0
Rathnayake et al. (2015)	Pannala and Negambo	100	20–45 ys.	2012	Women	38	34
FHB (2015)	National survey			2014	Pregnant mothers	20.2	
Jayawardena et al. (2016)		526	Adults	2012	Cardiac patients	22	33
DCS (2017)	National survey	16,806	15–49 ys.	2016	Urban women	35.8	20.1
					Rural women	31.9	12.4
					Estate women	17.3	6.1
					All women	31.9	13.3
Jayawardana et al. (2017)	Kandy	2469	16–72 ys.	2013–2014	Men presented for a routine medical exam at the National Transport Medical Institute	31.8	12.3

2.4.3.3 Determinants of Overweight and Obesity

There were eight studies, which have explored the association between overweight and obesity, and household income, physical activity level, leisure activities, and education level of adults (Arambepola et al. 2008; Katulanda et al. 2010; Jayatissa et al. 2012a, b, c; Perera and Ekanayake 2012; Rathnayake et al. 2014a, b; Jayawardana et al. 2017). Of these, five have reported a positive association between overweight and household income (Katulanda et al. 2010; Jayatissa et al. 2012a, b, c; Rathnayake et al. 2014a, b; Jayawardana et al. 2017), and three reported a negative association between physical activity level and overweight among adults (Rathnayake et al. 2014a, b; Jayawardena 2014). Two of these studies have found a positive relationship between urban living and overweight (Katulanda et al. 2010; Arambepola et al. 2008). The association between female gender and overweight and obesity was reported by Katulanda et al. (2010) and Perera and Ekanayake (2012). Surprisingly, three studies have reported a positive association between education level and the incidence of overweight, of which one has studied the overweight among preschool children (Jayatissa et al. 2012a, b, c), and the other has studied the overweight among adults (Katulanda et al. 2010). One study focusing on overweight and obesity among adolescents reported that lower fruit consumption, skipping of breakfast, and intake of high-energy dense food are positively associated with the overweight prevalence among adolescents (Rathnayake et al. 2014a, b; Table 2.11).

Table 2.11 A summary of studies reporting the determinants of overweight and obesity

Author and year	Association or causality	Outcome variable	Food-related factors	Income	Physical activity level	Other factors
Arambepola et al. (2008)		20–64 years	Meal size	Household income	Physical activity	Age, marital status
Katulanda et al. (2010)		>18 years old		Household income		Higher education, urban living, being female
Jayatissa et al. (2012a, b, c)		Women 15–49 overweight and obese		Household wealth		Husband's education
Perera and Ekanayake (2012)	Causal					Gender (female +), ethnicity (moor +)
Rathnayake et al. (2014a, b)			Lower fruit consumption, skipping breakfast, energy intake	Household income	Physical activity	First-born child, screen viewing hours (+), irregular menstruation
Jayawardana et al. (2017)		Men 16–72 years old		Household income	Physical activity	Age, being moor

2.5 Diagnosis of Nutrition Transition

2.5.1 Key Trends, Patterns in Malnutrition and Its Determinants

As Table 2.12 indicates, child's nutritional status has improved in Sri Lanka, while WRA anemia has declined over the last 40 years. However, there is a slight increase in obesity and overweight among adults. This contradicting sign indicates that Sri Lanka is facing double burden of malnutrition. In fact, more than one in ten preschool and school children in Sri Lanka suffer from at least one form of undernutrition. This proportion is even higher in the estate sector due to relatively higher rates of poverty and low level of maternal education, maternal nutrition level, and poor feeding practices.

As brought up in this review, income is one of the key determinants of undernutrition. People in the lowest income quartile consume monotonous cereal-based diets which lack some important macro and micronutrients. In contrast, people in the highest income quintile consume more diversified diets which are relatively high in animal-based foods. However, increase in income does not always lead to higher nutrient intake as part of the increased income is spent to buy nonnutritive attributes. Therefore, it is not the family income but the effective income allocated for the consumption of nutritious food that affect the nutritional outcome of the household. Any addiction that reduces the income allocated to food consumption

Table 2.12 Changes in the status of malnutrition and food supply in Sri Lanka during 1977–2016

Category	Indicator	1977–1986	1987–1996	1997–2006	2007–2016
Child anthropometry	Low birthweight[a]		23%	16.7%	15.7%[b]
	Stunting[c]	27.5%	23.8%	13.5%	17.3%
	Wasting[c]	12.9%	15.5%	14.0%	15.1%
	Underweight[c]	38.1%	37.7%	29.4%	20.5%
Adolescence Anthropometry	Overweight[d]			2.2%[d]	4.6%
	Obesity[e]	0.1%	0.26%	0.87%	1.4%
Adult anthropometry	Overweight: Men[f]			14.3%	
	Obesity: Men[g]	0.35%	0.6%	2.6%	
	Overweight: Women[e]			24.0%	31.9%
	Obesity: Women[c]		5.0% (2000)	7.2%	13.3%
	WRA thinness (BMI less than 18.5)[c]		22% (2000)	16%	9.1%
	WRA short stature (height less than 145 cm)[c]			10.6%	7.2%
Micronutrient deficiencies	WRA anemia		45%[h]	39%[i]	11.1%[j]
	Vitamin A deficiency (6–59 months)				
	Undernutrition[k]		33.9%[l] (1990–1992)	29.6%	22.1%
Food supply	Fruit and vegetable intake[k]	270.25 g per capita per day	192.37 g per capita per day	188.16 g per capita per day	205.93 g per capita per day
	% of calories from staples[j]			56.85%	56.5%

WRA Women of reproductive age (15–49 years)
[a]FHB (2013), MoH
[b]DCS (2016)
[c]DHS (various years)
[d]Jayatissa and Ranbanda (2006)
[e]WHO
[f]WHO
[g]Jayatissa et al. (2012a, b, c)
[h]Jayatissa (2009)
[i]Mudalige and Nestel (1996)
[j]Allen et al. (2017)
[k]FAO: Food balance sheet
[l]FAO (2010)

will negatively influence the nutrition outcome of the household. Addiction to alcohol and smoking habit of the father are two such addictions that have prevented the child achieving its desired level of growth in estate sector and fishing community in Sri Lanka. This has become a determinant in the estate sector because estate sector dwellers are more likely to consume alcohol compared to urban and rural dwellers (Jayathilaka et al. 2016).

This review suggests two opposite relationships between women's employment and child malnutrition. While many of the studies portray an inverse association between women's employment and child nutrition, one study suggests a positive association between the two variables, probably due to the interlinkage between women's income and caring. Though increase in women income can be used to improve the nutritional status of children, due to the increased time burden of women, growth of infants could be adversely affected by women employment (Johnston et al. 2018). However, older children are benefited by the income contribution of the mother. Similarly, Burroway (2017) also showed that although some jobs are capable of alleviating child malnutrition, malnutrition is high among people those who are engaged in agricultural jobs.

Income is a necessary but not a sufficient drive to alleviate child malnutrition. As indicated by the analysis of the available literature, nutritional status of the mother is a critical factor in determining the nutritional outcome of the child. Maternal height and BMI affect anthropometry of preschool children. This intergeneration vicious cycle needs to be broken down to improve child nutrition. Maternal height and BMI influence the birthweight and the nutritional outcome of the child. The chance of giving birth to a low weight baby is higher among malnourished mothers than other groups. The role of mother in determining the nutrition outcome goes beyond this intergenerational relationship. Mother's knowledge on health and growth of a child plays a significant role in improving the child health (Glewwe 1999).

With respect to the determinants of overweight and obesity, unlike in some developing countries where overweight and obesity are associated with lower income groups and least education level (Drewnowski and Specter 2004), the opposite has been observed in the Sri Lankan context where obesity rates are higher among high-income and highly educated groups than the low-income and less-educated groups. This may be partly due to the difference in food consumption basket and physical activity of these groups.

2.5.2 Results of the Econometric Analysis

The results of the meta-regression analysis were consistent with our expectations and previous literature (Table 2.13). The results of the analysis indicate that over the years, both stunting and underweight in Sri Lanka have significantly reduced (models 1–9), while wasting has increased (Model 8). Regression results further witnessed a geographical inequality in child malnutrition where urban children are less undernourished and estate sector children are more undernourished (Models 1, 7

Table 2.13 Results of the meta-regression analysis

		Model 1 (whole sample)	Model 2 (rural sample)	Model 3 (urban)	Model 4 (estate)	Model 5 (male)	Model 6 (female)	Model 7 (stunting)	Model 8 (wasting)	Model 9 (underweight)	Model 10 (overweight)
Predictor											
Year of survey		−0.1601[a] (0.0449)	−0.3316[a] (0.1130)	−0.1947 (00989)	−0.5506[a] (0.1756)	−0.2461 (0.1479)	−0.3650[b] (0.1570)	−0.2775[a] (0.0635)	0.1712[a] (0.0358)	−0.5108[a] (0.0638)	−3.6813[a] (0.2237)
Sector of residence	Rural	−1.0061 (2.0104)						−2.9375 (3.0600)	0.4617 (1.7237)	−2.1931 (2.7778)	−0.5836 (4.1828)
	Urban	−5.5408[a] (2.0104)						−7.5090[b] (3.0600)	−2.7953 (1.7237)	−9.3838[a] (2.7778)	−0.1836 (4.1828)
	Estate	9.2591[a] (2.0104)						22.6195[a] (3.0600)	−3.0667[c] (1.7237)	10.159[a] (2.7778)	−2.8836 (4.1828)
Gender of the child	Male	−0.7239 (2.044)						−1.3390 (3.2328)	0.2717 (1.8211)	−2.2747 (2.9142)	1.6875 (3.6727)
	Female	−0.8679 (2.0449)						−0.8390 (3.2328)	−0.7140 (1.8211)	−1.1461 (2.9142)	−0.9124 (3.6727)
Age of the child	< 6 mon.	−14.927[a] (2.4077)						−13.6574[a] (3.5533)	−6.4602[a] (2.0016)	−23.5728[a] (3.2050)	
	6–11 mon.	−1.3724 (2.0605)						−4.3278 (3.2003)	−0.6949 (1.8028)	−7.6817[b] (3.0182)	17.4542[a] (3.6727)
	12–23 mon.	6.0554[a] (2.0605)						7.8393[b] (3.2003)	5.2050[a] (1.8028)	4.8515 (3.0182)	9.8209[b] (3.6727)
	24–35 mon.	0.6354 (2.0605)						1.8278 (3.2003)	−0.3949 (1.8028)	2.3348 (3.0182)	−1.4457 (3.6727)
	36–47 mon.	0.0161 (2.1437)						1.7002 (3.3461)	−0.2231 (1.8849)	1.9613 (3.1997)	−4.5124 (3.6727)
	48–59 mon.	−0.7054 (1.5806)	−1.3180 (2.6563)	−3.8418 (2.3260)	−1.6305 (4.1272)			1.4997 (2.4297)	0.3271 (1.3686)	−0.3831 (2.2411)	−7.7457[b] (3.6727)

Income quartile	1st income quartile	4.2253[c] (2.5378)						7.7569[c] (4.2886)	2.4975 (2.4158)	4.8323 (3.8624)	2.2875 (3.6727)
	2nd income quartile	1.2715 (2.5378)						1.6903 (4.2886)	0.5642 (2.4158)	0.6323 (3.8624)	1.7209 (3.6727)
	3rd income quartile	−0.5284 (2.5378)						−1.1096 (4.2886)	−0.1691 (2.4158)	−2.7343 (3.8624)	0.6209 (3.6727)
	4th income quartile	−2.7822 (2.5378)						−3.7096 (4.2886)	−1.2357 (2.4158)	−7.2010[c] (3.8624)	−1.1790 (3.6727)
	5th income quartile	−5.3053[b] (2.5378)						−6.7096 (4.2886)	−5.1357 (2.4158)	−11.4010[a] (3.8624)	−0.8790 (3.6727)
Education level of the mother	No education	9.3710[a] (2.1132)						18.9491[a] (3.3457)	2.2603 (1.8847)	12.0848[a] (3.0182)	1.3875 (3.6727)
	Primary	5.8165[a] (2.1132)						7.9657[b] (3.3457)	2.2936 (1.8847)	8.2515[a] (3.0182)	4.7875 (3.6727)
	Secondary	0.3983 (2.1132)						−0.7675 (3.3457)	1.0270 (1.8847)	0.3015 (3.0182)	0.8875 (3.6727)
	Pass GCE O/L	−3.3704 (2.2152)						−5.9177[c] (3.5510)	−1.2224 (2.0003)	−6.6186[b] (3.1997)	0.8542 (3.6727)
	Pass GCE A/L or higher	−5.0951[b] (2.3517)						−5.7343 (3.8447)	−3.6147[c] (2.1658)	−9.6900[a] (3.4629)	−3.1790 (3.6727)
Form of malnutrition	Stunting	14.4175[a] (2.1274)	−12.4317[a] (3.9541)	−14.514[a] (3.4624)	16.6690[b] (6.1437)	−4.3276 (4.3789)	−1.7893 (4.6458)				
	Wasting	8.1787[a] (2.1274)	−16.2174[a] (3.9541)	−16.985[a] (3.4624)	−16.202[b] (6.1437)	−7.671[c] (4.3789)	−6.6179 (4.6458)				
	Underweight	21.1497[a] (2.1293)	−3.3317 (3.9541)	−8.0279[b] (3.4624)	12.5690[c] (6.1437)	3.9437 (4.3789)	7.1106 (4.6458)				
	Anemia	16.7519[a] (2.1142)									
	Overweight					−17.191[b] (6.8063)	−14.7993[c] (7.2212)				

(continued)

Table 2.13 (continued)

		Model 1 (whole sample)	Model 2 (rural sample)	Model 3 (urban)	Model 4 (estate)	Model 5 (male)	Model 6 (female)	Model 7 (stunting)	Model 8 (wasting)	Model 9 (underweight)	Model 10 (overweight)
Survey type	DHS	1.2223 (1.8573)	1.5373 (3.2499)	−2.2602 (2.8458)	−5.3300 (5.0496)	3.5502 (3.6136)	3.1120 (3.8339)	5.2598[a] (1.8054)	−2.6299[b] (1.0170)	0.3006 (1.6252)	
	Health ministry	−4.6830[b] (2.011)									
	NFSA					1.7886 (4.5450)	1.3045 (4.8221)	6.6169[b] (3.3011)	−6.1626[a] (1.8595)	−1.7268 (2.9743)	4.7022 (3.1517)
Constant		327.138[a] (90.3929)	694.033[a] (227.3511)	421.351[b] (199.0813	1134.788[a] (353.2449)	513.328 (298.002)	749.6973[b] (316.1677)	571.2411[a] (127.7966)	−325.300[a] (71.9895)	1052.479[a] (128.4357)	7417.52[a] (449.6836)
R^2		0.5667	0.7604	0.7603	0.8684	0.6555	0.6941	0.7327	0.5298	0.7705	0.9046
Number of observations		440	23	23	23	25	25	122	122	117	61

Standard errors are given within brackets
[a]Denotes significance at 99%
[b]Denotes significance at 95%
[c]Denotes significance at 90%

and 9). However, with respect to the prevalence of wasting, estate sector performs better than the other two sectors, where the prevalence of wasting is lower than the other two sectors. The results further suggest that all the sectors have been able to curtail the high prevalence rate of undernutrition, though the reduction is impressive in the estate sector as compared to other two sectors (Models 2–4).

High prevalence rate of malnutrition is associated with income quartile. As compared to the rate of child malnutrition in the first four income quartile, the rate of child malnutrition in the fifth income quartile is low. Though statistically insignificant at 10% level, the prevalence rate in the third and fourth income quartiles is low compared to the bottom two quintiles. The difference was statistically significant at 5% level only for the prevalence of wasting and underweight and not significant for stunting and overweight. Moreover, child malnutrition is inversely associated with the mother's education level. The children who are born to mothers with no education or only with primary education are at a higher risk of being stunted or underweight, whereas children born to mothers who have completed school education or with higher education qualifications are at a lower risk. The association is not significant at 10% level for wasting and overweight. Similarly, it was found that children less than 6 months old are at a lower risk of being stunted or wasted and higher risk of being underweight, compared to other age groups. In contrast to this, children above the age of 1 and less than 2 have a higher risk of being stunted and wasted. This reinforces the finding, which implies better care at first few months and improper introduction of complementary food at the age of 1. Indicating the double burden of malnutrition, children between the ages of 6 months and 2 years show a higher risk of being overweight. One of the salient finding of this study is that none of the estimated models provide empirical evidence for the presence of gender inequality of malnutrition.

2.6 Summary, Conclusions, and Way Forward

2.6.1 Summary and Conclusions

Based on the findings of the previous section, this review assessed the degree of nutrition transition in Sri Lanka. As stated in Sect. 5, over these years, Sri Lankan diet has transformed toward more animal-based products as compared to cereal-based monotonous diet consume in the past. However, still the share of food of animal origin to total intake of calorie, protein, and fat remains relatively small in Sri Lanka. These facts suggest that Sri Lanka has not reached the fourth stage of the nutrition transition in which the diet includes more fat that comes from animal products. The consumption of sugar and salt also has increased in Sri Lanka to a far high level to be classified into the third stage of the nutrition transition.

Considering the regional disparities in food consumption, in some geographical locations in the country, still the consumption of products of animal origin remains low while the consumption of cereal-based products is high. Contrast to this in urban areas, consumption of cereal-based products has declined in favor of products of animal origin. Thus, it can infer that even though the stage of nutrition transition Sri Lanka is not clear, in terms of diet transformation, certain geographical communities (e.g., estate sector) are still at the food insecurity stage[1], while urban communities are at the Degenerative disease stage.

The national-level statistics on undernutrition and overnutrition among preschool children and adult women are used in this review in the absence of those for overnutrition and undernutrition of the whole population. Accordingly, it was found that child malnutrition has declined in Sri Lanka, and in some geographical subgroups (children in the estate sector) it has declined by leaps and bounds. This reduction in undernutrition in the estate sector has been accompanied by reduction in anemia status as well. This reduction of nutritional deficiencies is a characteristic of a receding famine stage. Though the estate sector was the most vulnerable sector to food insecurity four decades ago, this result provides strong evidence for an existence of a nutrition transition from the famine stage (food insecurity) to the "receding famine stage."

The proportion of preschool children who are undernourished is higher than the proportion of preschool children who are overweight. However, in some geographical sub-groups in Colombo in the western province, overweight among school children is higher than that of undernourished implying that these communities (urban has already shifted to the stage 4 of the nutrition transition) are at different stages of nutrition transition.

The incidence of overweight and obesity among Sri Lankan adults are higher than South Asian average and almost same as the world average. As the DHS (2006/2007, 2016) indicates, overweight among women has increased over a 10-year period, the prevalence is higher among urban dwellers than the rest. By 2016, the major nutritional problem faced by women is overweight. This high prevalence rate of overweight among adult is one of the characteristics of the stage 4 of the nutrition transition (Degenerative disease). However, in the estate sector, the prevalence of thinness among adults is higher than the prevalence of overweight. Thus, it is obvious that estate sector has not reached the fourth stage of the nutrition transition process.

The burden of NCDs has increased in Sri Lanka. Hospitalization due to diabetes mellitus, hypertensive disease, and ischemic heart disease has more than doubled during the last three decades (MoH 2011, Ministry of Health, Nutrition and Indigenous Medicine 2017). Around 70% of the disease burden in Sri Lanka is due to NCDs. Nearly 30% of the total hospital deaths are due to cardiovascular diseases, which are the first leading cause of death in Sri Lanka for the past few years (WHO 2015). The hospitalization for NCDs has accelerated after 1990, while those due to infectious diseases has rapidly declined after 1990 (DCS 2009).

In summary, it is evident that Sri Lanka on average has swung between stages 3 and 4 of the nutrition transition, while some communities have already shifted to stage, some others are still struggling to move away from the third stage. It can be concluded that Sri Lanka, as in many developing countries, experiences the duel burden of malnutrition characterized by coexisting undernutrition and overnutrition.

---

[1] This condition mimics the stage as defined by Popkin (1999)

2.6.2 Way Forward

This review witnessed a nutrition transition in Sri Lanka similar to many developing and transitional economics where both undernutrition and overnutrition coexist. Poverty, low maternal nutritional status, and lack of maternal knowledge on nutrition hinder the success of alleviating malnutrition despite several attempts taken to combat the epidemics of undernutrition and micronutrient deficiencies among children and pregnant mothers in the country. Urbanization, change in food consumption pattern, and physical inactivity may have contributed to the emerging problem of overnutrition.

The findings from this review have important policy implications on several areas such as agriculture, nutrition and health, and education. Food policies, agricultural policies, and nutrition policies need to be integrated to combat the triple burden of malnutrition in Sri Lanka. Agricultural policies should be reoriented to diversify farm-level production. The existing policy framework of the country incentivizes paddy production at the expense of high-value agricultural products. To this end, the current land policy, which bans the use of paddy land for cultivation of other crops, could be relaxed. Agriculture research policies need to be directed to explore the possible ways of improving nutritive values of crop and animal-sourced food in addition to yield improvements.

Food policies of the country, government taxes, and subsides need to be reformulated to make nutritious food more affordable. Import taxes can be charged on unhealthy food item (i.e., energy dense and/or junk foods) to make them less affordable to the consumers. Legislations are required to transform the food processing industry to depict sugar and salt contents in processed foods through compulsory labeling requirements. Along with such changes, reformations in consumption pattern are required through attitudinal changes. Public awareness through campaigns on the food-based dietary guidelines is required. School children need to be educated and regulatory measures should be imposed on sales of food items within school premises.

As stated before, though there have been many attempts to combat undernutrition and micronutrient deficiencies, the need to prevent the rising tide of overnutrition has not been adequately emphasized in Sri Lanka. Thus, it is necessary to introduce programs and policies to curtail the rising epidemic of overnutrition in the country at this early phase.

More research is needed to provide more scientific evidence for in-depth understanding of the prevalence of malnutrition in Sri Lanka. As evident from the review, prevalence of certain forms of malnutrition is higher in certain communities. The studies conducted so far provide no evidence to explain such patterns. More research is needed to understand the determinants of multiple forms of malnutrition and their interconnections over the lifecycle. Further research is needed to synthesize the trends in physical activity and NCDs and their association with food consumption behavior and demographic changes in the society, which has not been covered in this review.

Given the trends in the nutritional outcomes, a sound national monitoring program to record the health and nutritional data of citizens of all age groups will be required to track the improvements in health and nutritional status.

References

Allen A, Allen S, Rodrigo R, Perera L, Shao W, Li C, Premawardhena A (2017) Iron status and anaemia in Sri Lankan secondary school children: a cross-sectional survey. PLoS One 12(11):e0188110

Amarasinghe GS, Naottunna NPGCR, Agampodi TC, Agampodi SB (2017) Factors associated with anemia among Sri Lankan primary school children in rural North Central Province. BMC Pediatr 17(1):87

Arambepola C, Allender S, Ekanayake R, Fernando D (2008) Urban living and obesity: is it independent of its population and lifestyle characteristics? Trop Med Int Health 13(4):448–457

Baker P, Friel S (2014) Processed foods and the nutrition transition: evidence from Asia. Obes Rev 15(7):564–577

Bandara S, Weerasinghe MC (2015) Health and socio-economic determinants of malnutrition in the plantation sector of Sri Lanka: a review, Working Paper Series No. 21. Institute of Policy Studies of Sri Lanka, Colombo

Bell AC, Ge K, Popkin BM (2001) Weight gain and its predictors in Chinese adults. Int J Obes 25(7):1079

Bhalla S (1991) Sri Lanka. In: Krueger AO, Schiff M, Valdes A (eds) The political economy of agricultural pricing policy, volume 2: Asia. Johns Hopkins, Baltimore, MD, pp 195–235

Bhargava A, Jamison DT, Lau LJ, Murray CJ (2001) Modeling the effects of health on economic growth. J Health Econ 20(3):423–440

Bloom DE, Canning D, Sevilla J (2004) The effect of health on economic growth: a production function approach. World Dev 32(1):1–13

Bogahawatte C, Kailasapathy K (1986) The consumption patterns of food in Sri Lanka with special reference to protein-calorie nutrition. Agric Syst 22(4):289–303

Burroway R (2017) Are all jobs created equal? A cross-national analysis of women's employment and child malnutrition in developing countries. Soc Sci Res 67:1–13

Central Bank (2017) Annual report, 2017. Central Bank, Colombo

Chathurani U, Dharshika I, Galgamuwa D, Wickramasinghe ND, Agampodi TC, Agampodi SB (2012) Anaemia in pregnancy in the district of Anuradhapura, Sri Lanka—need for updating prevalence data and screening strategies. Ceylon Med J 57(3):101–106

DCS (2002) Sri Lanka demographic and health survey 2000. Department of Census and Statistics in Collaboration with Ministry of Health, Nutrition and Welfare, Colombo

DCS (2009). Sri Lanka demographic and health survey 2006–07, Colombo: Department of Census and Statistics. http://www.statistics.gov.lk/samplesurvey/LFS_Annual%20Report_2016.pdf. Accessed 24 Jun 2018

DCS (2017) Sri Lanka demographic and health survey 2016. Department of Census and Statistics, Colombo

DCS (2018) Sri Lanka labour force survey annual report 2017. Department of Census and Statistics, Colombo

Department of Census and Statistics (1988) Sri Lanka demographic and health survey 1987. Ministry of Plan Implementation, Colombo

Department of Census and Statistics (1995) Sri Lanka demographic and health survey 1993. Department of Census and Statistics, Colombo

Department of Census and Statistics (2000) Sri Lanka demographic and health survey 2000. Department of Census and Statistics, Colombo

Department of Census and Statistics (2007) Sri Lanka demographic and health survey 2006–2007. Department of Census and Statistics, Colombo

Department of Census and Statistics (2016) Sri Lanka demographic and health survey 2016. Department of Census and Statistics, Colombo

De Silva K, Liyanage C, Wijesinghe C, Perera B (2017) Nutritional status of elders in Galle district, Sri Lanka. Galle Med J 22(1):1–5

Drewnowski A, Specter SE (2004) Poverty and obesity: the role of energy density and energy costs. Am J Clin Nutr 79(1):6–16

Edirisinghe N (1987) The food stamp scheme in Sri Lanka: costs, benefits, and options for modification, vol 58. International Food Policy Research Institute, Washington, DC

Ekanayake S, Weerahewa J, Ariyawardana A (2003) Role of mothers in alleviating child malnutrition: evidence from Sri Lanka. In: Research report presented at the 3rd Poverty and Economic Policy (PEP), 16-20th June 2004, pp 1–20

Ellis F, Senanayake P, Smith M (1997) Food price policy in Sri Lanka. Food Policy 22(1):81–96

Ellis P, Roberts M (2015) Leveraging urbanization in South Asia: managing spatial transformation for prosperity and livability. The World Bank

Family Health Bureau (2011). Annual Report 2010, Family Health Bureau, Ministry of Health, Sri Lanka. http://fhb.health.gov.lk/web/index.php?option=com_phocadownload&view=category&id=66:annual-reports&itemid=150&lang=en. Accessed 24 Jul 2018

Family Health Bureau (2012) Annual Report 2011, Family Health Bureau, Ministry of Health, Nutrition and Indigenous Medicine, Sri Lanka. http://fhb.health.gov.lk/web/index.php?option=com_phocadownload&view=category&id=66:annual-reports&itemid=150&lang=en. Accessed 24 Jul 2018

Family Health Bureau (2013). Annual Report 2012, Family Health Bureau, Ministry of Health, Nutrition and Indigenous Medicine, Sri Lanka. http://fhb.health.gov.lk/web/index.php?option=com_phocadownload&view=category&id=66:annual-reports&itemid=150&lang=en. Accessed 24 Jul 2018

Family Health Bureau (2014) Annual Report 2013, Family Health Bureau, Ministry of Health, Nutrition and Indigenous Medicine, Sri Lanka. http://fhb.health.gov.lk/web/index.php?option=com_phocadownload&view=category&id=66:annual-reports&itemid=150&lang=en. Accessed 24 Jul 2018

Family Health Bureau (2015) Annual Report 2014, Family Health Bureau, Ministry of Health, Nutrition and Indigenous Medicine, Sri Lanka. http://fhb.health.gov.lk/web/index.php?option=comphocadownload&view=category&id=66:annual-reports&itemid=150&lang=en. Accessed 24 Jul 2018

Family Health Bureau (2016) Annual Report 2015, Family Health Bureau, Ministry of Health, Nutrition and Indigenous Medicine, Sri Lanka. http://fhb.health.gov.lk/web/index.php?option=com_phocadownload&view=category&id=66:annual-reports&itemid=150&lang=en. Accessed 24 Jul 2018

FAO (2010) The state of food insecurity in the world 2004. Available at http://www.fao.org/3/y5650e/y5650e00.pdf

FAO (2016) City region food system situation analysis. Colombo, Sri Lanka. Working document

Fernando DJ, Siribaddana SH, Silva DR, Perera SD (1994) The prevalence of obesity and other coronary risk factors in a suburban Sri Lankan community. Asia Pac J Clin Nutr 3(4):155–159

Fernando WHKN, Wijesinghe DGNG (2010) Assessment of nutritional status and disease prevalence among elderly population in elderly homes in Kandy. Tropic Agri Res 21(3):229–237

FHB (2008) Family health bureau- annual report 2007. Ministry of Health, Colombo

FHB (2009) Family health bureau- annual report 2008. Ministry of Health, Colombo

FHB (2010) Family Health Bureau- Annual Report 2009. Ministry of Health, Colombo

FHB (2015) School health return. Family Health Bureau, Ministry of Health, Colombo

Gavan JD, Chandrasekera IS (1979) The impact of public food grain distribution on food consumption and welfare in Sri Lanka (No. 13). International Food Policy Research Institute, Washington, DC

Glewwe P (1999) Why does mother's schooling raise child health in developing countries? Evidence from Morocco. J Hum Resour:124–159

Hartwig J (2010) Is health capital formation good for long-term economic growth?—Panel Granger-causality evidence for OECD countries. J Macroecon 32(1):314–325

Hawkes C, Fanzo J (2017) Nourishing the SDGs: Global nutrition report 2017. Development Initiatives, Bristol. Available from https://openaccess.city.ac.uk/id/eprint/19322. Accessed 11 Nov 1999

Hettiarachchi M, Liyanage C, Wickremasinghe R, Hilmers DC, Abrams SA (2006) Prevalence and severity of micronutrient deficiency: a cross-sectional study among adolescents in Sri Lanka. Asia Pac J Clin Nutr 15(1):56

Jayasekera CR (2006) Nutritional status of children under five in three state foster care institutions in Sri Lanka. Ceylon Med J 51:63–65

Jayathilaka R, Selvanathan S, Bandaralage JS (2016) Is there a link between alcohol consumption and the level of poverty? Appl Econ 48(22):2054–2063

Jayathissa R, Mahamithawa S, Ranbanda JM (2002) Nutritional problems among Sri Lankan primary schoolchildren aged 5–9 years. Department of Nutrition, Medical Research Institute. Colombo, Sri Lanka

Jayatissa R (2009) Micronutrient deficiencies in Sri Lanka–Are we on track? Presentation made at National fortification meeting. https://www.researchgate.net/profile/Renuka_Jayatissa/publication/280659815_Micronutrient_Deficiencies_in_Sri_Lanka_-_Are_we_on_Track/links/55c0e28608aed621de150343/Micronutrient-Deficiencies-in-Sri-Lanka-Are-we-on-Track.pdf. Accessed 11 Nov 2019

Jayatissa R, Bekele A, Kethiswaran A, De Silva AH (2012a) Community-based management of severe and moderate acute malnutrition during emergencies in Sri Lanka: challenges of implementation. Food Nutr Bull 33(4):251–260

Jayatissa R, Gunathilaka MM, Fernando DN (2012b) Anaemia among children aged 6–59 month. National nutrition and micronutrient survey. Part I. Ministry of Health and United Nations International Children's Emergency Fund UNICEF, Colombo

Jayatissa R, Hossain SM, Gunawardana S, Ranbanda JM, Gunathilaka M, De Silva PC (2012c) Prevalence and associations of overweight among adult women in Sri Lanka: a national survey. Sri Lanka J Diabet Endocrinol Metab 2(2):61–68

Jayatissa R, Ranbanda RM (2006) Prevalence of challenging nutritional problems among adolescents in Sri Lanka. Food Nutr Bull 27(2):153–160

Jayatissa R, Hossaine SMM (2010) Nutrition and food security assessment in Sri Lanka 2009. Medical Research Institute, United Nations International Children's Emergency Fund (UNICEF)[T1] and World Food Programme, Colombo

Jayatissa R, Wickramage K (2016) What effect does international migration have on the nutritional status and child care practices of children left behind? Int J Environ Res Public Health 13(2):218

Jayawardana NWIA, Jayalath WATA, Madhujith WMT, Ralapanawa U, Jayasekera RS, Alagiyawanna SASB, Kalupahana NS (2017) Lifestyle factors associated with obesity in a cohort of males in the central province of Sri Lanka: a cross-sectional descriptive study. BMC Public Health 17(1):27

Jayawardena P (2012) Socio-economic determinants and inequalities in childhood malnutrition in Sri Lanka. Well-Being Soc Policy J 8(1):1–22

Jayawardena P (2014) Underlying causes of child and maternal malnutrition in the estate sector of Sri Lanka. J South Asian Stud 2(3):241–255

Jayawardena R, Byrne NM, Soares MJ, Katulanda P, Hills AP (2012) The obesity epidemic in Sri Lanka revisited. Asia Pacific J Public Health 27(2):NP1298–NP1299

Jayawardena R, Byrne NM, Soares MJ, Katulanda P, Hills AP (2013a) Food consumption of Sri Lankan adults: an appraisal of serving characteristics. Public Health Nutr 16(4):653–658

Jayawardena R, Byrne NM, Soares MJ, Katulanda P, Yadav B, Hills AP (2013b) High dietary diversity is associated with obesity in Sri Lankan adults: an evaluation of three dietary scores. BMC Public Health 13(1):314

Jayawardena R, Punchihewa P, Ranathunga I, Lokunarangoda N, Pathirana AK, Santharaj WS (2016) Body weight perception among Sri Lankan cardiac patients. BMC Obes 3(1):32
Johnston D, Stevano S, Malapit HJ, Hull E, Kadiyala S (2018) Time use as an explanation for the agri-nutrition disconnect? Evidence from rural areas in low and middle income countries. Food Policy 76:8–18
Kandeepan K, Balakumar S, Arasaratnam V (2016) Nutritional status and food insecurity among the children in northern Sri Lanka. Procedia Food Sci 6:220–224
Katulanda P, Jayawardena MAR, Sheriff MHR, Constantine GR, Matthews DR (2010) Prevalence of overweight and obesity in Sri Lankan adults. Obes Rev 11(11):751–756
Lee MJ, Popkin BM, Kim S (2002) The unique aspects of the nutrition transition in South Korea: the retention of healthful elements in their traditional diet. Public Health Nutr 5(1a):197–203
Malkanthi RLDK, Chandrasekera GAP, Jayasinghe JMUK (2007) High prevalence of malnutrition and household food insecurity in the rural subsistence Paddy farming sector. Trop Agri Res 19:136–149
Ministry of Health (2017) Annual health bulletin. Ministry of Health, Nutrition and Indigenous Medicine, Colombo
MoH (2011) Non-communicable disease status Sri Lanka. Ministry of Health, Colombo
Mudalige R, Nestel P (1996) Combating iron deficiency. Ceylon J Med Sci 39(1):9–16
Nanayakkara GJM, Silva KDRR (2011) Association among maternal nutrition, birth characteristics, feeding practices, nutritional status and body composition of term infants. The Proceedings of the Nutrition Society, 70(OCE4)
Naotunna NPGCR, Dayarathna M, Maheshi H, Amarasinghe GS, Kithmini VS, Rathnayaka M, Agampodi TC (2017) Nutritional status among primary school children in rural Sri Lanka; a public health challenge for a country with high child health standards. BMC Public Health 17(1):57
Narayan S, Narayan PK, Mishra S (2010) Investigating the relationship between health and economic growth: empirical evidence from a panel of 5 Asian countries. J Asian Econ 21(4):404–411
Pathmeswaran A, Jayatissa R, Samarasinghe S, Fernando A, De Silva RP, Thattil RO, De Silva NR (2005) Health status of primary schoolchildren in SriLanka. Ceylon Med J 50(2):46–50
Peiris TDR, Wijesinghe DJGNG (2011) Nutritional status of under 5 year-old children and its relationship with maternal nutrition knowledge in Weeraketiya DS division of Sri Lanka. Trop Agric Res 21(4):330–339
Perera R, Ekanayake L (2012) Relationship between nutritional status and tooth loss in an older population from Sri Lanka. Gerodontology 29(2):566–570
Perkins JM, Jayatissa R, Subramanian SV (2018) Assessment of nutritional status and disease prevalence among elderly population in elderly homes in Kandy. Nutrition 55–56:76–83
Piyasena C, Mahamithawa AMASB (2003) Assessment of anaemia status in Sri Lanka 2001 [survey report]. Ministry of Health, Nutrition and Welfare, Department of Health Services, Medical Research Institute, Colombo, pp 40–46
Popkin BM (1994) The nutrition transition in low-income countries: an emerging crisis. Nutr Rev 52(9):285–298
Popkin BM (1998) The nutrition transition and its health implications in lower-income countries. Public Health Nutr 1(1):5–21
Popkin BM (1999) Urbanization, lifestyle changes and the nutrition transition. World Dev 27(11):1905–1916
Popkin BM (2001) The nutrition transition and obesity in the developing world. J Nutr 131(3):871S–873S
Popkin BM (2002) The shift in stages of the nutrition transition in the developing world differs from past experiences. Public Health Nutr 5(1A):205–214
Popkin BM (2003) The nutrition transition in the developing world. Develop Policy Rev 21(5–6):581–597
Popkin BM (2004) The nutrition transition: an overview of world patterns of change. Nutr Rev 62(suppl 2):S140–S143

Prasanna IWK (2015) Nutritional status of preschool children and its associates: a Sri Lankan experience of a fishing community. Eur J Prevent Med 3(2–1):31–35
Rajapaksa LC, Arambepola C, Gunawardena N, Rosa C, Opatha S (2011) Nutritional status in Sri Lanka, determinants and interventions: a desk review. UNICEF, Colombo
Ranis G, Stewart F, Ramirez A (2000) Economic growth and human development. World Dev 28(2):197–219
Rannan-Eliya RP, Hossain SMM, Anuranga C, Wickramasinghe R, Jayatissa R, Abeykoon ATPL (2013) Trends and determinants of childhood stunting and underweight in Sri Lanka. Ceylon Med J 58(1):10–18
Rathnayake IM, Weerahewa J (2005a) Maternal employment and income affect dietary calorie adequacy in households in Sri Lanka. Food Nutr Bull 26(2):222–229
Rathnayake IM, Weerahewa J (2005b) Determinants of nutritional status among pre-school children in Sri Lanka. Trop Agri Res 17:148–161
Rathnayake KM, Roopasingam T, Dibley MJ (2014a) High carbohydrate diet and physical inactivity associated with central obesity among premenopausal housewives in Sri Lanka. BMC Res Notes 7(1):564
Rathnayake KM, Roopasingam T, Wickramasighe VP (2014b) Nutritional and behavioral determinants of adolescent obesity: a case–control study in Sri Lanka. BMC Public Health 14(1):1291
Rathnayake KM, Wimalathunga MPPM, Weech M, Jackson KG, Lovegrove JA (2015) High prevalence of undernutrition and low dietary diversity in institutionalised elderly living in Sri Lanka. Public Health Nutr 18(15):2874–2880
Sahn DE (1988) The effect of price and income changes on food-energy intake in Sri Lanka. Econ Dev Cult Chang 36(2):315–340
Sahn DE, Alderman H (1988) The effects of human capital on wages, and the determinants of labor supply in a developing country. J Dev Econ 29(2):157–183
Senadheera D, Goonewardene M, Mampitiya I (2017) Anaemia and iron deficiency in pregnant women attending an antenatal clinic in a Teaching Hospital in Southern Sri Lanka. Ceylon Med J 62(3):175–183
Shekar M, Somanathan A, Du L (2007) Malnutrition in Sri Lanka: scale, scope, causes, and potential response. World Bank report. Report No. 40906-LK. World Bank, Washington, DC
Smith LC, Haddad L (2002) How potent is economic growth in reducing undernutrition? What are the pathways of impact? New cross-country evidence. Econ Dev Cult Chang 51(1):55–76
Strauss J, Thomas D (1998) Health, nutrition, and economic development. J Econ Lit 36(2):766–817
Suri T, Boozer MA, Ranis G, Stewart F (2011) Paths to success: the relationship between human development and economic growth. World Dev 39(4):506–522
Thamilini J, Silva KDRR, Jayasinghe JMUK (2015) Prevalence of stunting among pre-school children in food insecure rural households in Sri Lanka. Trop Agri Res 26(2):390–394
Thilakarathne RMLR, Wijesinghe DGNG (2011) Association between nutritional status and life style practices of primary school children in the Colombo District: a pilot study. Trop Agri Res 22(4):392–401
Ubeysekara NH, Jayathissa R, Wijesinghe CJ (2015) Nutritional status and associated feeding practices among children aged 6-24 months in a selected community in Sri Lanka: a cross sectional study. Eur J Prevent Med 3(2–1):15–23
Weerahewa J, Korale-Gedara PK, Kanthilanka H (2017) The evolution of food policy in Sri Lanka: 1948-2017. Reference Module in Food Sciences, pp 1–19
Weerahewa J, Korale-Gedara P, Wijetunga CS (2018b) Nutrition transition in Sri Lanka: a diagnosis. Ann Nutr Food Sci 2(2):1020
Weerahewa J, Wijetunga CS, Babu SC, Atapattu N (2018a) Food policies and nutrition transition in Sri Lanka: historical trends, political regimes, and options for interventions. IFPRI Discussion Paper 1727. International Food Policy Research Institute (IFPRI), Washington, DC. http://ebrary.ifpri.org/cdm/ref/collection/p15738coll2/id/132683. Accessed 24 Jul 2018
Well DN (2007) Accounting for the effect of health on economic growth. Q J Econ 122(3):1265–1306

WHO (2012) Report of expert meeting on population sodium reduction strategies for prevention and control of non-communicable diseases in the South Asian region. World Health Organization (WHO), Geneva

WHO (2015) Non-communicable disease country profile. World Health Organization (WHO), Geneva. http://apps.who.int/iris/bitstream/handle/10665/128038/9789241507509_eng.pdf;jsessionid=FAF005500BC47F41343C767CC74284DF?sequence=1. Accessed 24 Jul 2018

Wijesekere G (2015) Changing patterns of food consumption in Sri Lanka: 1985–2009. The Australian National University, Australia South Asia Research Centre, Canberra

Wisniewski SL (2010) Child nutrition, health problems, and school achievement in Sri Lanka. World Dev 38(3):315–332

World Bank (2018) World Bank Open Data. https://data.worldbank.org/indicator/SP.URB.TOTL.IN.ZS. Accessed 24 Jul 2018

Cascaded Tank-Village System: Present Status and Prospects

3

P. B. Dharmasena

Abstract

Sri Lanka is well known worldwide for its ancient water civilization, and the development and evolution of irrigation systems in the country are well documented. Of them, the cascaded tank-village systems (CTVSs) are of greater significance. They exist in the country with their unique environment which comprises of tanks, paddy fields and the passages of surface water movement. Low cropping intensity, tank sedimentation, high tank water losses and low resource productivity have been identified as the major problems related to the CTVS. Various water management practices have been adopted in Sri Lanka to improve land productivity under small tanks. There is a need to understand the underlying principles and mechanisms of the evolution of the CTVSs with a focus on its each component. Building awareness among farmers on the ecological aspects of the components in the CTVSs would help to introduce novel cascade management systems towards improving its ecological conditions and land productivity.

Keywords

Village tanks · Cascade systems · Cascade ecology · Sustainability

3.1 Introduction

Over the history of mankind, a large number of communities have developed their own culture and lifestyle that are intricately tied to nature and the local landscapes. The ancient water civilization of Sri Lanka is not an exception, but its social, political and economic interactions with neighbouring countries have been interfered

P. B. Dharmasena (✉)
Rajarata University of Sri Lanka, Mihintale, Sri Lanka

B. Marambe et al. (eds.), *Agricultural Research for Sustainable Food Systems in Sri Lanka*, https://doi.org/10.1007/978-981-15-2152-2_3

during the entire process of the evolution (Dharmasena 2010). The earliest reservoir referred to in the *Mahavamsa*, the great chronicle of the history of the island, was the *Jayavapi* built in the reign of King Pandukabhaya (377–307 BC). The construction of larger-scale reservoirs initiated at the time of King Vasabha (65–109 AD) and then continued by the Kings Mahasena (276–303 AD), Dhatusena (455–473 AD) and Agbo II (575–608 AD). These were very advanced hydraulic structures, which would have required sound knowledge of key hydraulic principles pertaining to rainfall, runoff and storage volumes that had been designed, in their construction. This trend continued until the reign of King Parakramabahu I (1153–1186 AD). Subsequently, the evolutionary process of development established a widely spread mosaic system of reservoirs and canals regulating the water flow either extracted from natural streams or received from direct rainfall for agricultural and human utility purposes, as well as a means of flood management and soil conservation (Jayawardena 2010).

In view of the importance of reservoirs in the lives of the people, it is not surprising that much has been written on them; the historical aspects have been dealt with by Brohier (1934, 1937), De Camp (1974), Fernando (1979) and Parker (1981); sociological impact of reservoirs by Ellepola (1955), Brohier (1937) and Perera (1984); the engineering aspects by Brohier (1934, 1937), Van de Lippe (1951), Schnitter (1967), Smith (1971) and De Camp (1974); and the limnological and biological aspects by a host of authors, as reviewed by Fernando and Silva (1984). Perera (1984) traced six stages in the development of the irrigation systems in Sri Lanka (Table 3.1).

Fernando (1979) recognized 44 ancient, major irrigation reservoirs covering a surface area of approx. 39,000 ha. Almost all the major ancient irrigation works have now

Table 3.1 Development of the irrigation systems in Sri Lanka (Source: Perera 1984)

Stage	Description
First stage	Collection of rainwater in a pond or valley, lift irrigation using primitive implements to irrigate the surrounding paddies
Second stage	Development of low artificial embankments or weirs built across the valleys of small ephemeral rivulets, such a tank would have a depth of 2.5 m, tank followed downstream by yet another tank and paddy fields
Third stage	Improvement of the former type, the bunds strengthened; extent of irrigable land improved but still not part of a complex network of tanks
Fourth stage	Damming of the bed of comparatively large non-perennial rivers, e.g. Kala Oya (Dhatusena, 459 AD), water distribution capabilities increased, special channels constructed to transmit water, catchments linked
Fifth stage	Construction of reservoirs on large, ephemeral rivers and tributaries and linking these to anicuts built on rivers having catchment areas of perennial water supply in the wet zone, e.g. anicut at Elahera across the Amban Ganga (tributary of Mahaweli Ganga), built by Vasabha (65–109 AD), later enlarged and diversified by Mahasena (276–303 AD), feeds the Minneriyawewa. This stage was also characterized by building of spills and sluices
Sixth stage	Trans-basin transfer of water from a perennial catchment to an ephemeral catchment-based reservoir, e.g. Amban Ganga catchment linked with Kala Oya catchment, highest complexity by the eighth century AD

been rehabilitated, except perhaps for a few, which may have gone into disrepair and are covered by thick jungle. It is likely that most of the ancient works have been unearthed and their origins understood. Withanachchi (2015) has investigated the role of ancient canals in these networks of intrinsic reservoir canals operated in the dry (DZ) and intermediate (IZ) zones of Sri Lanka. There had been amalgamated irrigation systems linked by the canals in the DZ of Sri Lanka. At the operational level, the canals have behaved as veins transmitting water from one end to another end or many ends. It was not an isolated structure that existed on earth surface but a main part of a complex system, especially with inter- and intra-roles in cultural and ecological systems.

The early societies of water civilization have thrived on small irrigation systems with unique assemblages of land uses and agricultural attributes. Possibly, these systems have evolved from the early rainfed shifting agriculture into small-scale irrigation systems that in turn have led to major irrigation systems. The sedentary way of life facilitated by this hydraulic base has led to land tenure, property inheritance, and social organizations that persisted for centuries. Community leadership patterns had to be strong and effective with increasing size and complexity of irrigation systems (Witfogel 1957).

In Sri Lanka, the small irrigation systems have evolved in two forms referred to as tanks and anicuts, based on the landscape, rainfall and the people's need. There are 14,200 small tanks and 12,950 anicuts operational at present, though the numbers are higher with the irrigation structures abandoned or absorbed into the forest and wildlife sanctuaries. About 80% of the small tanks are organized in form of tank clusters referred to as tank cascades.

3.2 Cascaded Tanks

A tank cascade, according to the original definition made by Madduma Bandara (1985), is a 'connected series of tanks organized within the micro-catchments of the dry zone landscape, storing, conveying and utilizing water from an ephemeral rivulet'. Minor refinements in terminology to the above definition have been proposed by Panabokke (IIMI-SLFO 1994). With the advancement of understanding the perception of the DZ landscape, the tank cascade definition has shown limitations to describe the whole concept of cascade formation on the landscape and specific features evolved subsequently due to the human adaptation and nature's contribution to the system equilibrium. Accordingly, the need of a more comprehensive definition is felt in order to cover all aspects of tank cascade landscape leading to describe its whole 'cascade ecology' (Dharmasena 2017a). Thus, the tank cascade definition can be improved as 'an ecosystem, where land and water resources are origianized within the micro-catchments of the dry zone landscape, providing basic needs to human, floral and faunal communities through water, soil, air and vegetation with human intervention on sustainable basis'.

Tank cascades are found at the elevation range of 100–500 m amsl. The three main cascade zones are found in Sri Lanka (Table 3.2 and Fig. 3.1), which need to be understood and differentiated. The main cascade zones that could be identified in Sri Lanka are north and north-central, north-western and south and south-eastern

Table 3.2 Tank cascade zones in Sri Lanka (Source: DAD 2007)

Cascade zone	River basins	No. of cascades
North and north-central	Malwathu Oya, Mahaweli Ganga, Panna Oya, Pankulam Aru, Kunchikumban Aru, Yan Oya, Ma Oya, Mannal Aru, Per Aru, Kanakarayan Aru, Pali Aru, Paranki Aru, Nay Aru, Modaragan Aru, Kala Oya	617
North-western	Mi Oya, Rathambala Oya, Deduru Oya	255
South and south-eastern	Kirama Oya, Urubokka Oya, Walawe Ganga, Malala Oya, Kirindi Oya, Menik Ganga, Kumbukkan Oya, Karanda Oya, Maduru Oya	177
Total		1049

(Dharmasena 2017b). Tank cascade zoning is still a new aspect of the cascade ecology. The total number of tank cascade systems identified in Sri Lanka is 1162, of which about 90% could be included in these cascade zones, while others are scattered.

All large reservoirs constructed in the ancient Sri Lanka are located in the lower positions of the plains, just below the 150 m contour. The tank cascades are found in the catchment of these large reservoirs. Presence of these cascades slows down the inflow of large reservoirs, minimizing the risk of flood and drought in the river basins. The annual percentage loss of water to the sea is relatively lower in such river basins with intrinsic network of tanks in varying sizes (Malwathu, Deduru Yan, Kala) compared to others without many reservoirs such as Kelani, Kalu, Gin, etc. (Table 3.3).

3.3 Early Studies of Tank Cascades

The setting of the small tank village in its position in the landscape, and the principles of land and water use that had been understood and practiced by the early settlers are best described by Abeyratne (1956) including some of the basic features of traditional DZ agricultural systems. A shift in emphasis from the single small tank to the cascade took place following Madduma Bandara's (1985) approach to the study of catchment ecosystems and village tank cascades in the DZ of Sri Lanka. His approach emphasized the treatment of the total tank cascade rather than the individual tanks within a cascade as the more logical focus for any study of small tank systems.

Somasiri (1979) reported the important aspect of water balance at the Walagambahuwa village tank located in the Mahakanumulla cascade in the DZ of Sri Lanka. This tank has a water spread area of 30 ha at full supply level and a storage capacity of 36 ha.m. Its catchment area is 115 ha and the irrigated command area is 13 ha. The water balance of this small tank was studied for four consecutive seasons where the % catchment runoff during the *Maha* season (main rainy season; October to February) varies widely, with a high value of 25.4% of seasonal rainfall in a very wet *Maha* season compared to 5.2% in an average *Maha* season.

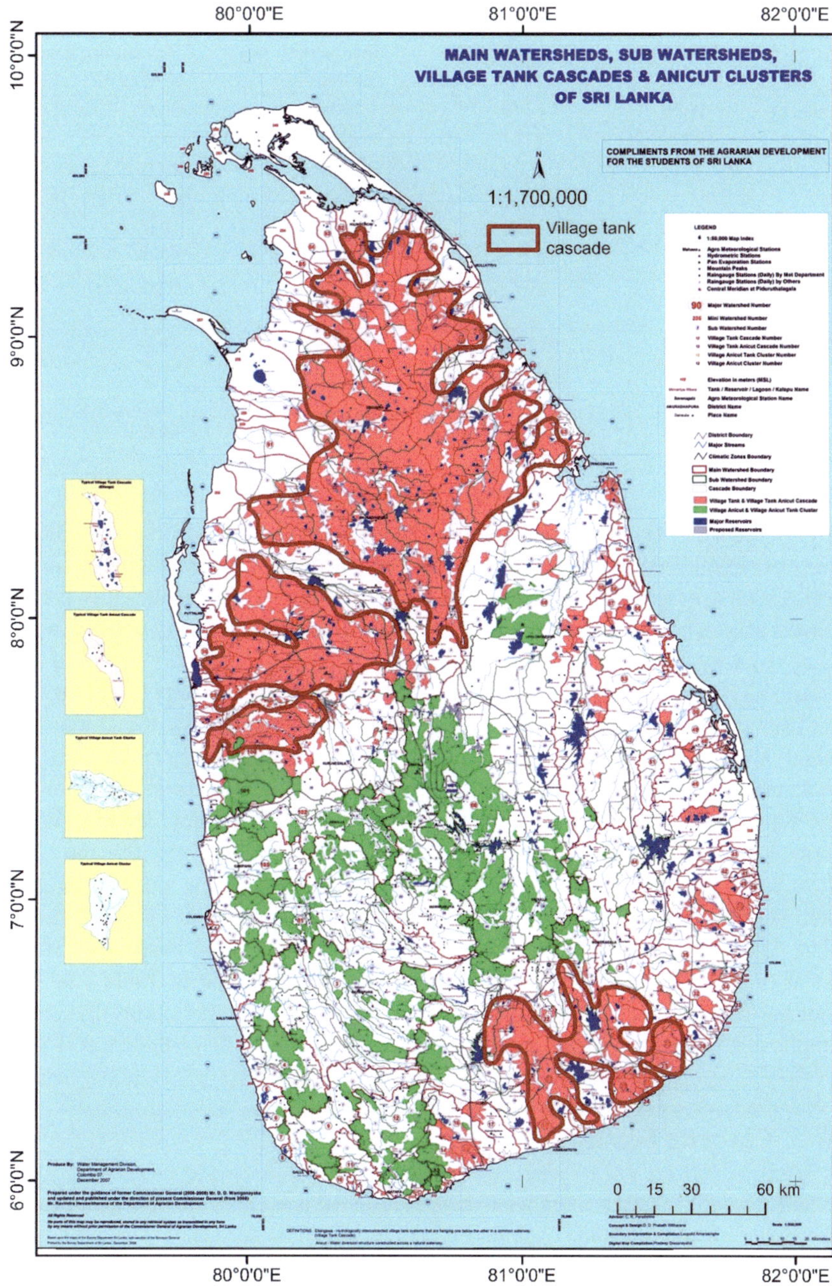

Fig. 3.1 Three main tank cascade zones of Sri Lanka (Source: Dharmasena 2017b)

Table 3.3 Water losses to the sea from some rivers (Source: Survey Department)

River basin	Total basin area (km²)	Total rainfall (million m³)	Drainage to sea (million m³)	Percent drainage to sea (%)
Deduru Oya	2616	4794	1608	34.0
Kala Oya	2772	4424	587	13.0
Malwathu Oya	3246	4592	568	12.0
Yan Oya	1520	2269	300	19.0
Gin Ganga	922	3039	1903	62.0
Kelani Ganga	2278	8692	5474	62.0
Kalu Ganga	2688	10,122	7862	77.0
Walawe Ganga	2442	9843	2165	22.0
Mahaweli Ganga	10,327	26,804	11,016	41.0

Dharmasena (1991) reported that the catchment area of a tank absorbs a significant amount of rainfall for initial soil saturation before it generates any productive or useful runoff. On an average, around 150 mm of rainfall is required during the early part of the *Maha* season before runoff commences. This value is in conformity with the water holding capacity of the reddish-brown earth soils that require around 150 mm of rain to moisten a 1.5 m depth of the soil profile to the field capacity moisture level. The first water balance study on a whole cascade was done by Itakura (1994) in the Thirappane cascade in the Anuradhapura District (DZ) of Sri Lanka located adjoining to Mahakanumulla cascade (Fig. 3.2). The Thirappane cascade is made up of four minor tanks along the main valley and two minor tanks on a side valley. Itakura (1994) reported that for two successive *Maha* seasons, the average runoff was 30% and 12%, while it was 10% and 2.5% for the two successive *Yala* (minor rainy season; March to September) seasons.

The conclusions made by Dharmasena and Goodwill (1999) based on a study done at the Siwalakulama tank cascade systems are presented in Table 3.4. The author has revisited this study after 20 years and provides interpretations that could be applicable to the present perception of the cascaded tank-village system (CTVS).

3.4 Cascade Ecology

The tank cascade systems exist with their unique environment. As defined initially by Madduma Bandara (1985), the cascade determines its surface water movement based on the sizes and positions of tanks and paddy fields. Further, the tank cascade system, once established, determines the various other aspects in addition to the surface water movement. For example, it manipulates the groundwater behaviour, availability and its spatial and temporal variations. Both these two aspects with human intervention determine the floral and faunal composition and dynamics.

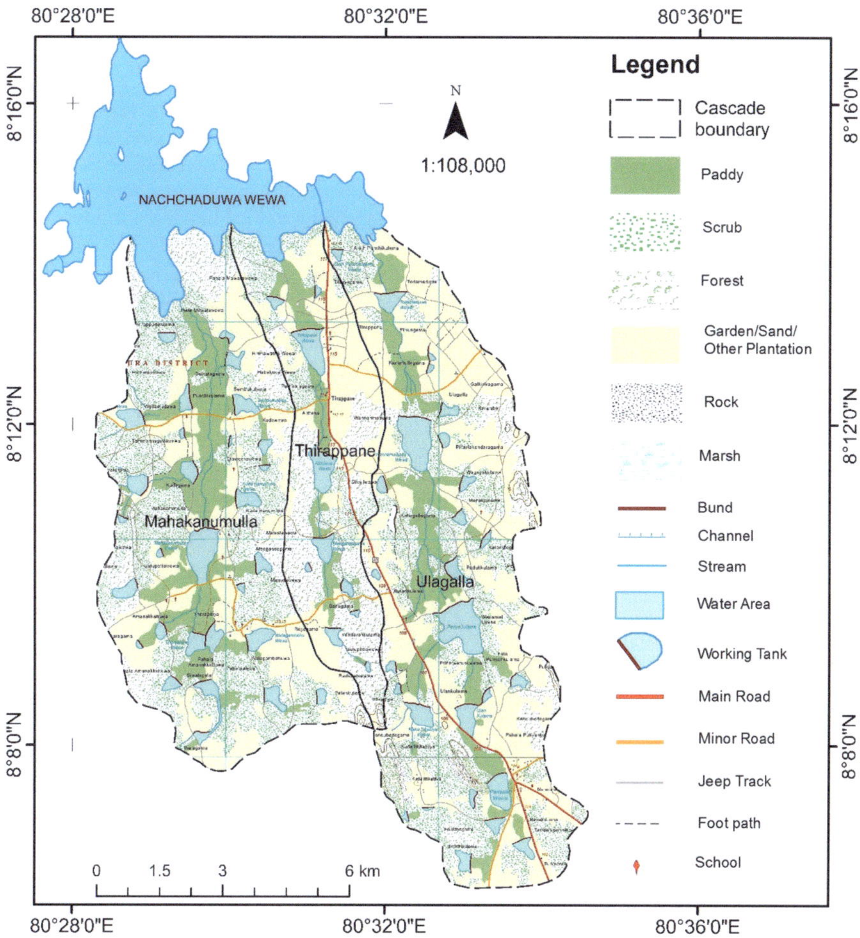

Fig. 3.2 Three tanks cascade systems in Nachchaduwa major watershed (Source: Kadupitiya 2018)

This means that the whole ecology is governed by the determinants of cascade formation especially rainfall, landscape and soil. Tank cascades are found (1) in the Provinces of northern, north-central, north-western, Uva and southern; (2) in the Districts of Mullaitivu, Vavuniya, Anuradhapura, Trincomalee, Puttalam, Kurunegala, Hambantota and Monaragala; (3) at the elevation range of 100–500 m amsl; (4) mainly in the agro-ecological regions of the low-country dry zone, where the isohyets of annual rainfall are between 800 and 2000 mm; (5) in the geologically in Wanni and Vijayan complexes and (6) mainly in reddish-brown earth and low humic gley soil associations. The small tank cascade ecology provides a great opportunity for researchers to explore the latent potential of enhancing the sustainable human health.

Table 3.4 Conclusions and interpretations based on a study done by Dharmasena and Goodwill (1999)

Conclusion	Interpretation
There is a wide variation in monthly rainfall over a small watershed, the extent of which is even below 4000 ha	The upper area of the cascade receives about 140 mm of rainfall more during the *Maha* season. The tanks in the upper landscape can store this excess water and release subsequently to the lower portions enhancing the total water use efficiency of the cascade
The months of October, November and December provide opportunities for replenishment of groundwater reserves and surface reservoirs in the dry zone	This enhances the availability of surface water and groundwater during the dry season ensuring a healthy landscape vegetation
Area with imperfectly and poorly drained soils, where groundwater is abundant occupies in about 40% of watershed	The tank ecosystems, paddy fields and traditional hamlets are occupying in these water regimes maintaining high biodiversity of the tank cascade
Agro-wells must be deepened to pass the shattered rock layer to obtain a satisfactory water yield	The shattered rock layer is the pathway for lateral groundwater movement, which contributes to storage of the lower tanks. This phenomenon ensures the water availability in lower tanks so that farmers in lower areas of the cascade cultivate paddy with confidence in the *Yala* season
Most of the catchments release runoff water at the rate of 5–10% of weekly rainfall	Slow release of rainwater leads to reduce the flood threat of lower areas of the cascade while maintaining a good water balance over the whole system. However, the land use plays a significant role in this water movement
Direct rainfall contribution to tanks is in the range of 25–30% of the annual total inflow	The amount of rainfall directly fallen on the tank is considerable for all tanks in a cascade. This is an immediate influence to increase the storage at the onset of seasonal rains, when most of the rains are absorbed by the catchment
Tank water is lost in a magnitude of 22–33% as evaporation and 27% as seepage and percolation	In a tank cascade system, water loss from an upper tank contributes to the water storage of the lower tank, hence not considered as a loss from the system
About 6% of the total annual rainfall percolate downward to replenish the groundwater reserve	The annual replenishment of the groundwater reserve is adequate to maintain the water table without causing permanent depletion. However, the use of groundwater through agro-wells needs careful planning over a tank cascade system
The aquifer specific yield is 1.5% and permeability is 5 m/day showing the limitation of groundwater use for agricultural purposes in the metamorphic hard rock basement areas in Sri Lanka	High permeability value is due to the presence of shattered layer found immediately above the hard rock. Although the lithology is a compacted complex, this shattered layer allows the water movement along the land catena

(continued)

Table 3.4 (continued)

Conclusion	Interpretation
Wells can be as closer as 100 m distance apart; however, it is the well density which should be emphasized more than the well spacing. Well density should not exceed the rate of 7–8 wells per 100 ha	The well density can be maintained with a firm control if the planning is done at cascade level. This again emphasizes the importance of cascade-level planning
It is safe to use 25% of potential storage of the aquifer for agriculture	About 50% of the total groundwater storage available after the *Maha* season moves laterally to lower areas of the tank cascade. Another 25% is a permanent storage and should not be allowed for extraction. Thus, cascade-level groundwater utilization planning will ensure the system sustainability
Imperfectly drained area, which spreads over 20% of the watershed, is the most suitable area for locating agro-wells	Limiting the area for locating agro-wells to the imperfectly drained land phase can be used as a precautionary measure to avoid occurrence of severe drought over the tank cascade systems

3.5 Issues in Tank Cascade Systems

Various water management practices have been adopted to improve land productivity under small tanks. Some of these practices are traditional while others are more recently introduced. The traditional *Bethma*[1] and *Kakulama*[2] have been in existence with minor irrigation since time immemorial. However, increase in *Akkarawela*[3] due to legal and illegal settlements has disturbed the water balance in small tanks, thus creating deficiencies in water during the *Yala* season even to cultivate a *Bethma*. The deteriorating village cohesiveness and traditional organizations have been attributed as reasons for the failure to implement a *Bethma*. The "*Bethma*" has been emphasized as a result of strong village customs and traditions. More recently, under minor tank rehabilitation programmes, crop diversification has been introduced as a measure of water management. However, in most attempts, this has not been successful due to storage, marketing and labour problems associated with minor tank agriculture. Location of minor tanks and preoccupation in *chena* cultivation have been deterrent factors to adopt more crop diversification.

Evidently there is a serious policy gap with respect to village tank irrigation in Sri Lanka. A national policy on minor irrigation should fill the vacuum created by the loss of ancient traditions and customs. There is a gap between the demand and the real need of the village society. It can only be filled by the bureaucracy, which

[1] *Bethma* is a practice that temporarily redistributes plots of land among shareholders (paddy landowners) in part of the command area (territory) of a tank (reservoir) during drought periods.

[2] *Kekulama* is a dry sowing of seed paddy in the asweddumized fields. Farmers advance the cultivation time in *Kekulama* using early seasonal rains whenever they feel that tanks would not get enough water to cultivate the command area. They have the experience that if September (second inter-monsoonal) rains are high, the total seasonal rainfall is not adequate to fill the tank.

[3] Subsequently developed upper land strip of the irrigable paddy land is known as *Akkarawela*.

has failed in this endeavour, due mainly to lack of reliable and enhanced database on natural resource management. To redress this situation, the Department of Agrarian Development is now in the possession of a database on village irrigation systems. This database consists of 76 main attributes and is capable of linking village irrigation systems and meso-catchment with the help of geographical information system mapping. Hence, it is now believed that the state bureaucracy will be in a better position to meet the gap between the demand and the real need of the village tank communities. However, since small tanks constitute a very important part of the rural landscape and its ecosystem, there is a strong rationale for ensuring the sustainability of these settlements for economic, social and environmental reasons.

It is not feasible now to adopt the same land use, which was in the traditional CTVSs. The challenge is to bring some elementary changes in planning, designing, and managing the CTVSs to sustain the ecosystem services provided. Importantly, we need to understand the underlying principles and mechanisms of the evolution of the CTVSs to treat them as systems. The role of each component needs better understanding, with probable modification of some with modern technologies with planning of entire agriculture and livelihood as a unit (as in the past). This is a must to bring the value of the CTVSs to the level of next generation. It is important to note that only the system will sustain not the components. Hence, collective tank-based sharing system should be strengthened with modern techniques and technologies without damaging them (Geekiyanage and Pushpakumara 2013).

The four main problems that have been identified in the tank cascade systems (IUCN 2015) are as follows:

1. Low cropping intensity:

   In a study carried out for the entire Anuradhapura District using rice cultivation statistics recorded from 1970 to 2003, it was observed that the cropping intensity[4] never exceeded one and fluctuated according to the rainfall received during the *Maha* season (Dharmasena 2005). Despite efforts made to renovate small tanks under various tank rehabilitation projects implemented during this period in Anuradhapura District, there has not been any significant improvement in cropping intensity. This calls into question the strategies currently being used to rehabilitate tanks, as well as their effect on the water storage efficiency of tanks.
2. Tank sedimentation:

   The area cultivable from the water in small tanks decreases gradually due to tank sedimentation and the subsequent high tank water losses. Dharmasena (1992) reported that three small tanks, namely Paindikulama, Siwalagala and Marikaragama, in the Nachchaduwa major watershed have been silted up by 35%, 30% and 23%, respectively, of their initial capacity. Siltation of tanks not only reduces storage capacity, but it also leads to alteration of the tank bed geometry. Subsequent rehabilitation work, where the capacity has been improved by raising the spill and the tank bund, has created shallow water bodies spreading

[4]Cropping intensity is the number of times a crop is planted per year in a given agricultural area.

over a larger surface area. This creates additional problems, including flooding of upstream paddy lands, increased water losses, upper areas becoming more saline and disappearance of the *Gasgommana*,[5] as well as the grass cover (*Perahana*)[6] underneath. Some indigenous fish species, which need deeper water to breed and live, also disappear.

3. High tank water losses:

   Water losses from small tanks are very high. Within 2–3 months of the cessation of the seasonal rains, most of the tanks appear as marshy lands infested with aquatic weeds. Previous studies conducted on hydrology of minor tanks indicated clearly that the total tank water loss through evaporation and percolation varies from 35 to 90% depending on the geometry of the water body (Dharmasena 2005). Water losses are higher from tanks with shallower water bodies, than those with deep water. Therefore, it is clear that the geometry of the tank bed is critical for the water storage efficiency of a tank. It follows that if the tank bed geometry is altered suitably, water loss can be reduced drastically.

4. Low resource productivity:

   The failure of previous projects to increase productivity could be attributed to a lack of focus on restoration of the tank bed and its surrounding ecosystem; confinement of programmes to tank and command area development, without addressing the problems of rainfed and homestead farming in the tank catchment; external interactions and socio-economic conditions; a poor social mobilisation process and the lack of a local institutional mechanism to continue activities, once the project ceases.

3.6 Restoration of Tank Ecosystem for Sustainable Use

In the recent past, efforts to restore tank ecosystems have been made through special projects developed and implemented in Sri Lanka, especially by selecting typical tank cascade systems in the DZ (Dharmasena 2016). Such efforts and technologies adopted would significantly increase paddy production under the new water regime; water level in many of the tanks could be augmented in the cascade system through tank rehabilitation works, and the close involvement of the Farmers' Organizations and the Department of Agrarian Development provide a sense of ownership of the rehabilitation work to these two entities (Mahindapala 2016). Further, Mahindapala (2016) reported that establishment of a Tank Cascade Management Committee could provide an oversight to the maintenance of the cascade system in the future. The government of Sri Lanka is currently planning to form federations of farmers' organizations as private companies through

[5] An area planted with large trees of the same species that acts as a wind break to minimize evaporation from the surface of the tank. It also provides a number of ecosystem services, including the provision of dry season fruits and habitats for wild animals such as nesting birds.

[6] A strip of grass and reeds on the periphery of the water body (dark green) that acts as sieve or filter to trap silt. A good habitat for herons and bitterns, as well as a breeding area for fish.

amendments to the Agrarian Development (Amendment) Act No. 46 of 2011. The proposed federations would take the responsibilities originally expected to be given to the Tank Cascade Management Committee.

A closer engagement of the lines, agencies, namely, the Department of Agriculture, the Forest Department and the Department of Animal Production and Health, would benefit the farming component of the managing tank cascade systems. A good resource base for cascade management is a need for future cascade management work (Mahindapala 2016). Ecological improvements to *Gasgommana* and *Kattakaduwa*,[7] during tank rehabilitation work, are needed for sustainability of the tank system. Mahindapala (2016) further reported that building awareness on the ecological aspects of the components in CTVSs to the farmers would help to introduce new and important facet of cascade management systems towards improved ecological conditions.

References

Abeyratne E (1956) Dry land farming in Ceylon. Trop Agric 112:191–229

Brohier RL (1934) Ancient irrigation works in Ceylon, vol I. Ceylon Government Press, Colombo, p 37

Brohier RL (1937) Ancient irrigation works in Ceylon, vol II. Ceylon Government Press, Colombo, p 43

DAD (2007) Main watersheds, sub-watersheds, village tank cascades and anicut clusters of Sri Lanka. Water Management Division, Department of Agrarian Development, Colombo

De Camp LS (1974) The ancient engineers. Ballany-De Books, New York, p 450

Dharmasena PB (1991) Present use of land and water resources in village tank farming. J Soil Sci Soc Sri Lanka 7:1–17

Dharmasena PB (1992) Magnitude of sedimentation in village tanks. Trop Agric 148:97–110

Dharmasena PB (2005) Small tank heritage and current problems. In: Aheeyar MMM (ed) Proceedings of the symposium on small tank settlements in Sri Lanka. HARTI, Colombo

Dharmasena PB (2010) Evolution of hydraulic societies in the ancient Anuradhapura Kingdom of Sri Lanka. In: Peter MI, Ward C (eds) Landscape and societies: selected cases. Springer, Dordrecht; New York, pp 341–352

Dharmasena PB (2016) Management of tank cascade systems for sustainable use. Paper presented at the World Water Day Workshop of the Department of Agriculture, Sri Lanka, 29 Mar 2016. Plant Genetic Resources Center, Peradeniya

Dharmasena PB (2017a) Overview of small tank cascades: evolution, present status and future scenarios. Presentation made at Policy Dialogue on Restoration and Management of Small Tank Cascade Systems, at Bandaranaike Centre for International Studies (BCIS), Colombo, 14 Feb 2017

Dharmasena PB (2017b) Ecosystem and human health issues. Presentation made at the Strategic Expert Consultation workshop: UNEP-GEF Project on Healthy Landscapes, Plant Genetic Resources Center (PGRC), Gannoruwa, Sri Lanka, 8 Sept 2017

Dharmasena PB, Goodwill IM (1999) Use of groundwater in minor tank irrigation schemes of Sri Lanka. In: Irrigation under conditions of water scarcity, 17th Congress on irrigation and

[7]A thick strip of vegetation located between tank bund and paddy fields. It also has a water hole called *yathuru wala* to retain saline water seeping from the tank. Various plants of salt absorbing features are found on *kattakaduwawa*, which reduce the salinity of the water seeping through the bund before it reaches the paddy fields.

drainage, water for agriculture in the next Millennium, transactions, vol 1D. ICID, Granada, pp 175–194
Ellepola DP (1955) Changing trends in village Ceylon. Proceedings of the Ceylon association for the advancement of sciences (Sec. 2), pp 121–46
Fernando ADN (1979) Major ancient irrigation works of Sri Lanka. J R Asiat Sco (Sri Lanka Branch) 22:1–24
Fernando CH, De Silva SS (1984) Man-made lakes; ancient heritage and modern biological resources. In: Fernando CH (ed) Ecology and biogeography in Sri Lanka. Junk Publishers, Hague, pp 431–451
Geekiyanage N, Pushpakumara DKNG (2013) Ecology of ancient tank cascade systems in island Sri Lanka. J Marine Island Cult 2(2):93–101
IIMI-SLFO (1994) Guidance package for water development of small tank cascade systems, International Irrigation Management Institute-Sri Lanka Field operations. IIM-SLFO, Colombo
Itakura J (1994) Water balance model for reform planning of tank cascade irrigation systems in Sri Lanka. Unpublished report. IIMI
IUCN (2015) Project implementation plan. IUCN programme on Restoring Traditional Cascading Tank Systems Technical Note No. 1. Colombo: IUCN, International Union for Conservation of Nature, Colombo, Sri Lanka & Government of Sri Lanka. ivi + 34 pp.
Jayawardena C (2010) Economic aspects of the ancient irrigation systems of Sri Lanka. Sri Lankan J Manag 15:128–136. Post Graduate Institute of Management, University of Sri Jayawardenapura, Sri Lanka
Kadupitiya HK (2018) Land use planning and governance—national/landscape level integrated spatial planning assessment and analysis report, UNEP/GEF Project on Healthy Landscape, Ministry of Mahaweli Development and Environment
Madduma Bandara CM (1985) Catchment ecosystems and village tank cascades in the dry zone of Sri Lanka. In: Lundqvist J, Lohm U, Falkenmark (eds) Strategies for river basin development. Reidel Publishing Company, Dordrecht, pp 99–113
Mahindapala R (2016) Project evaluation report of ecological restoration of Kapiriggama Tank Cascade System in Rambewa DS Division of Anuradhapura. IUCN, Colombo
Parker H (1981) Ancient Ceylon, 2nd edn. New Delhi, Asian Educational Services, p 695
Perera NP (1984) Natural resources, settlements and land use. In: Fernando CH (ed) Ecology and biogeography in Sri Lanka. W. Junk Publishers, The Hague, pp 453–492
Schnitter NJ (1967) A short history of dam engineering. Water Power 19:142–148
Smith NA (1971) A history of dams. Peter Davies, London, p 279
Somasiri S (1979) Village tank as an agricultural resource in the dry zone of Sri Lanka. Trop Agric 135:33–46
Van de Lippe P (1951) Ceylon restores its ancient irrigation works. Civil Engineering, London, pp 41–44
Witfogel KA (1957) Oriental despotism; a comparative study of total power. Yale University Press, New Haven, CT
Withanachchi CR (2015) Ancient irrigation system of dry and intermediate zones as revealed from ancient canals. Rajarata Univ J 3:4–16

Soil Survey, Classification and Mapping in Sri Lanka: Past, Present and Future

4

Udaya W. A. Vitharana and Ranjith B. Mapa

Abstract

Soil information is indispensable for sustainable use of the soil resource. Since its initiation in 1887, soil survey, classification and mapping activities in Sri Lanka have progressed through three successive stages. The latest soil maps of Sri Lanka developed through Sri Lanka-Canada soil resource project (1999–2004) are available at scales of 1:250,000–1:500,000. These choropleth soil maps delineate mapping units at soil series level, but due to the coarse nature of the mapping scale, majority of mapping units are either associations or complexes. These maps and the databases have primarily served for agricultural planning and management at its best. The state-of-the-art digital soil mapping techniques have been used in Sri Lanka mostly to generate soil information required for small-scale studies indicating a consistent development of expertise on advances in soil mapping. Presently, the need of soil information has expanded beyond its classical uses and has become a prerequisite for addressing environmental issues and food, energy and water security. These requirements are driving new demands for better soil maps at finer resolution. Catering the present soil information demand, the next national soil survey needs to be performed at a scale of 1:25,000 or larger using soil properties-based map legend. The multidisciplinary nature of the usage of soil information requires a collective decision of scientists on the soil properties to be considered for future mapping activities. Ultimately, the development of web-based national spatial soil information systems is a viable option to publish the next-generation digital soil information of Sri Lanka.

U. W. A. Vitharana (✉)
Department of Soil Science, Faculty of Agriculture, University of Peradeniya, Peradeniya, Sri Lanka

R. B. Mapa
University of Peradeniya, Peradeniya, Sri Lanka

B. Marambe et al. (eds.), *Agricultural Research for Sustainable Food Systems in Sri Lanka*, https://doi.org/10.1007/978-981-15-2152-2_4

Keywords
Traditional soil mapping · Digital soil mapping · National soil information · Detailed soil maps

4.1 Introduction

Sri Lanka consists of 64,454 km^2 of land area with a population of 21.7 million as of 2018. Further, about 1156 km^2 covers inland water bodies. When unsuitable terrain and wildlife reservations are added, about 35,000 km^2 of land become not arable. Therefore, the per capita arable land becomes as low as 0.14 ha indicating heavy pressure on land resource. Most of the rural people make a living from land-based activities and the per capita land area will further reduce with land degradation and increase in population. Therefore, there is a need to survey, classify and map the soils of Sri Lanka for sustainable management to ensure food security for future generations.

4.2 Diversity of Soils of Sri Lanka

The soil is a product of nature and it takes about 3000 years to form a 2.5 cm layer of soil under natural conditions. Soil is formed from parent material due to climate, topography, vegetation and organisms acting on the parent material. More than 90% of the rocks in Sri Lanka which weathers to form parent material are made up of pre-Cambrian rocks, while the rest is mainly sedimentary rocks towards the north-western and northern parts of the country. The variations of different rock types from Vijayan to Highland Complex and to Wanni Complex and then to more recent Miocene deposits are shown in Fig. 4.1. These different complexes are made out of different rock types which give rise to different parent materials resulting in different soils, and more details could be found elsewhere (Indraratne 2010; Panabokke 1996).

The climate of Sri Lanka shows a very high variability. Rainfall, the major climatic factor of soil formation, varies from 830 mm in the driest areas to 5750 mm in the wettest areas of the country. The rainfall zones (wet, intermediate and dry zone) with a combination of the elevation zones (low, mid and up country), which are related to variability of ambient temperature, give rise to the seven main agroclimatic zones. By considering many other factors, these zones are further divided to 46 agroecological regions (Punyawardena 2008) showing the ecological diversity. Topography, the next factor in soil formation, contributes to variation of soils in short range. The flat topography in the coastal area (0–2% slope) becomes undulating (2–8%), to rolling (8–16%) and subsequently to hilly (16–32%) when moving to the central highlands of the country.

The differences in the topography with the rise and fall of the water table create well-drained, imperfectly drained and poorly drained areas forming different soils along the centenary landscape with the same agroecological region and similar

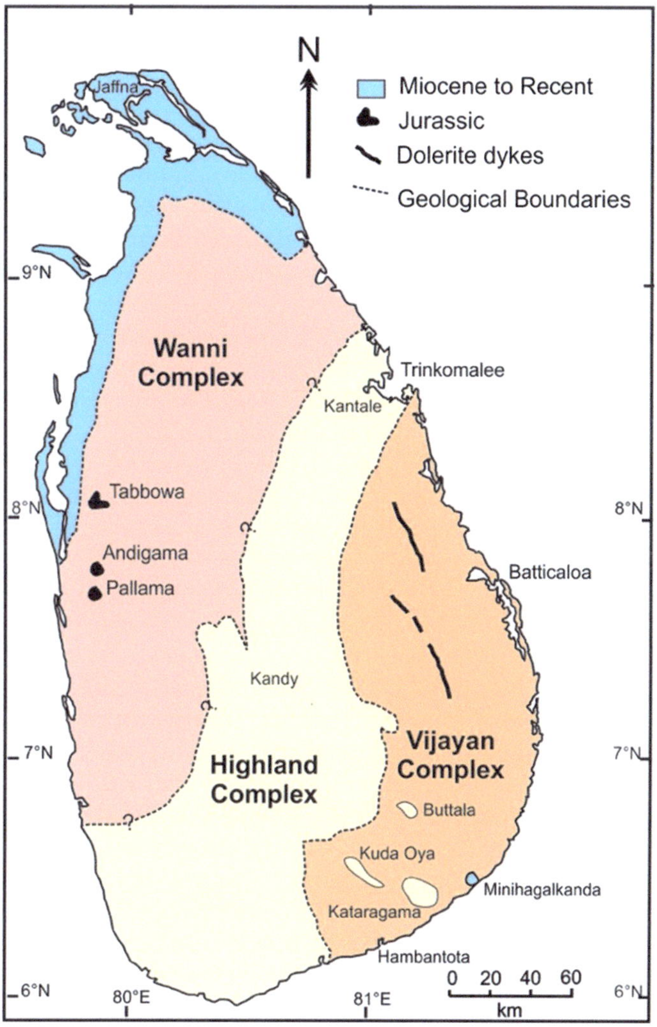

Fig. 4.1 Simplified geological map of Sri Lanka (modified after Cooray 1994)

parent materials. With the variation in soil-forming factors of parent material, climate and topography, Sri Lanka is very rich in soil diversity where more than 102 soil series have been identified by the Soil Science Society of Sri Lanka (Mapa 2014). This soil diversity shows the need for surveying, classifying and mapping the soil resource for sustainable management.

4.3 Soil Survey and Mapping in Sri Lanka Using Traditional Soil Survey Techniques

The need of systematic soil information inventories aroused as the perception and understanding of mankind on the non-renewable nature of soil resource and its crucial role in deciding their existence. The "Great Dust Bowl" of the 1930s in the United States of America was a triggering factor for many nations including Sri Lanka to initiate systematic soil surveys and production of maps to support the sustainable use of soil resource. Soil survey and mapping history of Sri Lanka could be divided into three distinct periods as the Pre-independence Era (before 1948), Post-independence Era (from 1948 to 1972) and the Present Era (from 1972 to today). All these mapping endeavours were relied on traditional mental pedologic models developed by soil mappers (Hudson 1992). These empirical or conceptual models qualitatively relate soil properties measured at a limited number of sites with landscape and environmental variables. The soil variability is presented as by dividing the survey area into polygons called individual delineations of map units. A map unit contains one or more soil types identified according to a chosen soil classification method.

The concept of zonal, azonal and intrazonal soils was the consideration in the early classification of soils of Sri Lanka and gradually shifted to great soil groups-based local classification and then to USDA soil taxonomy and FAO-WRB legend. The accompanying soil survey report with the soil map describes the soil types enclosed in mapping units. Among these mapping attempts, the most comprehensive survey report consisting of the description of 102 soil series with soil horizons-based analytical data was produced at the soil survey conducted during the period 1999–2007 under the SRICANSOL project. Burrough (1993) described two models representing spatial variability of soils, namely, discrete and continuous models (Fig. 4.2). The continuous model resembles the natural spatial variability of soil. The discrete model is a generalization of the continuous model depicting abrupt changes of soil properties at boundaries but uniform soil between boundaries. This model has been used in soil mapping endeavours conducted in Sri Lanka which resulted in area-class or choropleth soil maps.

4.3.1 Pre-independence Era of Soil Survey (Before 1948)

The earliest documented evidence on the soils of Sri Lanka (then Ceylon) could be traced to the publication by Hughes (1887) on "Ceylon Coffee Soils and Manures" during the time coffee was introduced to the country as a plantation crop by the British government. Once the coffee plantations were destroyed by the coffee rust disease, tea was introduced. The nutrient status and physical constituents of tea-grown soils were studied by Bamber (1900). Subsequently, Harbord (1913) studied the coconut-grown soils, and Bruce (1922) studied rice and forest soils in the country. Similar to many other countries, soils of Sri Lanka were then classified

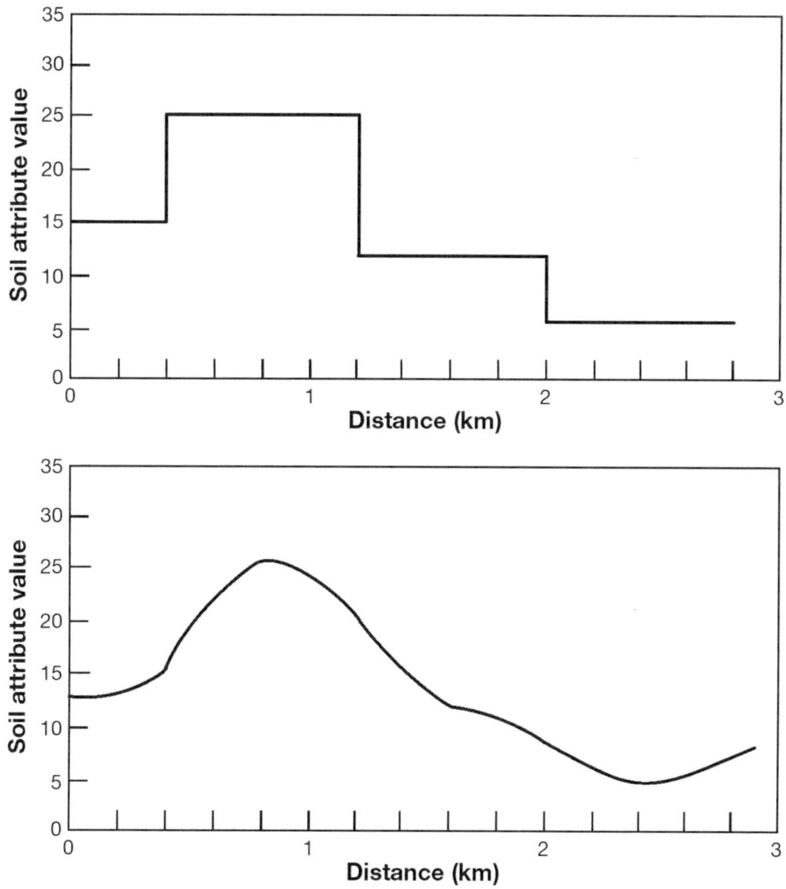

Fig. 4.2 Models for representing the spatial variation of soil properties, (top) discrete model and (bottom) continuous model (after Burrough 1993)

according to the crops grown as cinnamon soils, tea soils, rice soils, etc. Eden (1929) recognized the importance of the soil profile as the unit of study and attempted to classify the tea-grown soils and established the first factorial experiment at the St. Coombs Estate of the Tea Research Institute at Talawakelle.

The most prominent contribution to Soil Survey and Classification during this era was by Dr. A.W.R. Joachim who published 13 articles on "Studies of Ceylon soils" from 1935 to 1940 with his co-workers, S. Kandiah and D.G. Panditesekere (Joachim 1935a, b, 1937, 1955; Joachim and Kandiah 1935a, b, 1937, 1938, 1939, 1940; Joachim and Panditesekere 1935, 1937, 1938). In these, they explained adoption of modern methods for soil studies, classification of soils and their application to Sri Lankan (then Ceylon) soils. They described the typical soil groups of Sri Lanka consisting of red and yellow earths to forest soils of the wet and dry

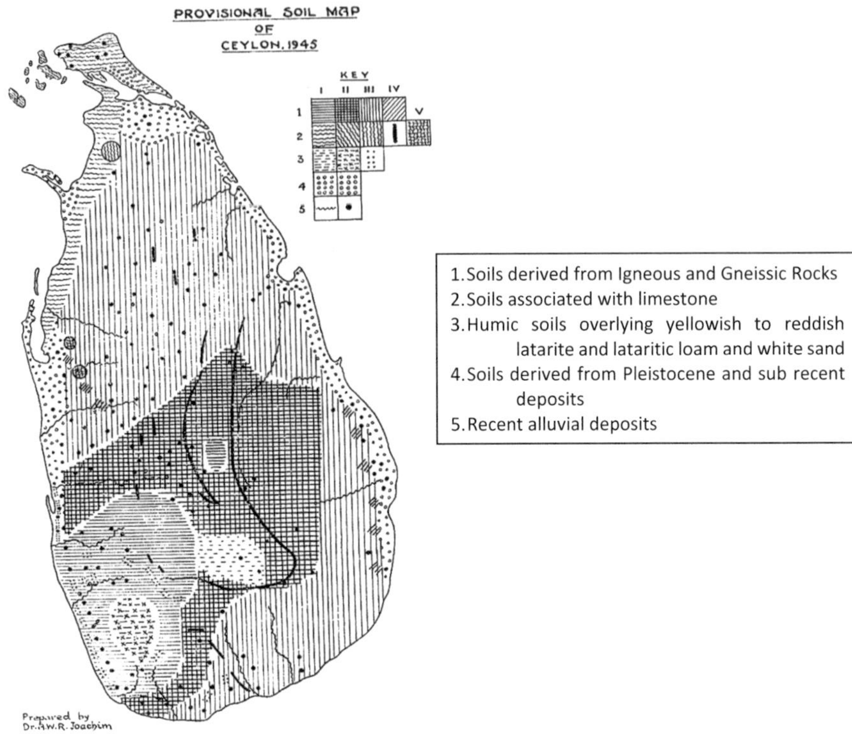

Fig. 4.3 The first provisional soil map of Sri Lanka (adopted from Joachim 1945—refer Table 4.1 for the description of the legend)

zones, wet and dry Patna soils, light sandy soils of the dry zone and soils associated with limestone in Jaffna Peninsula derived from Miocene limestone deposits. The soils of the Matale and Nalanda areas were named as red loams and chocolate red heavy loams overlying crystalline limestone, mainly dolomite.

To collect information related to pedology, they included two important aspects of modern soil concepts: the "unit of soil studies" to be the soil profile and the importance of characterizing the clay complex. Using these information, the first provisional soil map of Sri Lanka was published by Joachim and his co-workers in 1945 (Joachim 1945) which is shown in Fig. 4.3. They considered the soil-forming factors which are the nature of parent rock, climate, vegetation and topography in developing this soil map.

In his scholarly work, Joachim (1945) identified the incompleteness of the zonality-based classification to classify the diverse range of soils of Sri Lanka; thus, he included two additional categories (categories 3 and 4) in addition to zonal (category 1), intrazonal (category 2) and azonal (category 5) soils (Table 4.1).

Table 4.1 Soil classification used by Joachim (1945)

Categories	Units
1. Soils derived from igneous and gneissic rocks	1. Lateritic (cabook) and lateritic reddish yellow loams and gravelly loams of ultra-wet zone 2. Reddish to yellowish red lateritic loams of wet zone 3. Lateritic and non-lateritic red, reddish brown and dark grey loams of dry zone 4. Non-lateritic grey brown sandy loams
2. Soils associated with limestone	1. Red loams derived from Miocene limestone 2. Grey loams derived from Miocene limestone 3. Black heavy loams associated with Miocene and crystalline limestone 4. Brown red loams associated with crystalline limestone (mainly dolomite) 5. Red loams associated with Jurassic limestone
3. Humic soils overlying yellowish to reddish laterite and lateritic loams and white sands	1. Grassland *(Patna)* soils (wet and dry) 2. Fernland *(Kekilla)* soils 3. Low-lying peaty soils (paddy lands, etc.)
4. Soils derived from Pleistocene and sub-recent deposits	1. Red and brown sandy soils (coconut soils) over gravel 2. White sands (cinnamon soils, etc.)
5. Recent alluvial deposits	1. Reddish brown, grey brown and yellowish brown loams and clays 2. Existing paddy (gley) soils

4.3.2 Post-independence Era of Soil Survey (from 1948 to 1972)

During the beginning of this period, the government policy on agriculture gradually shifted to dry zone due to the initiation of irrigation development schemes. A methodological study on soil mineralogy was conducted by Panabokke (1956), which was helpful in future soil classification studies. Rice soils and the topo-sequence of the Hewagama and Hapitigama Korale were studied by Thenabadu and Fernando (1966) and Thenabadu et al. (1966). Subsequently, in collaboration with a Japanese soil scientist, a soil survey was conducted in four selected rice-grown areas, and a classification was proposed according to the differences in hydro-morphism as observed by gleying and mottling of submerged soils (Tokutome et al. 1971).

Much needed aerial photographs for soil survey work became available in 1963–1968 with the Hunting Survey Corporation of Canada, conducting resource surveys of 15 river basins in the north-western part of the country and in Kirindi Oya basins of the south of Sri Lanka. This provided the much-needed landscape and soil information (Hunting Survey Corporation 1963; Hunting Survey Corporation 1968) for future soil surveys. Using these information, and based on morphological characters, Mooman and Panabokke (1961) classified the soils of the country into 14 great soil groups. The great soil group (GSG) was defined as a soil unit with the same arrangement of genetic horizons in the soil profile immaterial of their thickness.

An important landmark during this time was the formation of the Soil Science Society of Sri Lanka (SSSSL) in 1969 whereby soil scientists in government departments such as agriculture and irrigation joined with the scientists from the plantation research institutes and private sector for the development of soil science. The mandate of the society was to promote the advancement of soil science in the country, to foster relationships among soil scientists and to disseminate the knowledge pertaining to soil science by publishing a journal and by other means. From then, most of the soil survey and classification work took a team approach through the leadership of the SSSSL.

Combining the data collected by Land and Water Use Division of the Department of Irrigation for the Mahaweli Project, the soil classification proposed by Moorman and Panabokke (1961) was revised by De Alwis and Panabokke (1972). This was done by extending the 14 great soil groups to 23 and proposing equivalent soil taxonomic names according to Seventh Approximation of USDA soil taxonomy, which was the international soil classification method used at that time (Soil Survey Staff 1960). A revised soil map at 1:500,000 scale with 31 mapping units consisting with combinations of different great soil groups was also produced by them (Fig. 4.4).

The mapping units were categorized to soils occurring in the dry zone and semidry intermediate zone and wet zone and semi-wet intermediate zone and miscellaneous land units. The soil map produced was at reconnaissance level at a scale of 1:500,000, where the mapping units consist of soil associations and complexes.

4.3.3 The Present Era (from 1972 to Date)

At the beginning of the present period, Dimantha (1977) surveyed the soils of the low-lying areas, bog soils, half-bog soils and alluvial soils. Panabokke and Somasiri (1980) classified the rice-grown soils of the wet zone of Sri Lanka from the highest elevation to the lowest elevation into four categories, namely, land system, land subsystem, land complex and land element. The next milestone of soil survey and classification was the initiation of the Sri Lanka-Canada Soil Resource Project (SRICANSOL Project) in 1999, which was a twinning project between the Soil Science Societies of Sri Lanka and Canada to classify the soils of Sri Lanka according to the international methods and map them. The soils of the wet, intermediate and dry zone were studied in detail, classified according to the international systems of soil taxonomy (Soil Survey Staff 2010) and the Food and Agriculture Organization method (FAO 1998, 2014). The soils were also mapped in detail than before where the wet zone was mapped at 1: 1:250,000 (Mapa et al. 1999, Fig. 4.5), the intermediate zone at 1:400,000 scale (Mapa et al. 2005) and the dry zone at 1:500,000 scale (Mapa et al. 2007).

These soil maps developed at soil series level (with map units as consociations, associations and complexes) overcame some of the weaknesses of the previous work. A soil series was defined as a soil having similar genetic horizons immaterial of the depth intervals. A benchmark site represented each soil series with the GPS location, description of the soil profile according to FAO method and colour photos

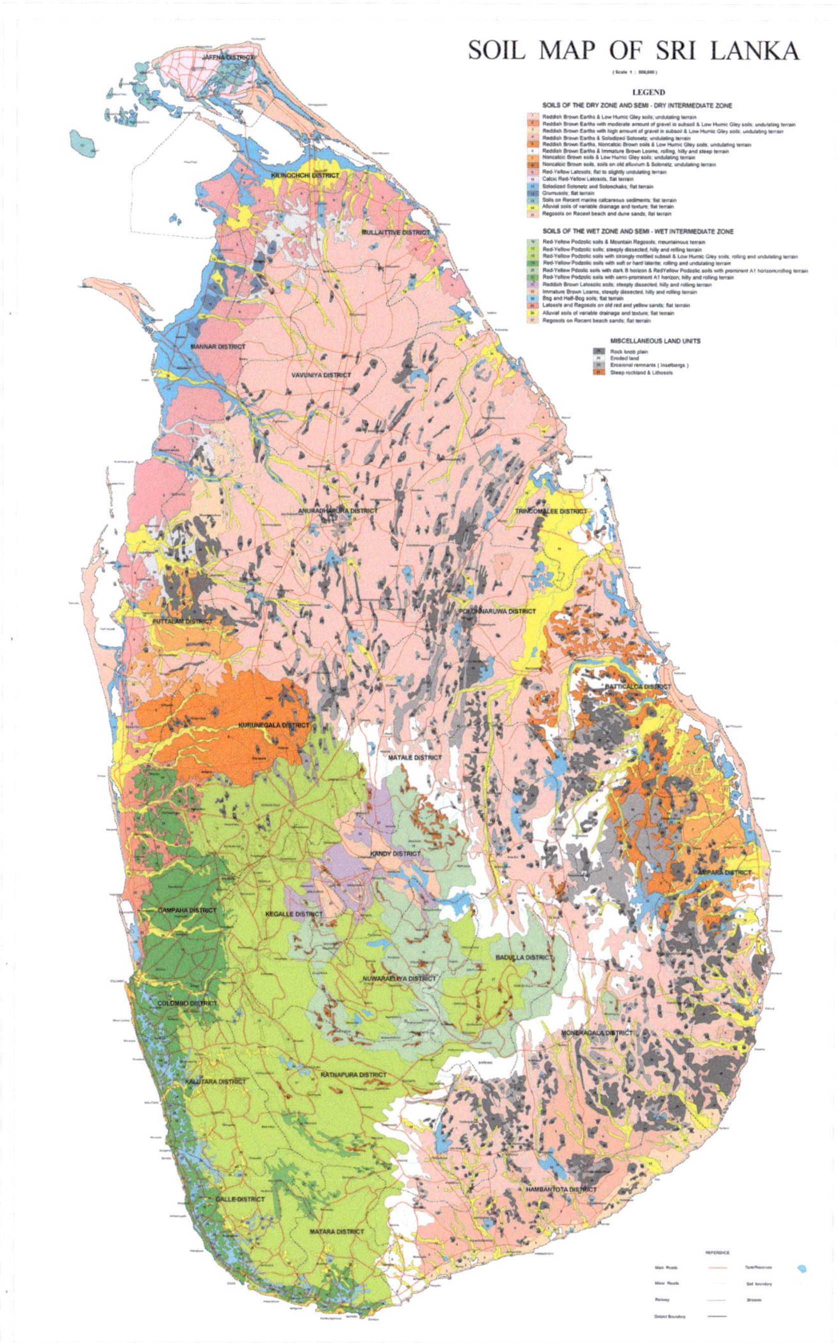

Fig. 4.4 Soil map showing the great soil groups of Sri Lanka (Adopted from De Alwis and Panabokke 1972)

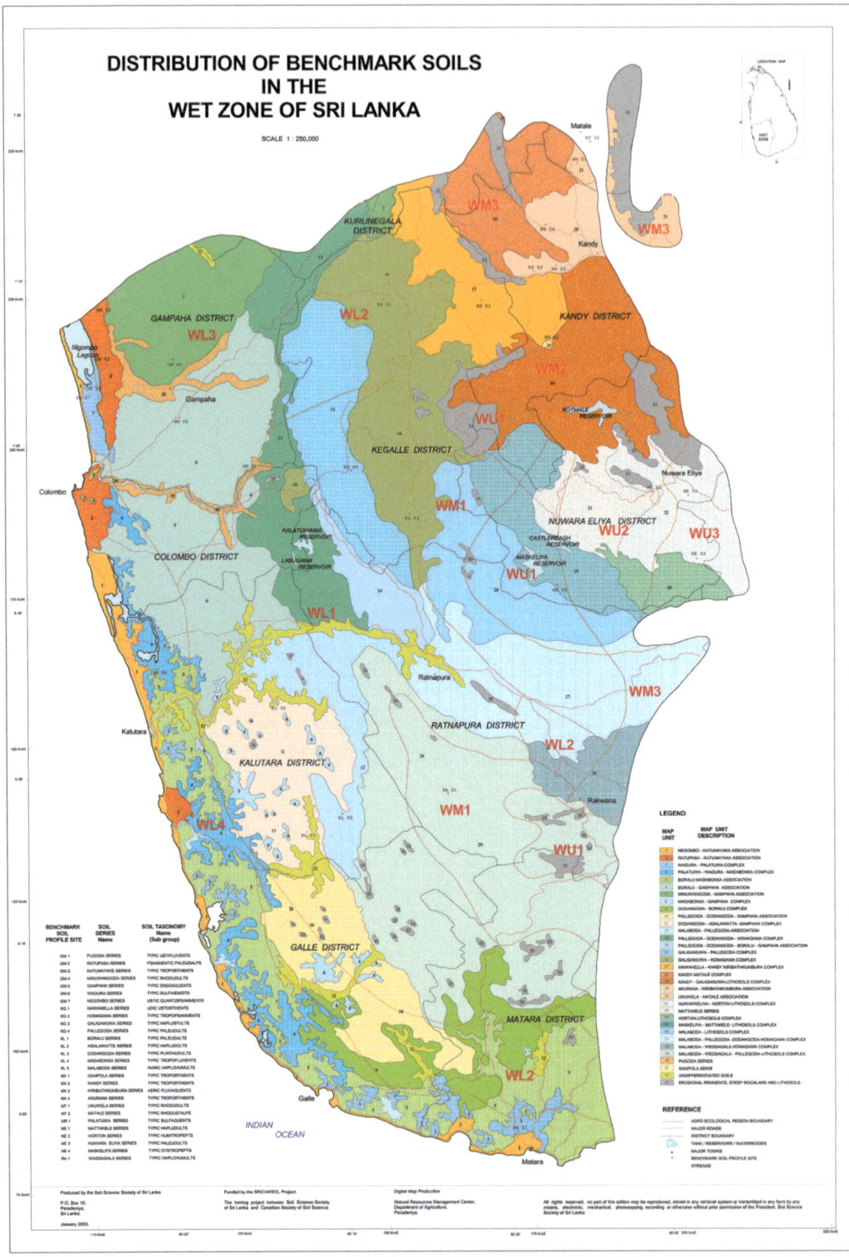

Fig. 4.5 Soil map of the wet zone of Sri Lanka showing the soil series (Adopted from Mapa et al. 1999)

of the soil profile and landscape. In addition, a database comprising of selected soil's physical and chemical properties for each horizon was also made available. The soils also were classified according to the international methods as soil taxonomy (Soil Survey Staff 1998) and FAO method (FAO 1998).

These maps have been used for practical applications as potential risks on groundwater contamination with herbicides (Adikaram and Kumaragamage 2002), for re-demarcation of conservation area in Sri Lanka based on soil erosion potential (Munasinghe and Pushpakumara 2002), to estimate organic carbon stocks in rubber (*Hevea brasiliensis*)-grown soils of Sri Lanka (Dharmakeerthi 2013) and to predict the soil organic carbon in soils of Sri Lanka (Vitharana et al. 2017a).

4.4 Digital Soil Mapping

Generation of detailed soil information for specific purposes such as for precision agriculture, process-based earth systems models, environmental studies and assessing climate feedbacks is a challenge faced by soil scientists (Lagacherie and McBratney 2007). The qualitative nature of the empirical predication models used in the traditional soil survey limits its suitability for detailed soil mapping (McBratney et al. 2003). Nonetheless, such conceptual models are not capable of utilizing geo-computational capabilities in geographic information systems (GIS). In the last two decades, digital soil mapping (DSM) evolved as an alternative for traditional soil survey-based mapping. Advances in information technology, spatial computation methods, proximal and remote sensing technologies and satellite-based navigation techniques have largely contributed to the evolution of digital soil mapping which is defined as "the creation and population of spatial soil information by the use of field and laboratory observational methods, coupled with spatial and non-spatial soil inference system" (Carré et al. 2007; Lagacherie and McBratney 2007). Digital soil mapping techniques are used to predict the spatial variability of soils either in the form of choropleth or continuous soil maps (Fig. 4.2). Geographic information systems play a key role in DSM by providing a platform to store, manage, analyse and display soil spatial information.

Jenny (1941) described the formation, thus the variability, of soils in relation to the "five factors of soil formation". This pedologic model is expressed as follows:

$$S = f(cl,o,r,p,t)$$

where properties of a soil (S) at a location is considered to be a function of climate (cl), organisms (o), relief (r) and parent material (p) interaction over time (t).

Addressing the present-day geo-computational computation options available for spatial data analysis, McBratney et al. (2003) modified Jenny's *clorpt* model to be used for quantitative prediction of soil variability:

$$S = f(s,c,o,r,p,a,n)$$

where soil (S) is a function of soil information (s) and its spatial relationship (n) with climate (c), organisms (o), relief (r), parent material (p) and age (a). This model highlights the need of georeferenced soil information and its relationship with environmental controllers of soil variation to predict the spatial distribution of soils.

4.4.1 Use of Ancillary Information for Digital Soil Mapping

Construction of spatially accurate detailed soil maps requires the collection of a large number of soil samples and subsequent laboratory analysis which are time-consuming and cost prohibitive. However, digital soil mapping makes use of the quantitative information of secondary variables to map the spatial variability of soil properties or classes. This information is referred to as exhaustive ancillary information. The need of lesser number of soil samples allows the creation of detailed soil maps in a cost-effective manner by combining with ancillary information. Different sampling approaches have been suggested for DSM. De Gruijter et al. (2006) recommended model-based sampling approaches such as centred grid sampling, spatial coverage sampling and geostatistical sampling as suitable soil sampling approaches for DSM. The conditioned Latin hypercube sampling (cLHS) is a novel approach developed by Minasny and McBratney (2006) as a method for soil sampling in the presence of ancillary information such as digital elevation models, remotely sensed images and proximally sensed soil information.

The use of cLHS allows choosing sampling locations adequately representing the parameter space of information represented by environmental covariates used for the prediction of soil properties. Remote sensing provides high-resolution information suitable for digital soil mapping. Kumar et al. (2016) combined land use/land cover information derived from AVNIR satellite data (10 m resolution) with other information such as slope, geomorphology and soil types to generate groundwater potential zones in Kilinochchi area. Muthuwatta et al. (2001) used NOAA-AVHRR sensor data to estimate soil moisture in Walawe basin of Sri Lanka at 1 km resolution. Rathnayake et al. (2016) reported that Landsat 8 satellite images can be used to predict soil organic carbon concentration in paddy-grown soils of the northern region of Sri Lanka. Vegetation indices such as the normalized difference vegetation index (NDVI) have widely been used for spatial characterization of soils in relation to the type of vegetation cover. Soil information obtained using invasive and noninvasive proximal soil sensors have proven to be useful for digital mapping of soils (Adamchuk et al. 2004), since these methods can detect the properties of materials below the soil surface. Compared to conventional techniques in the collection of soil information, proximal sensing methods are less expensive, more rapid and provide larger amounts of data for the area covered (Brevik et al. 2015). These methods can be applied in the field using portable devices (Doolittle and Brevik 2014). Soil sensors can be interphased with a global positioning system (GPS) receiver and manually operated to obtain georeferenced measurements on the go. Thus, a large number of observations can be taken with less field sampling effort. The potential of electromagnetic induction-based proximal soil sensing for digital mapping of soils of Sri Lanka has been well explored. Rathnayake et al.

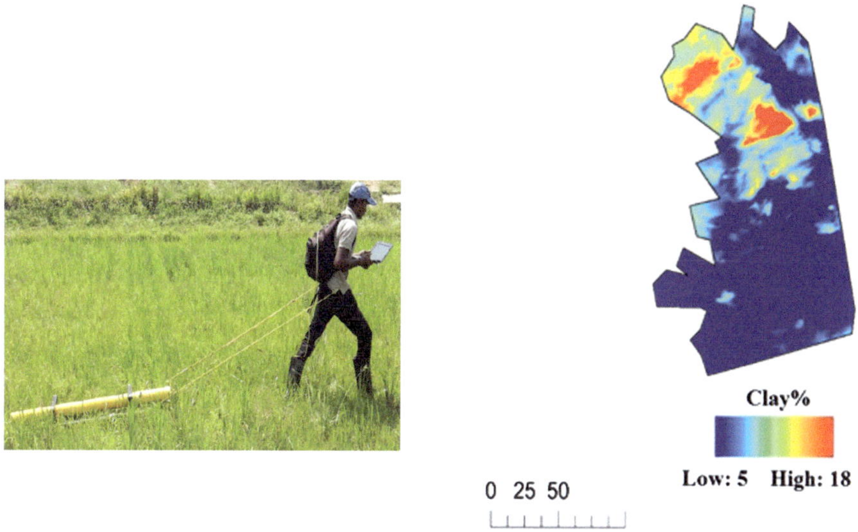

Fig. 4.6 Proximal sensing of paddy-grown soil using the DUALEM-1S sensor (left) and detailed clay content map constructed using ECa data (right)

(2017) used the DUALEM-1S sensor for detailed mapping (1 m resolution) of soil texture in a 2.5 ha paddy land (Fig. 4.6) belonging to Batalagoda soil series (Typic Endoaqualfs).

Hiranthika et al. (2015) elucidated sodic and saline soils using proximal soil sensing, and Perera et al. (2015) mapped salt-affected paddy-grown soils in Anuradhapura area. Further, Vitharana et al. (2013) reported differential behaviour of proximal soil sensing on red yellow podzolic soils (Ultisols) and reddish brown earth soils (Alfisols) and stated EMI-based proximal sensing is highly suitable to map soil texture in Alfisols of Sri Lanka. Vitharana et al. (2014) successfully elucidate soil texture and depth to the gravel layer using the DUALEM-1S sensor measurements. Other proximal soil sensors such as visible and near-infrared (Vis–NIR) spectroscopy, ground-penetrating radar (GPR), soil pH sensing and gamma-radiometric (γ) soil sensing have been successfully used in other countries to map soil physicochemical properties (Table 4.2). However, the potentials of these latest proximal soil sensors to characterize the soils of Sri Lanka are yet to be explored.

The availability of digital elevation data has opened up new possibilities for incorporating quantitative topographic information for supporting soil map preparation. Topographic attributes derived from square-grid digital elevation models (DEMs) have often been used as ancillary information for the spatial prediction of soil properties. Sources of elevation data for DEM creation can be from field measurement of elevations using theodolites, GPS receivers and digitization of contour lines on paper maps or data collected using airborne altimetry. Digitized contour data at 1:10,000 scale available with the Survey Department of Sri Lanka are excellent source of information for the development of DEMs.

Table 4.2 Available proximal soil sensing techniques for prediction of different soil properties (modified from Whelan and Taylor 2013)

Soil properties	Proximal soil sensing technique
Soil nutrients—N, P, K, S and trace elements	Visible (VIS)/near-infrared (NIR)/short-wavelength infrared (SWIR)/mid-infrared MIR spectroscopy Ion-selective electrodes Ion-selective field-effect transistor (ISFET)
Soil pH	Ion-selective electrodes ISFET VIS
Organic matter	VIS/NIR/SWIR/MIR spectroscopy
Soil sodicity	Electromagnetic induction (EMI) Resistivity
Soil salinity	EMI Resistivity Ground-penetrating radar
Soil texture/soil type	Gamma-radiometrics EMI Resistivity Visible/NIR/MIR Spectroscopy Ground-penetrating radar Tillage draft
Soil water storage capacity (PAWC)	EMI Resistivity VIS /NIR/MIR/thermal IR spectroscopy Radar
Rooting depth	EMI Resistivity Ground-penetrating radar

Moreover, scientists have the free access to global digital elevation data at a spatial resolution of 30 m through the NASA Shuttle Radar Topography Mission (SRTM) database (Farr et al. 2007). Primary topographic attributes such as slope and specific catchment area and secondary topographic attributes such as wetness index and stream power index derived from a DEM (Wilson and Gallent 2000) can be used as ancillary information for predicting soil properties (Vitharana et al. 2008). Jayakody et al. (2015) explored topographic-soil relationships to support spatial prediction of soil properties in a red-yellow podzolic soil scape in the wet zone of Sri Lanka. Vitharana et al. (2017a, b) showed the potential of the SRTM-based DEM as a covariate for digital mapping of soil K concentration in Sri Lanka (Fig. 4.7).

The development of national-level inventories of soil organic carbon stocks and monitoring are imperative for assuring the sustainability of the soil resource and minimize negative feedbacks on climate change. Vitharana et al. (2019) found topography of Sri Lanka modelled using the SRTM as a strong environmental controller of the spatial distribution of soil organic carbon stocks and used primary and secondary topographic attributes to produce detailed maps at a spatial resolution

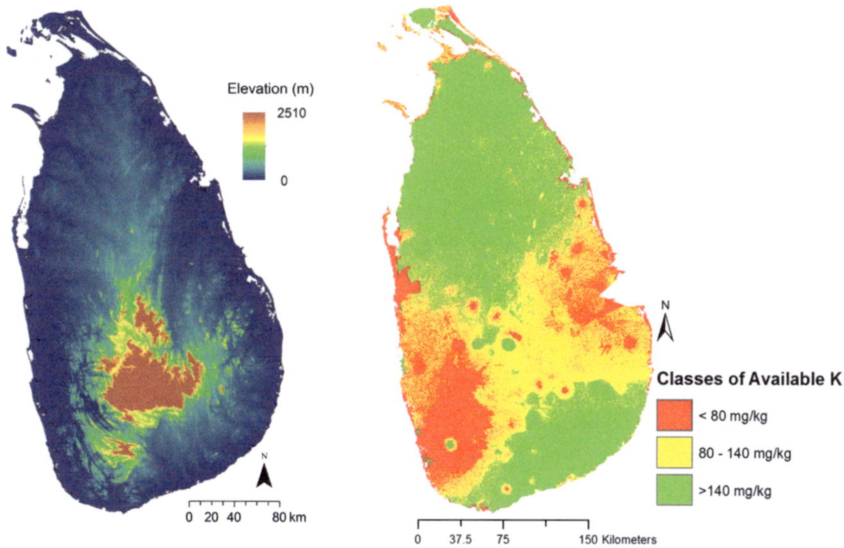

Fig. 4.7 SRTM data-derived DEM (left) and a digital maps of availability classes K (right) in Sri Lanka

of 30 m. They showed that information derived from legacy soil maps of Sri Lanka can improve soil organic carbon prediction models. Digital elevation models allow automated delineation of (sub) catchment areas. This intern is very useful to model soil formation process within a catchment scale and produce detailed soil maps. Vitharana et al. (2008) used digital elevation model to elucidate the soil variability, considering erosion and sedimentation processes at sub-catchment scale in a temperate soil-scape.

Such studies are yet to be performed under local conditions. Dobos et al. (2010) highlighted the potentials of original soil data collected in national soil survey and soil fertility assessment campaigns for digital soil mapping. These legacy data have been used in many countries for soil mapping. Vitharana (2017) fitted mass-preserving depth splines (Bishop et al. 1999) to soil profile data (n = 122) to extract soil organic carbon and bulk density data for soil depths 0–30 cm and 30–100 cm of soils of Sri Lanka. Digital maps of surface and subsurface soil carbon and K stocks (Fig. 4.7) of Sri Lanka were derived from these data by applying digital soil mapping techniques.

4.4.2 Use of Pedometric Models for Soil Mapping

Quantitative spatial prediction models play an essential role in digital soil mapping. In situations where ancillary information is available, these models need to be capable of combining sparsely measured soil properties with exhaustive ancillary information to produce digital soil maps. Pedometrics (McBratney et al. 2003), which is defined as "the application of mathematical and statistical methods for the quantitative modelling of soils, with the purpose of analysing its distribution, properties and behaviors", encompasses these spatial prediction models. Hiranthika et al. (2015) used multiple regression to map soil salinity of a paddy-growing soil, and Wijesinghe et al. (2018) applied the classical statistical method, Fuzzy *k*-means classification, to refine the soil boundaries of a choropleth soil map (1:10,000) of an agricultural experimental station in Sri Lanka. Geostatistical methods are developed on the basis of the theory of regionalized variables. These methods quantify the spatial structure of variables by calculating the experimental variogram and subsequent fitting of a variogram model. The variogram explains the nature, the scale and the magnitude of the spatial structure. Then, the variogram is used for interpolation (known as kriging) of soil information (Webster and Oliver 2007). Kriging provides the best unbiased linear estimate (BLUE) of a soil attribute. Rosemary et al. (2017) reported the presence of strong spatial structure of soil pH, EC, organic carbon and CEC within a soil catena in Sri Lanka. Further, they used kriging to produce detailed digital maps (spatial resolution = 1 m) of these properties within the catena (Fig. 4.8).

Rathnayake et al. (2015) studied the variability of soil chemical properties of rainfed lowland paddy fields using inverse distance to power interpolation technique, and they concluded that the variability of soil available P, K, pH and EC is structured enough to guide site-specific application of soil management practices. On the basis of maps of available P developed for the Polonnaruwa District, Rathnayake et al. (2015) reported that considerable reduction of usage of inorganic P fertilizers can be achieved by utilizing these maps.

Hybrid methods combine classical statistics and geostatistical methods in order to use ancillary information for soil property estimation (Webster and Oliver 2007). Cokriging considers the joint spatial variability (co-regionalization) between ancillary information and soil properties to estimate the latter (Webster and Oliver 2007).

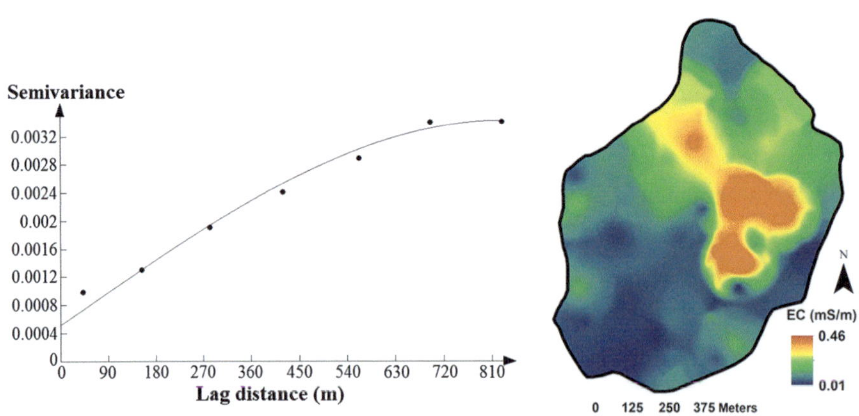

Fig. 4.8 Variogram showing highly structured variability of soil EC in a dry zone catena (left) and a digital soil map of soil EC (dS m^{-1}) within catena (right, source Rosemary et al. (2017))

Rathnayake et al. (2017) used cokriging to predict soil texture in a rice-grown soil using proximally sensed ECa as an ancillary information. They found a strong co-regionalization between soil textural fractions and ECa. Regression kriging combines linear regression models with ordinary or simple kriging of residuals (Hengle et al. 2007). Rathnayake et al. (2017) found greater accuracy of predicting the soil texture of paddy-grown soil by regression compared to cokriging. Vitharana et al. (2019) found geographically weighted regression kriging approach as a suitable method for digital mapping of soil organic carbon stocks across the island.

The recent studies on soil mapping in Sri Lanka reveal a gradual shifting towards nationwide digital soil mapping. It can be expected that comprehensive detailed digital soil information system should be developed in the future, assuring the sustainable use of soil resource amid the constraints of limited availability of lands, higher food demand due to the ever-increasing population growth and negative feedbacks on the national food production due to extreme climatic events.

4.5 Future Directions of Soil Survey and Mapping in Sri Lanka

The need of soil information has expanded beyond its classical uses, and presently soil information is vital for addressing environmental issues, food security, energy security, water security and human health (Brevik et al. 2015). According to Mermut and Eswaran (2000), this has led to the emergence of new areas of interest such as land and soil quality, recognition of problems of land degradation and desertification, cycling of biogeochemical and soil pollution assessment and monitoring. These requirements are driving new demands for better soil maps at finer resolution. Soil survey and mapping priorities of any country need to be focused on the present and future needs of soil information. Sri Lanka went through a successful era of soil survey, classification and mapping; these legacy soil maps have primarily served for agricultural purposes at its best. Lack of details resulted by the coarse cartographic scale and qualitative nature of the spatial model used limits the use of these legacy soil maps of Sri Lanka for many applications at present. According to classification proposed by Rossiter (2000), soil maps available in Sri Lanka belong to a very low intensity level (Table 4.3), thus less appropriate for moderately intensive uses such as farm-level planning and district-level planning and for intensive uses such as site planning, precision agriculture and urban land use planning. Therefore, there is a need of a soil survey and mapping in Sri Lanka at a more detailed scale. Improvements of technology pertaining to digital soil mapping have allowed making better soil maps. Catering the present soil information demand, the next national soil survey needs to be performed at a scale of 1:25,000 or larger.

The classification-based mapping legend that is presently used is merely useful for the present-day quantitative soil information needs. Thus, soil properties-based map legend will be more appropriate. The multidisciplinary nature of the usage of soil information requires a collective decision of scientists on the soil properties to be considered for future mapping activities. Further, with the progress of digital soil mapping, the soil survey and mapping have been shifted from a solitary activity carried out by small soil survey teams with similar level of skills to an endeavour

Table 4.3 Classification of soil survey types according to cartographic scale and sampling intensity (modified from Rossiter 2000)

Intensity level	NRCS 'Order'	Observation density	Scale of publication scale	Minimum legible delineation	Purposes
1. Very high (intensive)	1st	>4 per ha	1:2500	0.025 ha	Site specific soil management, detailed engineering, precision agriculture
2. High (intensive)	1st	1 per 0.8 ha to 4 ha	1:10 000	0.4 ha	Experimental site planning, intensive uses, urban land use planning
3. Moderately high (detailed)	2nd	1 per 5 ha to 25 ha	1:25 000	2.5 ha	General agriculture, detailed project planning, urban planning
4. Medium (semi-detailed)	3rd	1 per 20 ha to 100 ha	1:50 000	10 ha	Moderately intensive uses at farm level, community planning, district level planning
5. Low (semi-detailed)	4th	1 per 100 ha to 400 ha	1:100 000	40 ha	Extensive land uses, project feasibility, regional land inventory, district-level planning
7. Very low (reconnaissance)	5th	<1 per 100 ha	1:250 000	250 ha	National land inventory, regional planning, very extensive land use
Exploratory		Opportunistic	1:1 000 000 1:5 000 000	4000 ha 100,000 ha	General information on unmapped areas

requiring a number of professionals with different skills, including soil surveyors, pedomatricians, GIS specialists, computer programmers and remote and proximal soil sensing experts (Lagacherie and McBratney 2007). Thus, there is a need of establishing a consortium of such relevant professionals before undertaking a national-level soil mapping project in the future.

The limitations of available arable land resources exert heavy pressure on existing agricultural lands to increase the productivity at a minimum cost to the environment. Adoption of advances of precision agriculture technology is one of the viable options to achieve present agricultural production targets while minimizing the overutilization of finite natural resources and the detrimental environmental impacts of associated agrochemical pollutants. Precision agriculture (farming) is defined in the literature in different forms, and according to Pierce and Nowak (1999), the application of technologies and principles to manage spatial and temporal variability associated with all aspects of agricultural production for the purpose of improving crop performance and environmental quality. Site-specific crop management is a form of precision agriculture whereby decisions on resource application and agronomic practices are improved to better match soil and crop requirements as they vary in the field (Whelan and Taylor 2013; Vitharana 2008). Site-specific application of fertilizers, irrigation and soil improvement is relying on detailed and quantitative maps (Table 4.3) of soil attributes. Though isolated attempts have revealed (e.g. Rathnayake et al. 2017) the presence of highly structured short-scale spatial variability of soils of Sri Lanka that suit for site-specific soil management, extensive mapping of key soil properties required for site-specific soil management (Vitharana et al. 2008) needs to be conducted covering soils used for agricultural production. These maps could be used to delineate management zones to practice site-specific nutrient management, tillage practices and irrigation for major crops including rice. Presently, researchers from the Department of Soil Science of the Faculty of Agriculture, University of Peradeniya, Sri Lanka are in the process of calibrating a decision support tool for supporting site-specific management of nutrients for paddy, which requires detailed information of soil properties for site-specific fertilizer recommendations.

Global initiatives such as 4 per mille (Lal 2016; Minasny et al. 2017) and Land Degradation Neutrality (LDN) in 2030 (United Nations General Assembly 2015) targeted to combat climate change, and land degradation impacts are relying on accurate and detailed inventories of soil organic carbon at national level. Detailed spatial inventories of forest soil organic carbon stocks and establishing observational networks (Vitharana et al., 2017b) for monitoring are prerequisites for C credit initiatives. Vitharana et al. (2019) attempted to generate a map of soil organic carbon stocks using organic carbon data extracted from 122 soil profiles from SRICANSOL database and the recently inventoried database of benchmark soil profiles (Mapa 2016), representing the soils of the northern region of Sri Lanka. The use of advanced digital soil mapping techniques allowed them to generate an acceptable estimate of soil organic carbon stocks in surface (0–30 cm) and subsurface (30–100 cm). However, a considerable uncertainty of estimates has been reported which was resulted due to the limitation of soil observations, and it has been recommended for the nationwide digital soil mapping with larger soil observation network comprised of 1552 samples.

Soil acts as a source and a sink for many pollutants including trace elements (Helmke and Losco 2013). Interactions between soils and groundwater contamination and consequent environmental and health issues became a major concern of the latter stage of the twentieth century (Hartemink 2002). Thus, maps of baseline concentrations of soil pollutants are a requirement to assess the present status and establish monitoring programs to sustain the environmental integrity. Sanjeevani et al. (2015) attempted to establish baseline concentrations of some trace elements in a soil map unit of the dry zone of Sri Lanka. However, extension of such studies covering the entire nation with detailed digital maps is of urgent necessity.

Land degradation has become a severe problem of Sri Lanka (Nayakekorala 1998). Detailed assessment of soil erosion hazard is one of the major focus to avoid further degradation of the soil resource. Though many attempts have been undertaken to map soil erosion hazard, lack of information on soil erodibility causes greater uncertainty in hazard predictions. Therefore, generation of soil information, such as soil texture, structure and erodibility at a finer scale, would support accurate assessments. Further, the development of suitable soil attribute indices to recognize marginal lands and mapping of such lands are important to undertake proper rehabilitation activities.

Though numerous information acquisition methods, sources of secondary information and pedometric techniques representing quantitative pedologic models are presently available for soil mapping (Minasny and McBratney 2016), the requirement of sufficient soil sampling and analysis remains. This is often seen as a constraint in developing detailed soil maps due to cost and time involvement. Therefore, national-level policy decisions are required for allocating adequate funding for prospective soil survey and mapping projects. Further, the legacy soil information from existing soil maps and profile databases (e.g. SRICANSOL database) needs to be utilized for digital soil mapping. Soil map predicts the geographical distribution of soil properties or soil classes and involves an uncertainty. Present uses of soil information such as for process-based models require quantitative information on the uncertainty of soil maps. Therefore, future endeavours of soil mapping should include uncertainty assessment of survey products. Brevik et al. (2015) highlighted that soil information has to be more easily comprehended to raise awareness about the importance of soil for sustainable land use and be focused on important environmental problems. Advances of information technology can be used to present soil maps and information that can be accessible to all. The development of web-based spatial soil information system (SSINFOS) (Lagacherie and McBratney 2007) is one of the viable options to present digital soil information of Sri Lanka in the future.

4.6 Conclusion

The history of soil survey, classification and mapping in Sri Lanka exemplifies the gradual advancement of inventorying soil information for agricultural purposes. These maps and the databases have primarily served for agricultural planning and management at its best. However, the need of soil information has expanded beyond its classical agriculture-based uses and has become a prerequisite for addressing environmental issues, food security, energy security, water security and human health. These requirements have driven new demands for better national soil maps at a finer resolution. Addressing these demand, the next national soil survey needs to be performed at a scale of 1:25,000 or larger using soil properties-based map legend. The multidisciplinary nature of the usage of soil information requires a collective decision of scientists on the soil properties to be considered for future mapping activities. Ultimately, the development of web-based national spatial soil information systems is a viable option to publish the next-generation digital soil information of Sri Lanka.

References

Adamchuk VI, Hummel JW, Morgan MT, Upadhyaya SK (2004) On-the-go soil sensors for precision agriculture. Comput Electron Agric 44(1):71–91

Adikaram AMTMK, Kumaragamage D (2002) Potential risks on groundwater contamination with herbicides in the wet zone of Sri Lanka. Trop Agric Res 14:233–244

Bamber MK (1900) Report on tea soils. Tea Research Institute of Sri Lanka, Thalawakele

Bishop TFA, McBratney AB, Laslett GM (1999) Modelling soil attribute depth functions with equal-area quadratic smoothing splines. Geoderma 91(1–2):27–45

Brevik EC, Calzolari C, Miller BA, Pereira P, Kabala C, Baumgarten A et al (2015) Soil mapping, classification, and pedologic modeling: history and future directions. Geoderma 264:256–274

Bruce A (1922) A contribution to the study of the paddy soils of Ceylon and Eastern countries. Ceylon Depart Agric Bull 57:54

Burrough PA (1993) Soil variability: a late 20th century view. Soils Fertil 56:529–562

Carré F, McBratney AB, Mayr T, Montanarella L (2007) Digital soil assessments: beyond DSM. Geoderma 142(1–2):69–79

Cooray P (1994) The Precambrian of Sri Lanka: a historical review. Precambrian Res 66(1):3–18

De Alwis KAN, Panabokke CR (1972) Handbook of the soils of (Ceylon) Sri Lanka. J Soil Sci Soc Ceylon 2:1–97

De Gruijter J, Brus DJ, Bierkens MFP, Knotters M (2006) Sampling for natural resource monitoring. Springer, Berlin

Dharmakeerthi RS (2013) Organic carbon stocks in rubber (*Hevea brasiliensis*) growing soils of Sri Lanka and strategies to increase: a review. J Rubber Res Inst Sri Lanka 93:16–36

Dimantha S (1977) Soils of the low lying areas of west and southwest of Sri Lanka, their properties and management. Trop Agric 133:13–28

Dobos E, Bialkó T, Micheli E, Kobza J (2010) Legacy soil data harmonization and database development. In: Boettinger JL, Howell DW, Moore AC, Hartemink AE, Kienast-Brown S (eds) Digital soil mapping. Progress in soil science, vol 2. Springer, Dordrecht

Doolittle J, Brevik EC (2014) The use of electromagnetic induction techniques in soils studies. Geoderma 223–225:33–45

Eden T (1929) Soil acidity and base exchange. Tea Quar (Ceylon) 3:100

FAO (1998) World reference base for soil resources, by ISSS-ISRIC-FAO. World soil resource report, no. 84. Food and Agriculture Organization of the United Nations, Rome

FAO (2014) World reference base for soil resources. In: International soil classification system for naming soils and creating legends for soil maps. Food and Agriculture Organization of the United Nations, Rome

Farr T, Rosen P, Caro E, Crippen R, Duren R, Hensley S et al (2007) The shuttle radar topography mission. Rev Geophys 45:1–33

Harbord G (1913) Coconuts: experiments at Maha-Illuppallama, Bullatin No. 7, Department of Agriculture, Ceylon. H.C. Cottle, Government Printer, Ceylon. pp 95–103

Hartemink AE (2002) Soil science in tropical and temperate regions—some differences and similarities. Adv Agron 77:269–292

Helmke MF, Losco RL (2013) Soil's influence on water quality and human health. In: Brevik EC, Burgess LC (eds) Soils and human health. CRC Press, Boca Raton, pp 155–176

Hengl T, Heuvelink GBM, Rossiter DG (2007) About regression-kriging: from equations to case studies. Comput Geosci 33:1301–1315

Hiranthika HGK, Vitharana UWA, Perera RACJ (2015) Identification of salinity hazard zones using DUALEM-1S proximal soil sensor. Proceedings of the Peradeniya University International Research Sessions, Sri Lanka, 5th–6th November 2015, p 18

Hudson BD (1992) The soil survey as paradigm-based science. Soil Sci Soc Am J 56(3):836–841

Hughes, T. (1887). Ceylon coffee soils and manures

Hunting Survey Corporation (1963) Report of the survey of Kelani Aruvi Aru area of Ceylon. Hunting survey Cooperation Ltd. Toronto, Canada. Canada Ceylon Colombo Plan project, Colombo: Government Press

Hunting Survey Corporation (1968) Report of the survey of Walawe basin of Ceylon. Hunting Survey Cooperation Ltd. Toronto, Canada. Canada Ceylon Colombo Plan project, Colombo: Government Press

Indraratne SP (2010) Mineralogy of dry zone soils. In: Mapa RB, Somasiri S, Dassanayake AR (eds) Soils of the dry zone of Sri Lanka. Survodaya Press, Colombo, pp 59–79

Jayakody KP, Vitharana UWA, Gunatilake J (2015) Investigation of topographic-soil relationships to support spatial prediction of soil properties. In: PGIS research congress. Postgraduate Institute of Science, Peradeniya, p 13

Jenny H (1941) Factors of soil formation. McGraw-Hill, New York

Joachim AWR (1935a) Studies on Ceylon soils I—modem methods of soil study and classification and their application to Ceylon soils. Trop Agric 84:196–208

Joachim AWR (1935b) Studies on Ceylon soils. II—general characterization of Ceylon soils. Some typical soil groups of the Island and a tentative scheme of classification. Trop Agric 84:254–275

Joachim AWR (1937) Studies on Ceylon soils IX. Trop Agric 89:89–92

Joachim AWR (1945) Soils of Ceylon. Presidential address, Proceedings of Agriculture and Forestry Section of the Ceylon Association for the Advancement of Science

Joachim AWR (1955) The soils of Ceylon. Trop Agric 89:89–92

Joachim AWR, Kandiah S (1935a) Studies on Ceylon soils IJI—the red and yellow earths, and the wet and dry Patna soils. Trop Agric 111:161–173

Joachim AWR, Kandiah S (1935b) Studies on Ceylon soils V—soils associated with limestone. Trop Agric 85:67–77

Joachim AWR, Kandiah S (1937) Studies on Ceylon soils VIJ—the characteristics of further important soil groups. Trop Agric 88:12–25

Joachim AWR, Kandiah S (1938) Studies on Ceylon soils XI—some paddy soils of the island. Trop Agric 90:200–221

Joachim AWR, Kandiah S (1939) Studies on Ceylon soils XJI- some forest and other characteristic soil types of the wet and dry zones. Trop Agric 95:268–277

Joachim AWR, Kandiah S (1940) Studies on Ceylon soils XIII—some characteristics of less extensive soils of the dry zone. Trop Agric 92:16–22

Joachim AWR, Panditesekere DG (1935) Studies on Ceylon soils IV—the light Sandy soils and the red dry zone and semi humid soils. Trop Agric 85:21–29

Joachim AWR, Panditesekere DG (1937) Studies on Ceylon soils VIII—the fernland (Kekilla) and some dry and semi-dry zone soils. Trop Agric 85:146–156

Joachim AWR, Panditesekere DG (1938) Studies on Ceylon soils X—further characteristics of dry zone soils. Trop Agric 90:136–141

Kumar P, Herath S, Avtar R, Takeuchi K (2016) Mapping of groundwater potential zones in Killinochi area, Sri Lanka, using GIS and remote sensing techniques. Sust Water Res Manag 2(4):419–430

Lagacherie P, McBratney AB (2007) Spatial soil information systems and spatial soil inference systems: perspectives for digital soil mapping. In: Lagacherie P, McBratney AB, Voltz M (eds) Digital soil mapping: an introductory perspective. Elsevier, Amsterdam, pp 3–22

Lal R (2016) Beyond COP 21: potential and challenges of the "4 per thousand" initiative. J Soil Water Conserv 71:20A–25A

Mapa RB (2014) Developing a soil database and its applications. Invited Presentation–ASDA 2014-Annual Symposium of Department of Agriculture, pp 25–26

Mapa RB (2016) Characterization of soils in the northern region of Sri Lanka to develop a soil data base for land use planning and environmental applications. Final report submitted to National Research Council of Sri Lanka. Grant No. 12–122

Mapa RB, Dassanayake AR, Nayakekorale HB (2005) Soils of the intermediate zone of Sri Lanka. Morphology, characterization and classification. Special publication No. 4. Soil Science Society of Sri Lanka. Survodaya Publishers, Colombo, p 241

Mapa RB, Somasiri S, Dassanayake AR (2007) Soils of the dry zone of Sri Lanka. Morphology, characterization and classification. Special publication No. 7. Soil Science Society of Sri Lanka. Survodaya Publishers, Colombo, p 288

Mapa RB, Somasiri S, Nagarajah S (1999) Soils of the wet zone of Sri Lanka. Morphology, characterization and classification. Special publication No. 1. Soil Science Society of Sri Lanka. Survodaya Publishers, Colombo, p 288

McBratney AB, Santos ML, Minasny B (2003) On digital soil mapping. Geoderma 117(1–2):3–52

Mermut AR, Eswaran H (2000) Some major developments in soil science since the mid-1960s. Geoderma 100(3–4):403–426

Minasny B, McBratney AB (2006) A conditioned Latin hypercube method for sampling in the presence of ancillary information. Comput Geosci 32(9):1378–1388

Minasny B, Malone BP, McBratney AB, Angers DA, Arrouays D, Chambers A et al (2017) Soil carbon 4 per mille. Geoderma 292:59–86

Minasny B, McBratney AB (2016) Digital soil mapping: a brief history and some lessons. Geoderma 264(B):301–311

Mooman FR, Panabokke CR (1961) Soils of Ceylon. Trop Agric 117:5–69

Munasinghe MAK, Puspakumara SMV (2002) Re-demarcation of conservation area in Sri Lanka based on soil erosion potential. J Soil Sc Soc Sri Lanka 14:58–70

Muthuwatta L, Chemin Y, Chandrapala L (2001) Variability of soil moisture in the Walawe River Basin: a case study in Sri Lanka suing low-resolution satellite data. In: Proceedings of 22nd Asian Conference on Remote Sensing, 5–9 November 2001, Singapore

Nayakekorala HB (1998) Soil degradation status of Sri Lanka. J Soil Sci Soc Sri Lanka 10:1–35

Panabokke CR (1956) A study of some soils in the dry zone of Ceylon. Soil Sci 87:67–74

Panabokke CR (1996) Soils and Agro-ecological environments of Sri Lanka. Natural Resources Series No. 2. Colombo: Natural Resource, Energy and Science Authority of Sri Lanka. p 219

Panabokke CR, Somasiri S (1980) The identification and characterization of the rice lands of the wet zone of Sri Lanka and its application to rice research. In: Proceedings of Rice Symposium, Department of Agriculture

Perera RACJ, Vitharana UWA, Nawarathne NRA (2015) Elucidation of spatial variability of salinity in soils in a rice—other field crops cropping system using proximal soil sensing. In: Proceedings of the annual symposium of the Department of Agriculture, 3rd and 4th September 2015, PGRC, Gannoruwa. ASDA proceedings, 17, pp 33–36

Pierce FJ, Nowak P (1999) Aspects of precision agriculture. Adv Agron 67:1–85

Punyawardena BVR (2008) Rainfall and agro-ecological regions of Sri Lanka (in Sinhala). Department of Agriculture Publication, Peradeniya, p 29

Rathnayaka RAAS, Vitharana UWA, Balasooriya BLWK (2017) Detailed mapping of soil texture of a paddy growing soil using multivariate geostatistical approaches. Trop Agric Res 29(4):300–312

Rathnayake WMUK, De Silva RP, Dayawansa NDK (2015) Variability of some important chemical properties of rainfed low land paddy fields and its effect on land suitability for rice cultivation. Trop Agric Res 26(3):506–516

Ratnayake RR, Karunaratne SB, Lessels JS, Yogenthiran N, Rajapaksha RPSK, Gnanavelrajah N (2016) Digital soil mapping of organic carbon concentration in paddy growing soils of Northern Sri Lanka. Geoderma Regional 7:167–176
Rosemary F, Vitharana UWA, Indraratne SP, Weerasooriya R, Mishra U (2017) Exploring the spatial variability of soil properties in an Alfisol soil catena. Catena 150:53–61
Rossiter DG (2000) Methodology for soil resource inventories. In: Soil Science Division, 2nd edn. International Institute for Aerospace Survey and Earth Sciences, Enschede
Sanjeevani UKPS, Indraratne SP, Weerasooriya SVR, Vitharana UWA (2015) Baseline concentrations of some trace elements in tropical Alfisols of Sri Lanka. Geoderma Reg 4:73–78
Soil Survey Staff (1960) Soil classification: a comprehensive system. 7th approximation. USDA, SCS, US Government Printing Office, Washington DC
Soil Survey Staff (2010) Keys to soil taxonomy, 11th edn. USDA, Natural Resources Conservation Service, U.S. Government Printing Office, Washington, DC
Soil Survey Staff (1998) Keys to soil taxonomy, 8th edn. USDA, Natural Resources Conservation Service, U.S. Government Printing Office, Washington, DC
Thenabadu MW, Fernando BL (1966) Rice soils in Hewagam Korale. Trop Agric 122:71–82
Thenabadu MW, Fernando BL, Perera BL (1966) Rice soils in Hewagam Korale. Trop Agric 122:117–126
Tokutome S, Handawala J, De Silva SOJ (1971) A method of classification of paddy soils of Ceylon. Trop Agric 127:153–159
United Nations General Assembly (2015) Transforming our world: the 2030 agenda for sustainable development. Report A/RES/70/1
Vitharana UWA, Van Meirvenne M, Simpson D, Cockx L, De Baerdemaeker J (2008) Key soil and topographic properties to delineate potential management classes for precision agriculture in the European loess area. Geoderma 143:206–215
Vitharana UWA (2008) Spatial inventory techniques in support of site-specific soil management. PhD thesis. Ghent University, Belgium
Vitharana UWA, Van Meirvenne M, Verdoodt A (2013) Response of EMI based proximal soil sensor in two contrasting tropical landscapes. Proceedings of 3rd Global workshop on proximal soil sensing, pp 56–62
Vitharana UWA, Thilakarathna EMSK, Balasooriya WK, Verdoodt A Saey T, Van Meirvenne M (2014) Potential of soil proximal sensing for mapping of key soil features of an Alfisol in Sri Lanka. In Proceedings of 20th World Congress of Soil Science 20WCSS (532). Jeju, South Korea
Vitharana UWA, Mishra U, Mapa RB (2017a) Predicting high-resolution soil organic carbon stocks of Sri Lanka using environmental covariates and a harmonized profile database. In: Proceedings of 13th International Conference of the East and Southeast Asia Federation of Soil Science Societies. Pattaya, Thailand. p 82
Vitharana UWA, Mishra U, Jastrow JD, Matamala R, Fan Z (2017b) Observational needs for estimating Alaskan soil carbon stocks under current and future climate. J Geophys Res Biogeo 122:415–429
Vitharana UWA (2017) Digital mapping of soil available potassium levels in Sri Lanka. In: Proceedings of advances in potassium research for efficient soil and crop management. IPNI, New Delhi, pp 33–35
Vitharana UWA, Mishra U, Mapa RB (2019) National soil organic carbon estimates can improve global estimates. Geoderma 337:55–64
Webster R, Oliver MA (2007) Geostatistics for environmental scientists, 2nd edn. Wiley, Chichester
Whelan B, Taylor J (2013) Precision agriculture for grain production systems. CSIRO Publishing, Collingwood
Wijesinghe WHAR, Vitharana UWA, Kulasinghe HPGTN (2018) Upgrading soil map of Dodangolla experimental station using proximal soil sensing. In: Proceedings of the 4th Faculty of Agriculture Undergraduate Research Symposium. Peradeniya, Sri Lanka. p 119
Wilson JP, Gallant JC (2000) Digital terrain analysis. In: Wilson JP, Gallant JC (eds) Terrain analysis: principles and applications. John Wiley & Sons Inc, New York, pp 1–27

Milestones in the History of Rice Improvement in Sri Lanka

5

M. P. Dhanapala

Abstract

The recorded history of rice breeding in Sri Lanka (then Ceylon) dates back to the British colonial era (1815–1948). The Ceylon Agricultural Society (CAS) was involved in promoting rice production in the country until the early twentieth century, a responsibility that was taken over by the Department of Agriculture (DOA), since its inception in 1912. Though no specific details of breeding procedures adopted are available, a variety named 'improved *Hatiel*', which takes 7 months to mature from planting, appears in the literature in the early nineteenth century. This would have been developed by seed selection, which was the procedure adopted then to purify rice varieties. During the early twentieth century, the targeted national average yield was 0.75 t/ha. Pure-line selection began in the 1920s, and 16 pure lines were identified by the end of the 1940s for cultivation under different agro-ecological conditions. As pure-line paddies failed to fulfil the anticipations, a scheme for paddy production improvement was drafted in 1945 to alleviate the food situation as two-thirds of the country's rice requirement had to be imported. Exotic rice varieties were introduced to fulfil the gap. The rice breeding strategies took a new turn in the late 1950s where hybridization of different varieties was employed to create a new genetic variability for the selection of genotypes with better fertilizer responses and disease tolerance. A series of 'H' varieties of different growth durations was developed and recommended. Improved plant type was conceptualized in the 1960s for *Indica* rice varieties, and new improved semidwarf varieties emerged thereafter. Pest and disease tolerance appeared as an important breeding objective in the 1970s with the change of the microclimate surrounding the plant and increased use of fertilizer. Resistance to rice blast, bacterial blight, gall midge and brown planthopper became a major issue in rice varietal improvement. Resistant varieties were

M. P. Dhanapala (✉)
Rice Research and Development Institute (RRDI), Department of Agriculture, Bathalagoda, Sri Lanka

B. Marambe et al. (eds.), *Agricultural Research for Sustainable Food Systems in Sri Lanka*, https://doi.org/10.1007/978-981-15-2152-2_5

developed to combat these pest problems. Grain quality aspects were given prominence in varietal improvement since the 1990s as the country was reaching self-reliance.

Keywords
Rice breeding technologies · New improved varieties · Pest and disease resistance · Higher yields

5.1 Early Records of Rice Production

Status and the historical developments in rice production in Sri Lanka during the precolonial era were not well documented, though some facts are confined to legendary tales with full of exaggerations. During the colonial era, the Ceylon Agricultural Society (CAS) has taken the initiative to promote rice production in the country. Much of the related information is recorded in the scientific journal 'Tropical Agriculturist' published by the CAS since 1881. Pioneering work pertaining to rice production in this era was in the hands of the civil servants and recognized citizens in the society.

Elliott (1913) reported that rice production in Sri Lanka (then Ceylon) was at its lowest by the mid-nineteenth century. This was attributed to abolishing of forced labour (*Rajakariya* system) in 1832, which led to the general neglect of irrigation work and gradual decline of communal labour exchange (*Aththam*) system. Accordingly, the average annual production within the 10-year period ending in 1856 was reduced to 5.5 million bushels (1 bushel = 20.5 kg), though the harvested extent was not reported. Rice production statistics, prior to the establishment of the Department of Agriculture (DOA) in 1912, are summarized in Table 5.1 (Elliott 1913). These statistics are apparently estimated from returns from paddy in government records. Inorganic fertilizer usage in rice cultivation during this era was not recorded despite that they were used in crop cultivation in other countries.

Table 5.1 Annual rice production and productivity during 1851–1907 (Source: Elliott 1913)

Period	Extent, acres (hectares)	Production, million bushels (million tons)	Productivity, bushels/Ac (t/ha)
1851/1856[a]	401,000 (162,276)	5.8 (0.114)	14.2 (0.7)
1892	613,000 (248,068)	10.0 (0.2)	16.32 (0.81)
1893	589,000 (238,355)	10.7 (0.214)	18.17 (0.9)
1902	663,000 (268,302)	11.75 (0.235)	17.73 (0.89)
1903	713,000 (288,535)	14.5 (0.27)	18.93 (0.94)
1903/1907[a]	587,000 (278,014)	12.3 (0.246)	17.91 (0.89)

[a]Mean annual production for the period

5.2 Traditional Varieties/Land Races

Molegoda (1924a) reported a detailed account of traditional paddies, naming procedure and possibilities of duplication, etc. Accordingly, a collection of 300 named traditional rice varieties was exhibited in 1902 by Nugawela *Disava* in Kandy Agri-Horticultural and Industrial Exhibition. A few years later, an attempt has been made to fill the missing ones of the above collection, but efforts have failed. There had been no proper maintenance of varieties by farmers. In subsequent attempts to collect only the names of varieties grown in the years 1915 and 1916 through the office of CAS, however, only 168 names have been collected.

A collection of 150 varieties grown throughout the country has been sent to the British Empire Exhibition in the 1920s. Some of these varieties were replicas with slight changes in name. Molegoda (1924a) did not express belief in the existence of many different rice types in the then Ceylon. The procedure adopted by farmers in naming varieties had been the possible reason for the increase in the number of varieties with time. De Soyza (1944), too, observed the possibility of confusion in naming varieties.

Molegoda (1924b) provided a list of 567 names of traditional varieties but made no comments on their availability. Apparently, the listed varieties are lowland paddies as de Soyza (1944) provided another list of 42 upland types (*El Wee*) used for upland or slash and burn (*Chena*) cultivation systems in the country. Iliffe (1922) disclosed that any 'type of rice' named by the farmer constitutes of many different types. The traditional paddies maintained by farmers had been, therefore, heterogeneous due to the genetical and/or physical mixing. The Plant Genetic Resources Center (PGRC) of the DOA in 2018 had 5256 viable accessions including 2463 traditional accessions (Dr. Gamini Samarasinghe, PGRC—Personal Communication) in its medium-term storage (personal communication). Majority of these were descending down from the traditional varieties/land races grown in the past in Sri Lanka.

5.3 Selection and Introduction of Cultivars

5.3.1 Seed Selection

The responsibility of rice production in the country has been gradually vested with the DOA since its inception in 1912. Prior to this, CAS was the forum for discussion of investigations made in rice. Apparently, the first step in rice breeding was evident around 1914 with the appearance of improved paddy seed '*Hatiel*', (CAS 1914), where the seed selection and multiplication selected *Hatiel* was performed by the CAS. The improved paddy seed would have been developed by seed selection, the procedure adopted then to purify rice varieties.

Beven (1914) reported that the national average target of 15 bushels per acre (0.75 t/ha) during the early twentieth century was high under the prevailing cropping conditions. Beven (1914) predicted the possibility of increasing this up to 25 bushels per acre (1.25 t/ha) with some improvements such as seed selection, which is simply a selection of true-to-type better-looking panicles from the crop in the field, without elaborated testing of their progenies, to establish the next generation (Molegoda 1924b).

5.3.2 Pure-Line Selection

Iliffe (1922), the economic botanist of the DOA during the early twentieth century, dealt solely with paddy following the policies laid down by the institution.[1] He noticed heterogeneity among and within traditional varieties, different types of individuals within any given type of paddy and insisted on pure-line selection. Pure-line selection was initiated in 1919 with a representative collection of seed samples having 105 different paddies that mature in 6 months from planting (6-month age class) and 272 different paddies of 4–5-month age class, from the traditional varieties grown in Sri Lanka (then Ceylon). For convenience, Iliffe (1922) did arbitrary grouping of the original paddies in the 6-month age class paddies into '*Ma wee*' and '*Samba*'. The middle-aged paddy varieties (4–5-month age class), too, were categorized into nine main groups, namely, *Bala wi*, *Danahala* (probably *Dahanala*), *Ilankalayan, Murunkan, Heenati, Suwanda-el, Suduhatiel, Kalupannity* and *Madoluwa*. The basis of grouping was not clear. At least 20 pure lines were established from each variety, and selection among them was done at 5% selection intensity. Pure-line selection was concentrated at two locations, Anuradhapura (North Central Province; dry zone) and Peradeniya (Central Province; wet zone). Lord (1927) reported of a new series of variety selections of local interest, based on ear-heads (13 varieties in 1924 and 12 varieties in 1926). Selection was done in ear-to-row progenies (around 50 ear-heads per variety), at a selection intensity of 40%.

The importance of regional adaptability of varieties was realized in 1926, and 19 paddy stations were established in different parts of Sri Lanka for testing pure lines and in situ selection of the locally popular types. Production and distribution of improved seed became the main service of these stations (Lord 1927). Haigh (1932) reported that the pure lines selected at Anuradhapura and Peradeniya paddy stations demonstrated 15% yield increase over their ancestral types, probably in their test locations. The Amended Departmental Circular No. 156 (Rhind 1948) listed out 21 pure lines during this period (Table 5.2). Their purity maintenance was done at four different paddy stations, namely, Ambalantota (9 lines), Mahailluppallama (8 lines), Madampe (2 lines) and Batalagoda (2 lines) (DOA 1948).

Gunawardena and Wickramasekera (1947) published the botanical characteristics of these paddy pure lines where names of few varieties in this list are not in conformity with the varieties listed in the Departmental Circular No 156. The names in botanical descriptors are authentic as they appear consistently in the literature.

[1] Mr. F. Summers, the predecessor of Iliffe (1922) in the Division of Botany at the DOA, was responsible in developing the policies related to paddy breeding.

Table 5.2 Purity maintenance list of pure-line paddies adopted from amended (Source: DOA 1948)

Age class	Varieties
5–6 months	*Podiwee* A8[a,b]
	Kohumawi B11
	Molagusamba 1 18 6
	Kurulutuduwi b 13[a]
4–4.5 months	*Dewaredderi* 26031[b]
	Balamawi 31009
	Oddavalan 2449/20
	Vellai illankalayan 28061
	Perillanel 26014
	Honderawala 868
3–3.5 months	*Madael* 38 MY 137
	Sudheenati Hf 9
	Sulai 27614
	Kaluheeneti 39 YM 3254
	Dahanala 37 YM 2014
	Vellai perumal 28724
	Suduheeneti CPYN 19[b]
	Sinnanayan 38 Y 2203[b]
	Pachchaiperumal 2452/11[b]
	Rathkarayal 38 YM 3753
	Murunga 39 YM 137

[a]White-pericarped
[b]Gunawardena and Wickramasekera (1947)

Table 5.3 Annual rice production data in the 1940s (Source: DOA 1948)

Period	Extent ac (ha)	Production bu. (t)	Productivity bu./ac (t/ha)
1944	1,030,232 (416,912)	13,566,205 (275,631)	13.17 (0.66)
1945	895,600 (362,430)	1,080,470 (219,438)	12.06 (0.61)
1946	9127,078 (275,168)	11,350,510 (230,613)	12.24 (0.61)
1947	926,683 (375,008)	11,520,330 (234,064)	12.43 (0.62)
1948 (Yala)[a]	343,055 (138,827)	4,344,562 (88,270)	12.66 (0.64)
1949	1,072,803 (434,140)	15,235,247 (309,542)	14.2 (0.71)
1950[b]	1,081,616 (437,706)	15,000,430 (304,771)	13.87 (0.7)

[a]Data from one season (minor season)
[b]Estimated figures

Despite the efforts made, pure-line paddies virtually have failed to have the anticipated impact on rice production in the country. The average rice yields remained around 13 bushels/acre (approx. 0.65 t/ha) as evident from rice production statistics in 1940s (Table 5.3).

This failure may be partly due to the large number of selections released without making an attempt to recommend the most desirable pure lines. As a result, the subsequent pure-line selections in the 1940s were restricted to the best lines from the earlier selections and several others selected from popular indigenous varieties.

The best of these were *Podiwi* a-8 (5–6-month age class, photoperiod sensitive), *Murugakayan* 302 and *Vellai illankalayan* 28061 (4–4.5-month age class), *Vellai perumal* 28724 (3.5-month age class) and *Pachchaiperumal* 2462/11 (3-month age class) (Senadhira et al. 1980). Irrespective of the consequences, the pure-line selection paved the way to group rice varieties into different age classes and identifying the key varieties that are adapted to different agro-ecological conditions.

5.3.3 Introduction of Exotic Varieties

As pure-line paddies failed to fulfil the anticipations, a scheme for paddy production improvement was drafted in 1945 by the Paddy Advisory Board (established in 1943) to improve the food situation in the country. Two-thirds of the country's rice requirement to feed around 6 million people had to be imported during this period (Anonymous 1945).

Application of inorganic fertilizer to pure lines, which was the quickest way to increase yields, made them to succumb to blast disease (Peiris 1966). The blast-resistant pure lines, namely, *Vellai illankalayan* 28061 and *Murungakayn* 302 (4–4.5-month age class), did not respond to fertilizer adequately or grew excessively leafy, thus will a high tendency for premature logging causing yield losses (Senadhira et al. 1980). With these developments, exotic rice varieties were introduced to improve the productivity of rice in Sri Lanka in the 1950s.

Three tropical varieties, namely, Ptb 16 (*Riyan wee*; 5–6-month age class) from India and *Mas* and *Sigadis* (4–4.5 month class) from Indonesia, were selected for release in the country, among many varieties screened from the tropical region. Only *Sigadis* was found resistant to blast disease but needed a longer duration exceeding 4.5 months in the field causing other problems. Eventually, the option was left with rice hybridization to create new genetic variability (Senadhira et al. 1980).

5.3.4 Hybridization and New Genetic Variability

Rice hybridization in the world was first employed by Japanese scientists in the late 1920s to improve *Japonica* varieties. Based on the Japanese experience, the rice breeders in Sri Lanka adopted rice hybridization[2] in the wake of the failure of the earlier strategies adopted to attain self-sufficiency. Niles (1951) and Weeraratne (1954) have described the hybridization techniques adopted locally. The new program was initiated at the Dry Zone Agricultural Research Station (probably the Anuradhapura Paddy Station as identified in early literature) located at Mahailuppallama.[3] As there was no assured supply of water to cultivate both seasons of the year, the main program was shifted to Batalagoda Farm in 1952, which

[2] Dr. M. F. Chandraratne, the then botanist of the Department of Agriculture, was instrumental in launching the rice hybridization program in the early 1950s.

[3] During this period, Dr. E. Abeyratne has managed the Dry Zone Agriculture Research Station.

was eventually named as the Central Rice Breeding Station (CRBS). The development of rice varieties through hybridization was initiated, and the first such variety 'H4' was released in 1957.[4] This was a red pericarped variety in the 4.5-month age class. A series of 'H' varieties differing in growth duration was developed and recommended thereafter, namely, H9 (5–6 months), H8 (4.5 months), H7 (3.5 months) and H10 (3 months). The varieties H9 and H7 were white pericarped, H8 was short round-grained (samba) and H10 was red pericarped. There were two other red pericarped varieties, namely, H105 and H501 (4.5 months) developed elsewhere. Variety H501 may have originated from Labuduwa Rice Experiment Station as its sister lines named LH1 and LH2 (Labuduwa hybrids 1 and 2) were known to exist at Labuduwa in the Southern Province of Sri Lanka. The H105 may have originated from Ambalantota Rice Experimental Station located in the Southern Province. The Coordinated Rice Varietal Testing (CRVT) program initiated in 1968[5] expedited the adaptability testing and release of 'H' varieties. By this time, the agro-ecological regions and age groups of the varieties were well defined, and varieties in the 4.5-month age class were the major group cultivated in the country.

Senadhira et al. (1980) reported that H varieties (identified as old improved varieties or OIVs) made a tangible breakthrough in rice cultivation. Their built-in resistance to blast minimized crop losses, H4 and H9 had a sensational increase in the yield potential and the single cropping pattern changed to double cropping with H4 being cultivated in *Maha* season and H7 or H10 in *Yala* season. Furthermore, the moderate fertilizer response of these varieties resulted in an increase in fertilizer consumption by 350%, and the national average yield shot up to 2.5 t/ha (50 bu/ha). However, with conceptualization of the improved plant type, the breeding process of H varieties ended abruptly.

5.3.5 Improved Plant Type and Modern Cultivars

Improved plant type was conceptualized in the 1960s for *Indica* rice varieties by the International Rice Research Institute (IRRI). However, the traditional *Japonica* rice grown in Japan prior to the 1920s were tall and leafy resembling the traditional *Indica* paddies (Dr. Y. Yamakawa, Professor of Tropical Agriculture at the Saga University of Japan—personal communication). A mutant, *Jikkoku*, carrying dwarf gene/s changed the plant architecture and thereby the yield potential of Japonica varieties. This was done through rice hybridization initiated in the late 1920s. A hybridization program between *Japonica* and *Indica* varieties, supported by the Food and Agriculture Organization (FAO) of the United Nations, was executed in Cuttack (probably the Central Rice Research Institute) in India to transfer the

[4] Dr. H. Weeraratne, the pioneer breeder in charge of this program, was credited for developing and releasing H4.

[5] The CRVT program was initiated by Dr. J. W. L. Peiris, the then deputy director (Research) at the DOA.

Japonica plant type to *Indica* rice. However, this was abandoned without success due to lack of affinity between the two groups.

The modern plant type in *Indica* rice originated with the isolation of dwarf rice mutant Dee-geo-woo-gen (DGWG) in Taiwan. Two modern cultivars, namely, Taichung Native 1 (TN1) in Taiwan (1960) and IR8 in the IRRI (1970), emerged carrying DGWG dwarf gene/s. These varieties were tested in Sri Lanka in the late 1960s and early 1970s but failed due to their susceptibility to diseases, especially to bacterial leaf blight (BLB), and the intensive crop management demanded by them (Senadhira et al. 1980). Furthermore, TN1 and IR8 did not reach the anticipated plant height (1 m) under different agro-ecological conditions in the country. As an immediate response, CRBS used a semidwarf gene from *Engkatek*, an Indonesian variety, to introduce improved plant type. Varieties Bg11-11 (4–4.5-month age class), Bg3-5 and Bg38 (5–6-month age class) are varieties of *Engkatek* origin. Variety Bg11-11 was the first improved variety released by Sri Lankan breeders in the early 1970s. Subsequently, the dwarf gene/s of DGWG were introduced when the farmers learned to manage improved plant type. These varieties are identified presently as the new improved varieties (NIVs).

The varieties in the 4–4.5-month age class were in the major as well as the popular age group at the initial stages of the NIV development. In 1972, a decision was made by the DOA[6] to improve the short-duration varieties (3–3.5-month age class) for cultivation in two seasons (*Maha* and *Yala* seasons). As a result, the rice breeders placed more emphasis on short-duration varieties, which eventually became the major age group of rice varieties in the country; especially for rainfed cultivation in the *Maha* season, there was a necessity for a blast-resistant 100-day variety. Varieties Bg300 and Bg352 were released in the early 1990s fulfilling this requirement. These two varieties are grown in more than 50% of the annually cultivated extent even today.

In the early 1980s, the CRBS at Batalagoda has experienced difficulties in establishing legume crops during the *Yala* season in rainfed paddy fields of 'Rice-Based Cropping System Project' conducted in the intermediate zone (Katupotha). The legume crops established succumbed to stagnating water during the peak rainfall in *Yala* season despite the rainfall distribution in *Yala* was inadequate for a short-duration (3–3.5-month age class) crop of rice. Hence, the requirement of an ultrashort-duration paddy variety was strongly felt for *Yala* cultivation in rainfed lands in the intermediate zone. As a result, the 2.5-month age group emerged to fulfil this requirement. Variety Bg750 was released in 1981 and was the first variety in this group. As Bg750 was tall, susceptible to lodging and poor yielding, Bg250 with semidwarf plant type and high-yielding ability was developed later. Subsequently, two other varieties, Bg251 and Ld 252, were developed and released in this age group.

Sri Lanka is specific about the age groups of rice varieties due to the variability in agro-ecological regions and especially the bimodal rainfall pattern. In general, varieties in the 5–6-month age class are photoperiod sensitive and are for cultivation

[6]An initiative taken by Mr. D. V. W. Abeygunawardena, the then assistant director (research) of the DOA

during the *Maha* season in the low-lying paddy tracts in the coastal belt of the wet zone. The majority of presently abandoned paddy lands are in Colombo, Kalutara and Gampaha Districts in the Western Province which was formerly cultivated to varieties in the 5–6-month age class. These varieties were identified as date-fixed varieties as their heading takes place in the end of December or beginning of January (short days) irrespective of the date of sowing, provided that the basic vegetative phase (3-month age class) is completed by mid-December. The crops are usually established from July to mid-September depending on the availability of rainfall and harvested in February. These lands are inundated by October, the beginning of regular *Maha* season, and inundation continues until the dry-spell sets in in February.

The varieties in the 4–4.5-month age class are for general cultivation in *Maha* season. The short-duration varieties (3–3.5-month age class) are for cultivation in rainfed and irrigated lands in the *Maha* season in general and *Yala* season in rainfed lands in the wet zone and irrigated lands in the dry zone. Varieties in the 2.5-month age class are for rainfed cultivation in the *Yala* season in the intermediate zone. Even under major irrigation schemes, these varieties can be cultivated to save irrigation water. They can also fit into short cropping season after flood recedes in flood-prone rice lands in the low country wet zone (Abeysiriwardena et al. 2011). A shift from long- to short-duration varieties is a common practice at times when rains get delayed.

The NIVs are dwarf, less leafy, non-lodging and photosynthetically efficient. They responded to fertilizer with comparatively higher harvest index (HI) producing more grains and less straw. The source sink relationship of these varieties was modified to translocate more photosynthates into grain formation. Consequently, the farmers began to adopt the NIVs rapidly replacing the traditional and the OIVs (Fig. 5.1). Among the NIVs, varieties were developed in research stations, namely, Bombuwela (Bw), Ambalantota (At) and Labuduwa (Ld), to fulfil the demand specifically for regional requirements. Some of these varieties have high-yield potential and wide adaptability (e.g. Bw367) to become major cultivars in the country.

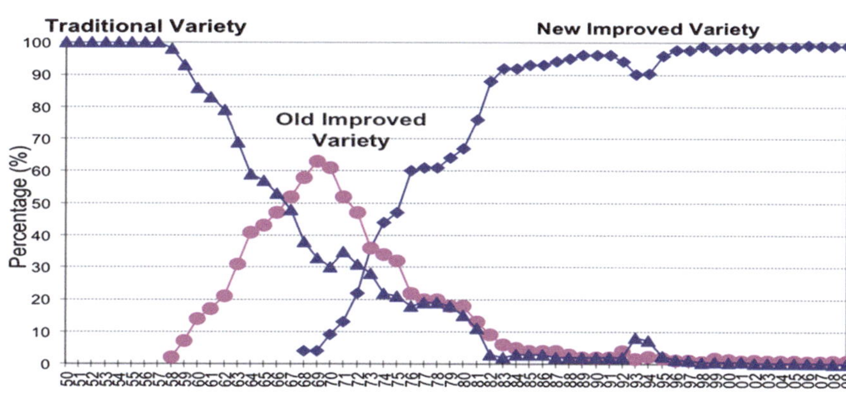

Fig. 5.1 Percentage distribution of rice varietal categories in Sri Lanka, 1950–2009 (Source: RRDI, Batalagoda)

Grain quality aspects were given prominence in the varietal improvement since the 1990s as the country was reaching self-reliance. Accordingly, Bg360 (*keeri Samba*) and At405 (*Lanka Samurdhi*) were introduced in the mid-1990s and, At373 (*Ambalantota Suwanda Samba*) and At311 (*Neeroga)* were introduced in the mid-2010s as quality rice varieties by the DOA for cultivation. The private sector initiated its rice improvement program for superior grain quality development in 2006 for the first time in Sri Lanka and has been successful in developing varieties with Basmati-type grain qualities.

With the introduction of NIVs, supported by other technological innovations, the rice production exceeded the requirement of the country in the years 2013 (4.6 mill t) and 2015 (4.8 mill t) (Fig. 5.2). The national average yield was increased up to 4.6 t/ha (92 bu/ac). At present, Sri Lanka is self-sufficient in rice except in drought years. However, sustenance of self-sufficiency in rice depends on government policies.

5.4 Breeding for Resistance to Biotic and Abiotic Stresses

The plant breeders have to encounter the dynamic changes in the environment and the evolutionary processes in nature. A few major biotic stresses, e.g. brown planthopper, gall midge, etc., have affected rice cultivars in the process of varietal improvement. Some experts believed that this was due to the changes of microclimate below the canopy of the plant and the increased use of chemical fertilizer. Tolerance to these biotic stresses appeared as an important breeding objective from the mid-1970s, especially with the spread of the NIVs. Some biotic stresses had sources of resistance but some did not.

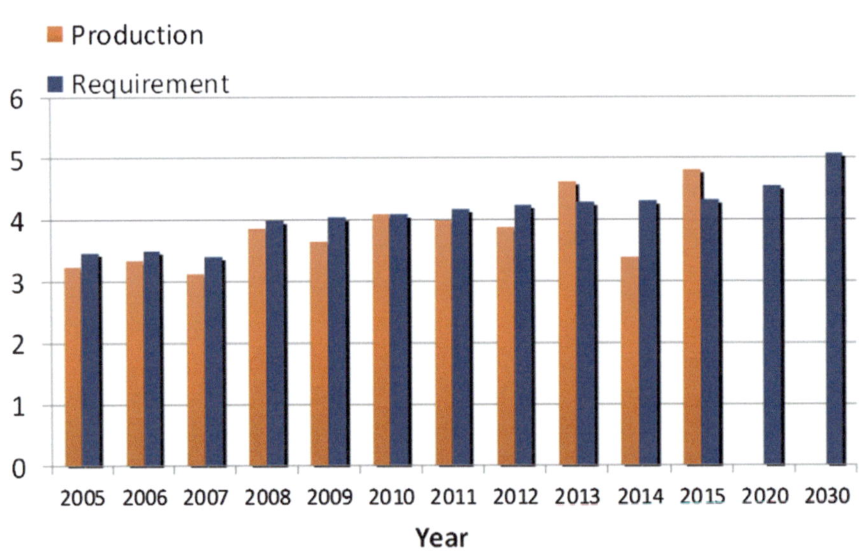

Fig. 5.2 Recent trend in annual rice production and requirement in the country (Source: RRDI, Batalagoda)

5.4.1 Disease-Resistance Breeding

5.4.1.1 Rice Blast Disease (Caused by the Fungus *Magnaporthe grisea*)

Breakdown of blast resistance was first reported in H4 grown in potato-rice cropping system in the Uva-Paranagama area in the late 1970s. Further, Bg34-8, a major breakthrough in short-duration varieties (3-month age class), was found comparatively more susceptible to blast a few years of its release. Introduction of *Tetep* and *Thadukan* as resistance donors in parental nurseries and the change of having bombardment rows to blast resistance screening nurseries have helped containing the problem. Rice breeders of CRBS at Batalagoda strongly believed that having a highly susceptible variety (SLO 15) in bombardment rows of the blast screening nursery would encourage selective multiplication of less virulent strains of the pathogen that would mislead the breeder in the selection process. This has led to the loss of unnecessary virulence in the pathogen when the host plant is highly susceptible (Dr. H. Weeraratne—personal communication). This can possibly be attributed to selective multiplication of less virulent strains in the pathogen complex due to the reproductive advantage they possess.

5.4.1.2 Bacterial Leaf Blight (Caused by Bacteria *Xanthomonas oryzae*)

Bacterial leaf blight (BLB) was a serious disease from the late 1960s in the country. An epidemic of BLB, including Kresek phase, was reported from the Polonnaruwa District in 1972. Watanabe et al. (1971) reported a high degree of resistance to BLB in varieties *Nakashin 120* and *Wase Aikoku 3*, after screening many traditional and exotic varieties at the Central Agricultural Research Institute (CARI) at Gannoruwa, Sri Lanka. However, none of these varieties were successful as donors in hybridization as they were *Japonicas*. The resultant hybrid progenies with *indicas* were sterile. Further, the grain shattering of *Wase Aikoku 3* progenies was abnormally low.

Among the varieties screened for resistance at CRBS at Batalagoda, the variety IR20 (an IRRI variety) was found to be tolerant to BLB (field resistance). This variety apparently inherited BLB resistance from TKM6 (an Indian variety). Introduction of IR20 as a hybrid parent helped to control the BLB-spread in Sri Lanka. Rice variety Bg400-1 released in Sri Lanka inherited BLB-resistance from IR20.

Resistance to Blast and BLB was considered a varietal requirement to release for cultivation from the year 1974.

5.4.2 Insect-Pest Resistance Breeding

Many insect pests are prevalent in Sri Lanka but not always in epidemic proportions. Some important insect pests; paddy bug, leaf folder, stem borer etc. do not have effective sources of resistance.

5.4.2.1 Brown Planthopper (*Nilaparvata lugens*)

A few seasons after release of Bg11-11, brown planthopper (BPH) attack (hopper burn) emerged in epidemic proportions, especially in the Ampara District of Sri Lanka. The variety IR36, developed by the IRRI is resistant variety and resistant to BPH, was introduced. However, this variety succumbed to the pest apparently due to differences in the biotype of the pest. Later, a resistant genes identified in Ptb33, an Indian variety, was employed at the CRBS to combat thc situation successfully. Rice variety Bg379–2 was the first to be released in Sri Lanka in 1980 with the built-in BPH resistance.

5.4.2.2 Gall Midge (*Didymomyia tiliacea*)

Gall midge (GM) became a serious pest, especially in the wet zone rice cultivation with the release of Bg94-1. Many varieties and two immune progenies of a double cross (Warangal 1263/IR 8//Eswarakora/IR8) were identified for GM resistance.[7] They were registered at the CRBS as observation lines (Ob) 677 and 678. With these sources of resistance, GM-immune varieties, Bg400, Bg380, Bg276-5, etc., were developed and released in the early 1980s. A few years later, galls were reported in Bg400 from the Matale District and subsequently from other districts in Sri Lanka. The 'resistant varieties' became more susceptible than the susceptible ones (similar to *Vertifolia effect* in potatoes). This was apparently the first incidence of breakdown of major genetic resistance in rice in Sri Lanka.

The BPH and GM tolerance/resistance eventually became a prerequisite to release a variety for cultivation from the year 1975.

5.4.3 Breeding for Resistance to Abiotic Stresses

Tolerance to iron toxicity, salinity (coastal and inland), flood and drought is the major issue in abiotic stresses. Breeding for abiotic stress resistance was a part of the responsibilities of regional breeding centres. These targets were difficult to achieve as the breeders could not surpass biological limits the rice plant can tolerate.

At the early stages, iron toxicity and flood tolerance breeding was confined to the Regional Agricultural Research Center (RARC) at Bombuwela and salinity tolerance to the RARC at Ambalantota. The CRBS was involved in research drought tolerance. After the establishment of the Rice Research and Development Institute (RRDI) in 1996, collaboration among the centres was strengthen, and collective efforts were made for stress tolerance breeding. Among the traditional varieties, a few varieties as *Pokkali*, SR 26B (SR—salt resistant, of which the origin is unknown) for salt tolerance and *Dewaredderi* 26081 for flash flood tolerance were used by farmers. Among the NIVs, a few varieties with appreciable degree of resistance to

---

[7] Based on studies carried out by Dr. Nalini Wickramasinghe, a research officer in entomology, through screening trials

biotic stresses are Bw367, Bw372 (iron toxicity), Bg455 (flood), Bg251 (drought), At354, At401, Bg310 and Bg369 (salinity).

The IRRI has identified Sri Lanka as the most important site to screen rice varieties for iron toxicity and zinc deficiency.

5.5 Innovative Strategies for Future Breeding

Conventional breeding through hybridization will have many obstacles to overcome. Even if the donor sources were identified, some varieties, particularly wild types, do not have affinity with others for cross-pollination. Some did not combine well to produce progenies having desirable combinations of characters (poor combining ability). Even among the indigenous varieties, incompatible reactions were experienced, e.g. progenies of the crosses using *Dahanala* and *Heenati* varieties with other *Indica* types, behaving like *Japonica*/*Indica* hybrids.

Innovative breeding techniques such as bridging varieties, embryo rescue, double haploids, chromosomal substitution, protoplast fusion and genetic transformation are used as options; however, these are all expensive, laborious and time-consuming, with no assurance of satisfactory end results. Sri Lanka does not have the critical mass of trained researchers and the infrastructure facilities to undertake basic research of this nature. These should be left to the international institutes while providing necessary collaboration at the country level.

5.5.1 Mutation Breeding

Among the possible innovative opportunities, emphasis should be placed on mutation breeding, especially to improve productivity of so-called traditional quality paddies. The basic argument here is the effect of background genome on gene expression. The exact grain qualities of any cultivar would not be retained in the process of cross-breeding as the background genome is affected during the process. One can argue that by backcross-breeding, the nuclear genome could be transferred, but desirable point mutations would retain the basic genome almost unaffected. Mutation is the ultimate source of genetic variability. We should not forget the fact that genetic variability of all living organisms on earth has resulted from mutations.

Many indigenous varieties possess specific grain qualities, nutritional or medicinal. Also, there are less productive, indigenous traditional quality rice such as *Suduru samba*, *Suwanda samba*, *Puwakmal-eta samba* and *Heenati*. These should be the target varieties for mutation breeding.

5.5.2 Hybrid Rice

Hybrid rice technology is prospering in China. Super hybrid rice can reap yields of 18 t/ha.[8] In the same agroecology, a conventionally bred high potential variety would also yield 16 t/ha. This marginal increase in yield is due to hybrid vigour. Tropical soil and climate do not have the potential to produce high yields, but research on hybrid rice technology to exploit hybrid vigour should continue as we do not have an alternative to proceed beyond the NIV yield ceiling. Precautions should be taken not to release hybrid technology until it is proven appropriate for the local farmer.

5.5.3 New Plant Type

The IRRI scientists conceptualized a new plant type[9] in the 1980s, having basic features of the improved plant type with reduced tiller number and heavy panicles. In this scenario, photosynthates are conserved by elimination of ineffective tillering in the vegetative phase. Constructing a plant having these features would be ideal for direct seeding. It may permit to exploit uni-culm advantage by adjusting the seed rate. Sri Lankans are the pioneers in direct seeding, and no other country has the experience of crop stand establishment by direct seeding. The concept of new plant type is worthwhile attempting.

5.5.4 Biotechnological Tools

Maker-assisted selection involves molecular-biological techniques to tag or mark the genes and monitor their presence in individuals in the selection process. The breeder needs not wait for gene expression. This is rather a breeding tool helpful in expediting the breeding process. Further, it can be used for pyramiding genes into a variety (e.g. pyramiding resistance genes specific for different strains of blast fungus without elaborate testing with differential set of varieties). However, the background genome of the ultimate variety may influence the penetrance of the gene. Regular verification of gene effect in the process of introducing the genes is, therefore, needed to ascertain the character expression.

[8] A value claimed by Dr. Yuan Long Ping, the father of hybrid rice. The author has personally witnessed this claim in the IRRI in the early 2000.

[9] The concept was presented by Dr. G. S. Khush, the principal rice breeder at the IRRI.

References

Abeysiriwardena DS d Z, Balasooriya BANK, Somaratne NS, Dissanayake DMN (2011) Minimizing water supply duration in the field with no reduction of yield in the rice variety Bg250 of the ultra-short maturity class. J Nat Sci Found Sri Lanka 39(1):43–52

Anonymous (1945) A draft scheme for the development of the paddy industry in Ceylon. Trop Agric 101(8):191–195

Beven AW (1914) The extension of paddy cultivation. Trop Agric 43(6):421–424

CAS (1914) Paddy (rice)–Ceylon agriculture society. Trop Agric 43(4):296–298

de Soyza DJ (1944) Hill paddy cultivation in Ceylon. Trop Agric 100(4):211–218

DOA (1948) Departmental circular no. 156, Department of Agriculture. Trop Agric 104(2):97–98

Elliott E (1913) Paddy cultivation in Ceylon during the 19th century (summary). Trop Agric 41(6):465–467

Gunawardena KJ, Wickramasekera GV (1947) Botanical characters of paddy. Trop Agric 103(2):105–112

Haigh JC (1932) A short review of paddy work done by the Economic Botanist. Trop Agric 77(6):384–385

Iliffe RO (1922) The improvement of Ceylon paddies. Trop Agric 58(2):27–70

Lord L (1927) Land crabs in paddy fields. Trop Agric 69:141–144

Molegoda W (1924a) Varieties of Ceylon paddy. Trop Agric 62(4):218–224

Molegoda W (1924b) Selection of seed. Trop Agric 62(1):16–17

Niles JJ (1951) Hybridization methods with paddy. Trop Agric 107(1):25–29

Peiris JWL (1966) Problems and organization of rice research. In: Abeygunawardana DVW (ed), Proceedings of Symposium on Research and Production of Rice in Ceylon, pp 19–29

Rhind D (1948) Pure-line paddies maintenance of purity. Amended Departmental Circular No. 156. Trop Agric 104(2):97–98

Senadhira D, Dhanapala MP, Sandanayake CA (1980) Progress of rice varietal improvement in the dry and intermediate zones of Sri Lanka. Rice Symposium, BMICH. Colombo: Department of Agriculture. pp 15–31

Watanabe S, Hemachandra S, Dhanapala MP (1971) Project report on studies on breeding of rice varieties resistant to Bacterial Leaf Blight disease. p 244

Weeraratne H (1954) Hybridization technique in rice. Trop Agric 110(2):93–97

Vegetable Breeding in Sri Lanka in Retrospect

6

H. Hemal Fonseka

Abstract

Plant breeding started when hunter-gatherers changed their lifestyle to the producers of selected plants and animals. During this gradual process, independent primitive forms of plants were transformed into fully human-dependent domesticated varieties through plant breeding technique "selection", which even today remains the primary strategy for crop improvement. Though it is difficult to establish as to when plant improvement started in Sri Lanka, it can be safely assumed that it is as old as civilization itself which is dating back to 29,000–30,000 years before present. The principles of genetics have become essential tools in plant breeding after the rediscovery of Mendel's laws in 1900, and during this era, Ceylon (renamed as Sri Lanka during 1978) was under the British Empire. In the activities of tropical agriculture, Ceylon has played a pioneering role, and the scientific workers of the Botanical Department in Ceylon had equally been pioneers in their work affecting tropical vegetation. The first pre-independence work related to crop varieties appeared in Volume I (1) of the *Magazine of the School of Agriculture, Colombo*, published in July 1889. Since then up to date, this review compares the development of regional plant breeding activities with the development of plant breeding in the country from prehistoric time to present day with special reference to vegetable breeding. The objective is to help in the identification of constraints and prospects for vegetable breeding in Sri Lanka.

Keywords

Germplasm · Breeding · Breeder seed · Post-independence · Pre-independence vegetable · Resistance breeding · Recommended varieties · Released varieties · Resources

---

H. H. Fonseka (✉)
Onesh Agri Pvt Ltd, Colombo, Sri Lanka

B. Marambe et al. (eds.), *Agricultural Research for Sustainable Food Systems in Sri Lanka*, https://doi.org/10.1007/978-981-15-2152-2_6

6.1 Introduction

Vegetable sub-sector plays an important role in the Sri Lankan economy. It provides livelihood to a large proportion of poor people, contributes to increasing the level of national income and export revenue, generates new employment opportunities, increases farm income and indirectly enhances the nutrition and health of the nation. In South Asia, 56% of the vegetables consumed are fruit types, followed by root and stem (22%) and leafy types (22%). There is a potential to increase the vegetable production by several folds with the increase of per capita income and expansion of exports and tourism industry in the country. Consumers tend to have health-safe and quality foods with the increase of income, thus enhancing the consumption levels of vegetables. This increasing demand should be met with an increased supply of quality vegetables. Even though prevailing diverse agroclimatic conditions in Sri Lanka make it possible to grow a wide variety of vegetable crops (about 40 species) throughout the year in different parts of the country, the per capita consumption of vegetables is only about 112 g/day (SAARC Report 2017), which is far below the dietary requirement (200 g/day) recommended by WHO (World Food Programme 2018). Thus, it is essential to increase vegetable production at least twofold to meet the recommended per capita vegetable intake.

Technological innovations can not only boost vegetable production and consumption in the country but also help to generate farm employment and off-farm employment opportunities and increase income and resource use efficiency of poor farmers. This review compares the development of regional plant breeding activities with the development of plant breeding in the country from prehistoric time to present day with special reference to vegetable breeding. The objective is to help in the identification of constraints and prospects for vegetable breeding in Sri Lanka.

6.2 Vegetable Sector Outlook

The position of Sri Lanka with regard to the area of cultivation, production and productivity of vegetables in comparison with its neighbours and the rest of the world is shown in Table 6.1. The average productivity of main vegetable crops, in comparison with India and the leading producers of the world, is given in Table 6.2. These statistics clearly indicate the need for further improvements in vegetable variety development to gain higher productivity to win the market competition at least within the South Asian neighbourhood. Data presented in Table 6.2 indicates that the productivity of cabbage, of which seeds of adaptable varieties are imported annually to meet the demand, has reached to a satisfactory level of productivity, while the other three crops, of which local breeding programmes have released some open pollinated varieties and continue the efforts to produce local F1 hybrids, are far below the recorded potential productivity.

In addition to the productivity issues in vegetables, in the last several decades, the carrying capacity of our land resources has surpassed by the constant decline in cultivable land due to other development requirements coupled with a booming

Table 6.1 The mean extent of vegetable cultivation (production and productivity in Sri Lanka and selected countries in comparison to the global average based on statistics from 2012 to 2016) (FAOSTAT 2017)

Country	Area (ha × 10^6)	% area (share of global)	Production (million t)	% production (share of global)	Productivity (t/ha)
Sri Lanka	**0.081**	**0.15**	**0.92**	**0.1**	**11.4**
India	8.26	14.9	119.0	11.6	14.4
Bangladesh	5.33	0.96	4.5	0.4	8.4
Japan	0.38	0.68	9.9	0.96	26.4
USA	10.2	1.8	34.4	3.35	33.6
China	22.7	41	513.7	0.1	23.0
World	55.4	100	1026.7	100	18.6

Table 6.2 Comparison of the average yield of selected vegetable crops grown in Sri Lanka (Source: IASRI 2018)

Crops	Average yield (t/ha)			Potential productivity (t/ha)	Max. productivity reported (t/ha)
	Sri Lanka	India	World		
Tomato	**18.9**	20.7	33.8	60-80	70.5 (USA)
Eggplant	**12.8**	18.6	26.5	40-50	34.7 (Japan)
Okra	**11.1**	12.0	7.8	15-20	17.8 (Jordan)
Cabbage	**25.0**	22.9	29.4	30-40	42.6 (Japan)

population with the ever-increasing demand for food. To feed the increasing population, it is necessary to utilize agricultural lands to the best of its capacities. One way of achieving higher productivity is to increase the per plant yield through genetic means.

6.2.1 Genetic Improvement of Vegetable Crops

Plant breeding is an endless challenge to improve crop plants by taming and training to suit human needs. To develop a new variety, one has to improve an already present trait or to add an entirely new trait. Typically, these traits include high yield, resistance to certain biotic and abiotic stresses or specific quality characters. The improvement of plant species, though insensible, started during the Stone Age and may last continuously. History of “insensible” plant breeding in Sri Lanka runs into Neolithic Ages. Among crop plants, the most notable crop, i.e. rice, is mentioned in chronicles of ancient Sri Lanka. Since rice culture in this country is an essential part of its customs and norms, a large number of detailed records covering different aspects of rice cultivation including breeding can be found. Though the agricultural crops were domesticated in the Asian region (“old world”) 10,000–9000 years before (Hancock 2004), such records on vegetable production and its improvement over the ages are scanty.

6.3 History of Plant Breeding

Plant breeding is a branch of agriculture which involves planting and harvesting of plants for genetic manipulation to develop improved types for the use of mankind (Acquaah 2007). Plant breeding started when our ancestors changed their lifestyle from hunter-gatherers to the producers of selected plants and animals. This lifestyle change was a gradual process during which independent progenitors of plants were transformed into fully human-dependent domesticated varieties. During this period, primitive people also discovered the most basic plant breeding technique—selection, the art of discriminating variation in a plant population to identify and pick desirable variants which even today remains the primary strategy for crop improvement.

Variability is the key to selection. During the very early stages of plant breeding, the exploited variability was confined to the naturally occurring variants and wild relatives of crop species. This form of selection is practised to date by some farmers, where they select most healthy plants or most desirable fruits and save them for planting in the next season. Though it is difficult to identify the exact beginnings of modern plant breeding, certain early observations by many individuals helped to lay the foundation for the discovery of the modern principles of plant breeding. Some selected milestones in plant breeding are presented in Table 6.3.

The above table clearly indicates that the "insensible" plant improvement has ancient origins, but the trait-specific and science-based plant breeding started little more than a hundred years ago.

6.4 Agriculture and Plant Breeding During the Prehistoric Time in Sri Lanka

Though it is difficult to establish as to when vegetable improvement started in Sri Lanka, it can be safely assumed that it is as old as civilization itself. Excavated material evidence tested at Cornell University, USA, has confirmed that the civilization of Sri Lanka dates back to 29,000–30,000 years. Discovery of oats and barley on the Horton Plains dating to about 15,000 BC suggest that Sri Lankans have been engaged in agriculture from prehistoric times.

Long before the recorded plant breeding activities in the "new world", these processes were already underway as early as 9000 BC (Table 6.3) in the "old world" including Sri Lanka. According to *Mahawansa* (the chronicle of ancestors of the country), backed by advanced irrigation technology, a number of rice varieties with specific qualities and a large number of indigenous vegetable varieties were grown in this country as far as 2500 years. The great historian and archaeologist, Dr. S. Paranavitana, describing the agricultural practices in the early Anuradhapura period had said: "Agriculture in ancient days, as it is today, was not confined to irrigated lands. Crops were raised during the rainy season on un-irrigated land. Rice grown on such lands was more sought after than that from irrigated fields".

Other varieties of subsidiary food-grains, beans and pulses were grown on un-irrigated fields; these products comprise an important and essential part of the

Table 6.3 Some selected global milestones in plant breeding (Acquaah 2007)

Year/era	Event
9000 BC	The first evidence of plant domestication in the Tigris river
3000 BC	Domestication of all important food crops in the old world completed
1000 BC	Domestication of all important food crops in the new world completed
700 BC	Hand pollination of date palm by Assyrians and Babylonians
1694	R.J. Camerarius demonstrated sex in plants
1716	Mather observed natural crossing in maize
1719	Fairchild created the first artificial interspecific cross between carnation (*Dianthus caryophyllus*) and sweet William (*D. barbatus*)
1727	Introduces pedigree breeding
1753	Linnaeus introduced binomial nomenclature
1761	Koelreuter produced the first hybrid in tobacco
1866	Mendel published his discoveries
1899	Hopkins describes ear-to-row selection in maize
1900	Rediscovery of Mendel's laws
1903	W. Johannsen developed a pure-line theory of selection
1904	Nilsson-Ehle multiple factor inheritance of colour in the wheat pericarp
1908–1909	Hardy-Weinberg law put forward
1909	Shull developed inbreds to produce maize hybrids
1909	Jones developed first commercial hybrid maize
1917	Pioneer establishes as a first seed company
1926	Colchicines discovered
1934	Harlan used a bulk breeding selection method
1940	DNA discovered to be heredity material
1944 1945	Hull proposed recurrent selection Selection of wilt-resistant tobacco variety "Oxford 26"
1950	McClintock discovered transposable elements
1953	Watson and Crick proposed a model for DNA structure
1970	Norman Borlaug received Nobel Prize for the green revolution/Berg
1994	Cohen and Boyer introduced the recombinant DNA technology
1995	First genetically modified tomato produced
1996	*Bt* corn developed
2004	Roundup ready soybean introduced Roundup ready wheat developed

people's diet. The main edible oil was that extracted from sesame ("*Thala*") also grown on highlands. Naturally, the maintenance of these varieties was done by saving seeds from the best-looking plants or most desirable fruit for planting the next season.

The needs of clothing of the population were satisfied with the cotton grown locally. Varieties of fruit trees were grown, the mango receiving special mention. Coconut groves are mentioned in an early inscription as well as in literary works, and its nut was eaten when tender. Table 6.4 depicts the important milestones of agriculture and crop improvement in Sri Lanka.

Table 6.4 Developments in agriculture and plant breeding activities with reference to vegetables in Sri Lanka

Year/era	Event
2500 ybp[a]	Numerous rice varieties and vegetables were grown
1884	Establishment of the School of Agriculture in Colombo and experimental stations at Toppur, Panapitiya, Minuwangoda, Galle, Mullaitivu, Akmimana, etc. started
1889	The *Magazine of the School of Agriculture, Colombo*, started
1900	A laboratory for the promotion of pure science opened at Peradeniya
1901	Closure of the School of Agriculture in Colombo
1902	First experiment station established at Gangaruwa (Gannoruwa) Coffee Estate near Peradeniya. Plant breeding experiments on peas and Indian corn started
1904	Ceylon Agricultural Society founded and the publication of *Tropical Agriculturist* initiated. First scientific paper titled Plant Breeding appeared in the *Circulars* and agriculture journal of the RBG, *Ceylon*
1912	Establishment of the Department of Agriculture
1916	Establishment of the school of Tropical Agriculture at Peradeniya
1921	The first record on the importation of vegetable and flower seeds
1935	Economic botanist took over the departmental seed stores
1937	Three vegetable seed stations (Matale, Tabbowa and Karadiyanaru) were established to multiply promising selections of local vegetables
1938	Two departmental promising cowpea pure lines found. Cross made between Marglobe x local varieties to develop a variety with low fruit cracking
1939	Local tomato variety and a superior brinjal variety found to be resistant to bacterial wilt. Seeds of these varieties multiplied for distribution to vegetable seed stations
1940	Wilt-resistant brinjal variety released. First-ever consumer survey to formulate objectives for vegetable breeding
1941	Interim or final selections of vegetable varieties (cowpea, V248 and V443; okra, H10; brinjal, SM 81; bitter gourd, MC 42; snake gourd, TA 77; luffa, LA33; capsicum, CA8) made available to the public
1944	Another seed station (10 acre) at Ambewela started for exotic vegetable seed production. Seed set was recorded to be successful in cauliflower, knol khol and cabbage and commercial scale seed production was started for peas, lettuce, cauliflower, radish, rhubarb and dwarf and runner beans
1948	Floral biology of cassava was studied prior to the breeding programme
1949	Wilt resistance of brinjal variety SM164 tested against eight local varieties
1950	Cultivation of improved selection of vegetables was encouraged through Rural Development Societies
1951	Announcement of the selection of cassava variety MU51. Vegetable seed station at Katugastota issued 3070 lbs (1394 kg) of vegetable seeds
1952	3000 lbs (1362 kg) of botanist's vegetable seeds were produced at Katugastota. Hybrid tomato variety development began (local variety × Greater Baltimore)
1953	4900 lbs (2225 kg) of botanist's vegetable selections were raised at Katugastota
1955	Results of the vegetable varietal trials published. Bush bean variety Top Crop and pole bean variety Kentucky Wonder Wax out yielded the others

(continued)

Table 6.4 (continued)

Year/ era	Event
1957	Back cross-breeding of okra (H10; high-yielding selection × Red *bandakka*; resistant to mosaic virus) for virus resistance reported. Wild tomato (*L. pimpinellifolium*) used to develop bacterial wilt resistance tomato
1967	Central Agricultural Research Institute (CARI) opened
1977	Based on 22 agro-ecological zones, eight major zones were identified for practical purpose and regional research stations established in these zones
1988	Plant Genetic Resource Centre (PGRC) established
1994	Horticultural Crop Research and Development Institute (HORDI) and other commodity-based Agricultural Research Institutes established

[a]*ybp* years before present

6.4.1 Vegetable Breeding During the Pre-independence Era

The present-day breeding is based on several well-known scientific principles introduced by Mendel in 1866. The role of genetics as a discipline and principles of genetics have become essential tools in plant breeding after the rediscovery of Mendel's laws in 1900. During this era, Ceylon (renamed as Sri Lanka during 1978) has been looked upon as the botanical centre of the British Empire's tropical possessions. In the activities of tropical agriculture, Ceylon has in the past played the part of a pioneer, and the scientific workers of the Botanical (later Agricultural) Department had equally been pioneers in their work affecting tropical vegetation (Stockdale 1921).

The first pre-independence work related to crop varieties appeared in Volume I (1) of the *Magazine of the School of Agriculture, Colombo*, published in July 1889. It carries notes from the experimental stations at Bandaragama, Galle, Kegalle and Marapana. Agricultural instructors worked at these stations reported the cultivation of rice, dhal (*Cajanus indicus*), arrowroot (*Maranta arundinacea*) and exotic and native vegetables in these experimental stations. It is not clear the purpose of these cultivations, but there were no indications whatsoever given in the reports on breeding activities carried out using these crops. A detailed account on brinjal varieties and cultivation practices was published in No. 12 of the same magazine (Kumaravelu 1889). De Silva (1890) reported three foreign tomato varieties, namely "The large Red", "The Yellow" and "Red Currant" as the most popular both among farmers and the consumers. In 1887 another garden was opened in Matara, which is neither a market nor an experimental station, but later (1890) developed into a combination of both (Anon 1890a, b). Native vegetables such as okra, brinjal, chillies and dhal and some common exotic vegetables such as lettuce, radish and beetroot were successfully cultivated in this garden too.

Whitaker (1979) described the milestones in vegetable breeding in other countries from 1900 to 1970. In the early twentieth century, many plant pathologists worked on vegetable breeding, and their work is mostly based on the objectives to produce high-yielding, highly nutritious, high-quality and pest-free, adapted cultivars of a specific crop which withstand distance transportation. The rulers of this country may definitely be influenced by the work conducted in the "new world in comparison to India, Java and Egypt where experiments in plant breeding had already been in progress for some time, Ceylon was far behind due to the scarce interest shown by the '*inhabitants of Ceylon*' in the improvement of their own crops though some experiments were carried out in 1903–1904. Mr. R.H. Lock was appointed as a specialist in plant breeding cum assistant director of the Botanic Gardens of Peradeniya in 1902 who undertook some research on maize improvement which were said to be first-ever experiments in plant breeding in a tropical country and predicted that "the colony will soon overtake its rivals" with regard to plant improvement and might have directed the relevant parties to conduct experiments of the same nature. The introduction of "nearly all best kinds of plants" as yet existing was also reported by Lock in Lock 1904. Without a doubt, these introductions would have definitely included best vegetable varieties existed during that period.

By 1908, rulers of Ceylon were in the opinion that the future success in agriculture will lie in the hands of the skilled scientific breeders of plants (Anon 1908). Mac Millan (1905a) described 25 types of yams and identified 6 best local varieties. He reported that the efforts were being made at Peradeniya to collect all best yams. By 1905, British were trying to utilize enormous plant diversity that exists in Ceylon and to introduce commercially valuable species from elsewhere. This was evident in the paper of Mac Millan (1905b) on fruit cultivation where he describes: "Whatever else may be said on to the condition of fruit culture here (Ceylon) the introduction and acclimatization of species and varieties from the other countries is a striking feature and probably no other country of equal area can show as great an assortment of fruits as may be seen in Ceylon".

In May 4, 1908, just 8 years after the rediscovery of the Mendel laws, Mr. R.H. Lock made a speech at a meeting of the Board of Agriculture of Ceylon on plant breeding and tropical agriculture (Lock 1908) where four stages of crop development of most cultivated species were highlighted:

1. The simple cultivation of the natural wild species
2. Unconscious recognition of differences in plants by the primitive growers which extended over a great period of time
3. Plant breeding as a trade
4. Application of definite scientific knowledge to the problem of plant improvement

Lock (1908) recognized that many crops in Ceylon like paddy, which have been in cultivation for thousands of years, have passed stages 1 and 2 and have a static condition at the time with regard to the improvements. He proposed to apply scientific methods of breeding, which have never yet been applied to the crops of this country. Lock (1908) recapitulated his speech as "In my mind, there is no kind of doubt that the most promising branch of scientific agriculture at the present time consists in raising improved varieties of existing native and introduced products; and doing it as far as possible in the place where these products are intended to be grown". These improvements of course Lock envisaged were for paddy and revenue-generating crops like tea, rubber, coffee, tobacco and coconut. Vegetables were not considered for improvement at all by this time. Administrative reports of 1909 and in 1910 contain the details of experimental vegetable cultivation (cauliflower, Chinese cabbage and other exotic vegetables) in Nuwara Eliya gardens. The responsibilities of the agriculture sector of Ceylon were vested on Mr. R. N. Lyne, the first director of the newly formed Agriculture Department of Ceylon in 1912.

After the establishment of the Department of Agriculture, records started to appear on the cultivation of dhal varieties, selection of manioc (cassava) varieties, sweet potato, field beans, yams, chillies, onions, eggplant, okra, spinach and lettuce (Administrative Reports of the Department of Agriculture 1918 and 1920).

Stockdale (1921), director of Agriculture submitting a report on "Scientific Research and investigation in regard to the agricultural industries of Ceylon" differentiated the terms of research, investigation, developmental experimentation and exploration. He divided research into two divisions, namely applied and pure, and categorized the application of Mendel's laws of heredity and mutation and evolution under pure research, whereas diseases of economic plants, studies on insect pests of main agricultural crops and paddy breeding and selection and composition of soils were categorized under applied research. Acclimatization work of the botanic gardens, varietal trials, experimental trials of imported types of crops and various cultural methods of different agricultural crops were classified under investigation. Developmental experimentation included the trials in economic crops (tea, rubber, coconut, cocoa, oil palm, fibres, coffee and castor) on a sufficiently large scale, while under exploration category, investigations were grouped into the flora and fauna and natural resources of Ceylon.

Vegetables were not specifically mentioned in this report, but the distribution of improved seed from seed farms and preparation of dried products from sweet potato, manioc, plantain, etc., were suggested under investigations. Due to increasing demand from the public for vegetables and flower seeds in 1922, vegetable seed importation for public cultivation was reported for the first time. For this purpose, a quantity of 19 1/4 lbs. (42,919 packets) of vegetables and flower seeds were imported from various countries including England, France, America, Australia, Japan and India (Administrative Report 1922).

Although the consumption and cultivation of vegetables have expanded by the 1930s, there was no organization for the distribution of pedigreed seeds to growers. In 1937, three vegetable seed stations were established at Matale, Tabbowa and Karadiyanaru for the purpose of providing vegetable growers with seeds carrying a departmental guarantee of quality, purity and viability. Variety trials were started using eight vegetable crop (okra, brinjal, capsicum, cowpea, cucumber, gourds, luffa and tomato) varieties, originated as botanist's selections. These trials were conducted from 1937 to 1940, and the performances of these varieties were reported (Anon 1940b). In 1939, bacterial wilt-resistant tomato and eggplant varieties were identified even though the disease was first recorded in Ceylon during the 1920s.

The list of crop varieties selected for multiplication and of the stations at which it was proposed to grow is presented in Table 6.5. Later Tabbowa and Matale seed stations were further developed to select best strains of local vegetable varieties and to replace mixed inferior varieties used in villages by the farmers for free distribution of seeds on large scale.

Matale seed station produced bacterial wilt-resistant eggplant variety and a strain of bitter gourd that had some degree of resistance to the fruit fly. At Peradeniya, an experiment was conducted during 1939/1940 *maha* season to find out bacterial wilt resistance in tomato using five single plant selections from Talathuoya and mass selection from Talathuoya along with imported varieties. Local single plant selection of tomato showed marked resistance.

These multiplied varieties of local vegetables were considered by botanist of the Department of Agriculture as ideal materials for intensive selection and hybridization. Prelude to selection and hybridization, a questionnaire was circulated among members of the various district agricultural committees and among agricultural instructors of the Department of Agriculture in 1940 to formulate breeding objectives other than "indisputably desirable and urgently demanded qualities like high yield, resistance to drought and diseases and adaptability to a wide range of climatic conditions". The survey was used particularly to answer the most difficult question of defining criteria of quality.

Table 6.6 presents an exemplary mass of data of considerable interest and value generated as a result of the first-ever survey of its kind in this country to perceive needs of the vegetable consumer before embarking on trait-specific scientific breeding. Ever since, the survey of such calibre has not been done to gather information from the end users to formulate breeding objectives for vegetable breeding.

The results of the above survey must have used to decide the characteristics of the varieties demanded by the farmers, the traders and the consumers and to formulate future breeding strategies accordingly. In the modern market-led (demand-led)

Table 6.5 List of crop varieties selected for multiplication and the stations selected for future multiplication

Crop	Varieties	Station selected
Okra	H10 and NWD[a]	Tabbowa
Brinjal	CD[a]	Matale and Tabbowa
Bitter gourd	4 CD varieties	Tabbowa, Karadiyanaru and Matale
Bottle gourd	CD	Karadiyanaru
Capsicum	NWD (Elephant trunk)[b]	Tabbowa
Cowpeas	V4, CD and Trinidad	Tabbowa
Cucumber	CD	Tabbowa
Luffa	NWD ribbed	Tabbowa and Karadiyanaru
Snake gourd	CD	Tabbowa and Karadiyanaru
Tomatoes	CD and Marglobe	Tabbowa

[a]*Varieties from central division and north-western division referred as CD and NWD, respectively*

[b]This is one of the varietal characters of CA8 which was later released as a national variety

Table 6.6 Data of the vegetable consumer survey conducted by district agricultural committees in 1940 (Anon 1940a)

Question	Consumer response
A. *The market*	
1. Percentage of the total production of vegetables handled by markets in towns and village fairs	58% estimated; over 50% is handled by town markets. 26% estimated; less than 50% is handled by town markets. 17% estimated; more or less 50% is handled by town markets
2. Percentage of customers buying for own consumption and of a middle man buying for the town market	50% buy for own consumption. 50% middle man buying
B. *Quality*	
3. Have the consuming public well-defined standard of quality and what emphasis consumers place on flavour and appearance	50% maintained that high and well-defined quality std. by consuming public, 50% answered the reverse. Over 75% replied that appearance is more important than flavour especially in the villages
4. Preference over quality and quantity	84% of the consuming public demand quantity rather than quality
A. *Size*	
5. Should vegetable breeder in this country go all out for size?	53% answered in the affirmative, while 47% in the negative
6. Does large size simplify handling and minimize costs of production?	59% answered yes, while 41% said no
7. Is there a general belief in the existence of a negative correlation between good flavour and large size?	69% replies maintained that flavour varies inversely with the size
B. *Uniformity*	
8. Does the market demand that growers conform to high std. of uniformity in size, shape and colour?	70% replied in the affirmative
C. *Keeping quality*	
9. What fraction of the vegetables consumed in large towns supplied from the immediate neighbourhood and what fraction from long distance?	Depend on the size of the town in which they are marketed
10. Is there an extensive and urgent demand for vegetables of good keeping quality?	76% concede the importance of good keeping quality. The rest consider this as of little importance
D. *Age*	
11. What is the extent and nature of the demand for early varieties?	83% emphasized the importance of earliness in vegetable varieties
12. Are short-age varieties ever grown with a view to catching the early market?	76% answered affirmatively

(continued)

Table 6.6 (continued)

Question	Consumer response
13. What are the other reasons for the preference for short-aged varieties?	(a) Poverty: 64% (b) To escape bad weather: 36%
14. Does the grower prefer varieties that spread their fruiting over a considerable period or short-season varieties?	67% replies in favour of long-season varieties
15. Do growers attempt producing vegetables for the out-of-season market?	A few growers attempt
E. *Growth habit*	
16. In what varieties perennial habit of growth desired?	Brinjals, chillies, beans and gourds
17. In what varieties would a bush habit be preferred?	Bush habit is generally preferred in beans, cowpea and tomatoes
F. *Diseases*	
18. Cite instances where diseases caused complete abandoning of crops	Tomatoes at Ampitiya, Dandagamuwa, Dambadeniya, Galle, Kumbaloluwa, Matale, Rathnapura and Peradeniya due to bacterial wilt, okra at Matale due to mosaic and beans and snake gourd at Mulgampola due to *Agromyza* and fruit fly, respectively
19. Demand for varieties of vegetables resistant to fungus diseases and insect pests	Nearly all replies indicated a general desire for pest- and disease-resistant varieties including bacterial wilt-resistant brinjal, tomatoes and chillies resistant to leaf curl, cucurbit resistant to fruit fly and mildew beans resistant to *Agromyza* and bandakka resistant to mosaic
G. *Individual crops*	
20. Evaluate the importance of the characters mentioned against each crop (a) Tomatoes: size shape, colour and uniformity of fruits, resistant to wilt and fruit crack, keeping quality and earliness (b) Brinjals: size shape, colour and uniformity of fruits resistant to wilt and *Phytophthora* spp., high-branching habit and earliness (c) Cucumber: resistant to mildew and fruit fly (d) Capsicum: thickness and succulence of fruit wall, reduction of pungency and resistant to leaf curl	The characters mentioned against individual crops are all considered sufficiently important to warrant their inclusion in the projected breeding programme

variety design programmes formulated by the leading private breeding companies, data of the above nature are collected for "product profiling" and "trait prioritization", and both of them are identified as vital information to produce varieties with greater acceptance by the end users. However, public breeding programmes in Sri Lanka are generally lack of such information during the formulation of breeding objectives, thereby the whole process of variety development is observed to be supply driven rather than demand driven. Vegetable production during pre-independence era was categorized into two groups:

1. *Truck crops*: those subjected to long-distance transportation before it reached to the markets in big towns.
2. *Market garden crops*: those produced in villages and sold in markets close by.

The consumers of these two types showed sharp division in their opinion regarding the quality of the product. The majority of consumers insisted on quantity rather than quality gave the signal for breeders to go all out for yield, which is still the prime objective of the present-day breeding programmes. Though seven decades has passed since the first-ever survey to feel the aspirations of the vegetable growers, traders and consumers of this country, it is disappointing to note that there are no objectively planned breeding programmes to improve specific quality traits of vegetables at present. Though there are many publications on the quality improvement in rice, this less attention paid to improve the quality traits of vegetables even at present may be well attributed to the reason given in the departmental notes (Anon 1940a), i.e. "If quality is considered to be of little importance, many of the difficulties inherent in vegetable breeding disappear".

Taking into consideration the survey results, economic botanist of the Department of Agriculture announced the release of final or interim selections of vegetable varieties (cowpea, V248 and V443; okra, H10; brinjal, SM 81; bitter gourd, MC 42; snake gourd, TA 77; luffa, LA33; capsicum, CA8) to the growers of Ceylon in 1941 (Table 6.2). Amazingly, even after 78 years, some of these varieties, namely CA8 (capsicum), LA33 (luffa), TA2 (snake gourd), SM164 (eggplant) and MC 43 (bitter gourd), are still popular among farmers and are performing better than recently developed varieties. Major events in vegetable breeding during the period of 1940–1948 are summarized in Table 6.4.

During the period 1941–2017, the Department of Agriculture (DOA) of Sri Lanka has released and recommended 88 different vegetable varieties belonging to 17 vegetable crops. During the last 15 years, Horticulture Crops Research and Development Institute (HORDI) and its regional stations have released 28 varieties of different vegetables. Breeding strategies adopted in developing these varieties, calculated on the basis of information given during variety release procedure, are given in Table 6.3.

This is in clear contrast to the All India Coordinated Research Project (Vegetable Crops) [AICRP (VC)] initiated in 1970 which has developed 407 varieties in 22 vegetable crops during the past 29 years. These crops have been identified for cultivation in different climatic zones of India. Among these, 246 are high-yielding open-pollinated cultivars, 107 are F1 hybrids, and 43 are cultivars resistant to pest and disease. The use of "old"/outdated varieties in Sri Lanka shows in one hand their persistence in adaptability for changing climate and acceptability by the farmers and the consumers over time, while on the other hand, it shows the slow rate of variety development and low replacement rate of old varieties by new high-yielding varieties to cater the current needs. This is well supported by the massive quantities of imported vegetables, 98.7% of the total demand in 2012, to Sri Lanka as recommended varieties (Agstat 2013). The age classes of varieties released by DOA from 1941 to 2019 are presented in Fig. 6.1.

It is important to note that there are two categories of vegetable varieties introduced by DOA. One such category is released varieties ("tailor-made varieties") that are developed through selection and hybridization procedures by the breeders of DOA, and the breeder seed production and further seed multiplication are responsibilities of the respective commodity research institutes and Seed and Planting Material Development Center (SPMDC) of DOA. List of released varieties is given in Tables 6.5, 6.6 and 6.7. The second category is recommended varieties ("ready-made varieties") which include the varieties tested extensively island wide by DOA for their adaptability and recommended for cultivation in Sri Lanka. As there is no responsibility of DOA to produce and multiply the seeds locally, seeds of almost all of the recommended varieties are imported annually, e.g. seeds of almost all upcountry vegetables, capsicum variety (Hungarian yellow wax – HYW); bean varieties such as Kentucky Wonder Green, Wade, Top Crop and Cora Black; low-country vegetables such as Naga (luffa) and tomato varieties such as Padma, Platinum, Ceres and Big Beef.

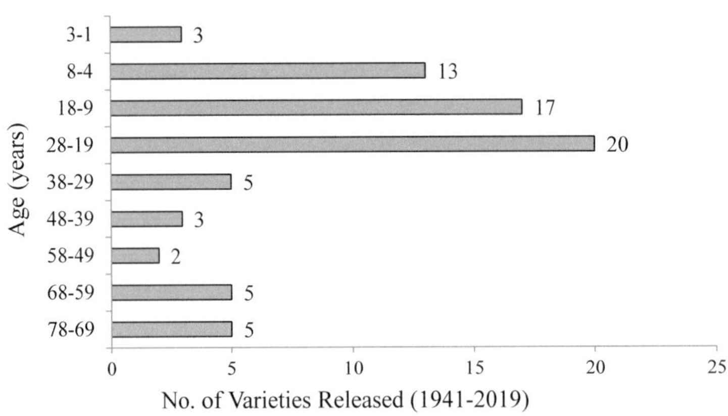

Fig. 6.1 Age classes of vegetable varieties released calculated from the year of release (1941–2019)

Table 6.7 Vegetable varieties released during 1950–1970

Crop	Variety	Year of release	Station released
Snake gourd	TA2	1950	Peradeniya
Cucumber	LY58	1950	Peradeniya
Bean	Top Crop	No records	Peradeniya
Radish	Lanka Ball Beeralu	1960 1960	Gannoruwa Gannoruwa
Capsicum	HYW	1960	Gannoruwa
Snake gourd	MI short	1960	Mahailuppallama
Tomato	Katugastota Wilt Resistant (KWR)	1960	Gannoruwa
Bitter gourd	Thinnavely white	1970	Thirunelvelly
Snake gourd	Thinnavely long	1970	Thirunelvelly

Table 6.8 Vegetable varieties released during 1971–1990

Crop	Variety	Year of release	Station released
Okra	MI 5	1975	Mahalluppallama
	MI 7	1975	
Vegetable cowpea	BS-1	1978	CARI
	Bushita	1987	
	BS 3	1987	
Tomato	T 146	1981	CARI
	T 246	1981	
	Bienz	1982	

6.4.2 Vegetable Breeding in Post-independence Era

Soon after independence, work carried out by the Department of Agriculture with respect to vegetable breeding seemed not deviated much from the activities during pre-independence. By the time there were quite a large number of pure-line selections of vegetables were released, and the seed production of those released varieties at Katugastota seed station and their distribution among local growers were given high priority. This was evident by the threefold increase in the quantity of seed produced in 1953 (3300 kg) compared to 1300 kg of seeds produced during 1951.

Three years after initiation of floral studies of cassava, in 1951, a famous cassava variety MU 51 which is popular even to date was released. Until 1950, botanist's selections of local vegetable varieties were released, and names were given abbreviating the first two letters of the botanical name of a particular crop. The period of 1950–1960 marks a turning point by decentralizing (Schokman 1981) vegetable breeding, where regional research stations are too embarked on their own vegetable variety development activities resulting in new varieties named on the basis of the place of origin, e.g. MI (Maha Iluppallama), Thirunelvelly and Katugastota (Tables 6.7 and 6.8). As a result of decentralization and initiation of individual breeding programmes at different research institutes with specific objectives to cater regional needs, the centralized breeding programme conducted by the botanist with national interest lost its coordination and organized nature after the 1950s. The Central

Agricultural Research Institute (CARI) inaugurated in 1967 at Gannoruwa had the administrative structure similar to pre-independence period to uphold the key role played by the pathologist, entomologist and botanist in the variety development process. However, great success was achieved with root and tuber crop development programme, while vegetable variety release was almost confined to tomato (Table 6.7).

The inception of Horticultural Crop Research and Development Institute (HORDI), the commodity-based research institute, in 1994 marked the beginning of a research approach exclusively for horticultural crops similar to rice. During the last 15 years, 28 varieties of different vegetables were released by HORDI and the regional stations under its purview (Table 6.9). Breeding strategies adopted in developing these varieties are given in Table 6.9.

6.5 Resource Allocation on Vegetable Breeding

In the National Agriculture Research Systems (NARS) of Sri Lanka, crop breeding is handled by the four Institutes, namely Horticultural Crops Research and Development Institute (HORDI), Field Crops Research and Development Institute (FCRDI), Fruit Crops Research and Development Institute (FRDI) and Rice Research and Development Institute (RRDI). The Department of Agriculture was mandated to undertake the programmes on vegetable improvement and other related disciplines.

Vegetable breeding is carried out solely as a public venture. Therefore, resource allocation to HORDI (has seven research centres and stations at different agro-ecological zones of Sri Lanka) in comparison to other research institutes of DOA to undertake vegetable variety development activities will be discussed here.

6.5.1 Financial Resources

During 2005, the government has allocated 25% of the recurrent budget to DOA, highest among other institutes, for agricultural research activities. From the total recurrent budget (692.21 million Sri Lanka rupees), 19% is allocated for the research projects in plant breeding and genetics, while agronomy and soil science-related studies were given 16% and 11%, respectively. Among 18 commodity groups, vegetables were allocated 6% of the total recurrent budget, while 8% was for cereals. Percentage of research projects and % officers involved in plant breeding activities of different commodity groups of DOA is shown in Table 6.10. During 2017, a sum of 873 million rupees was allocated for four research institutes, and around 357 million rupees (41%) was allocated to HORDI (Performance Report 2017).

Table 6.9 Vegetable varieties released during 1991–2017

Crop	Variety	Year of release	Station released
Tomato	Vihara F1	1993	CARI
	T 245	1993	CARI
	Thilina	1999	HORDI
	Tharindu	1999	HORDI
	Ravi	1999	HORDI
	Rajitha	2001	HORDI
	Rashmi	2001	HORDI
	Maheshi F1	2005	HORDI
	Lanka Sour	2006	HORDI
	Bhatiya F1	2008	HORDI
	KC1	2008	HORDI
	Lanka Cherry	2010	HORDI
Brinjal	Padagoda (BW 11)	1996	Bombuwela
	Amanda F1	2005	HORDI
	Anjalee F1	2005	HORDI
	HORDI Leanairi 1-F1	2011	HORDI
Tibbatu *(S. torvum)*	Bindu	2004	Aralaganwila
Luffa	Asiri	2002	Bombuwela
	Gannoruwa Ari	2013	HORDI
Okra	Haritha	1993	Mahailuppallama
Bean	Keppetipola Nil	2002	HORDI
	Balangoda Nil	2002	HORDI
	Lanka Butter	2002	HORDI
	Premel	2005	HORDI
	Sanjaya	2005	HORDI
	Gannoruwa Green	2011	HORDI
	Bandarawela Green	2011	Bandarawela
	Gannoruwa Bil	2013	HORDI
	Gannoruwa Kekulu	2017	Bandarawela
Pumpkin	Ruhunu	1993	Angunakolapelessa
	Padma	2016	HORDI
Vegetable cowpea	Sena	1997	Bombuwela
	Panduru polon Mae	2003	Bombuwela
Winged bean	SLS 44	1999	HORDI
	UPS 122	2004	HORDI
	Krishna	2006	HORDI
Leafy vegetables	Mukunuwenna (Piliyandala)	1999	HORDI
	Gotukola (Meerigama)	1999	HORDI
Capsicum	Lanka yellow wax	2004	HORDI
	Parthana F1	2015	HORDI
Carrot	Lanka carrot	2012	Bandarawela

(continued)

Table 6.9 (continued)

Crop	Variety	Year of release	Station released
Cucumber	Champion/ kalpitiya white	2004 2004	HORDI HORDI
	HORDI Green F1	2012	HORDI
	Gannoruwa White F1	2013	HORDI
Bitter gourd	Matale Green	2006	HORDI
	Neeroga F1	2015	HORDI
Thumbakarawila (spine gourd)	Thumbika (female)	2004	Aralaganwila
	Golika (female)	2004	Aralaganwila
	Parakum (male)	2004	Aralaganwila
	Visal (female)	2015	Angunakolapelessa
	Kesara (female)	2015	Angunakolapelessa
	Chandu F1	2015	Angunakolapelessa
	Visma (male)	2015	Angunakolapelessa
Yard long bean	Polon Mae	–	Bombuwela
	Gannoruwa Hawari	2011	HORDI
	A9	2016	HORDI

Table 6.10 Comparison of percent funds allocated, number of research projects (%) and number of scientists involved (%) in plant breeding and genetics, and cereal and vegetable breeding activities in DOA (Based on INFORM 2005)

	Breeding and genetics	Vegetable breeding	Cereal breeding
Funds allocated (%)	36.4	6	8
Number of research projects (%)	35	10	18
Number of scientists involved (%)	24	6	9

6.5.2 Human Resources

An overwhelming majority (43%) of the agricultural scientists in NARS is attached to the DOA under different disciplines of agriculture (INFORM 2005). Out of the scientists working in DOA, 15% are engaged in plant breeding and genetic research, while 19% and 9% are involved in agronomy and soil research, respectively. Commodity-wise research engagement of the scientists of DOA shows that 6% are working with research related to vegetables, while 9% are engaged in research related to cereals. The total number of scientists and number working on plant breeding at four research institutes of Sri Lanka during 2017 are given in Fig. 6.2.

Academic qualifications of the officers attached to DOA are encouraging. Among these officers, 31% possess a basic degree, 41% have master degrees, and 23% hold PhD degrees, while 5% do not hold a basic degree (Performance Report of DOA 2017). Though the academic qualifications of the officers of the DOA are very high, the majority of the officers who earned postgraduate degrees in other discipline are engaged in plant breeding research. Only a very few officers have postgraduate

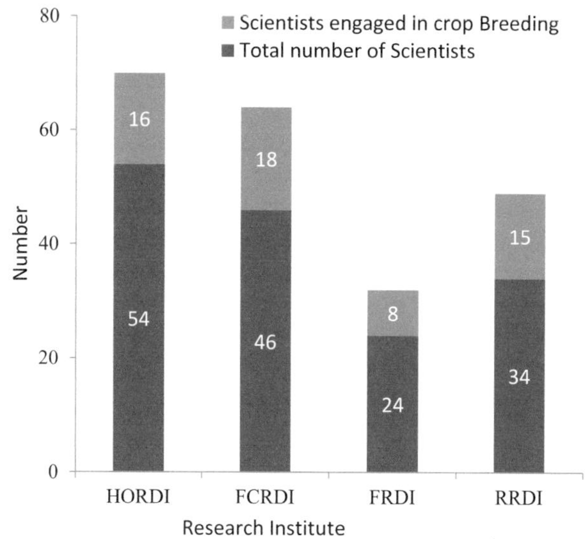

Fig. 6.2 Total number of scientists and number working on plant breeding at four research institutes of Sri Lanka (based on data of Performance Report of DOA 2017). *HORDI* Horticultural Crops Research and Development Institute, *FCRDI* Field Crops Research and Development Institute, *FRDI* Fruits Crops research and Development Institute, *RRDI* Rice Research and Development Institute

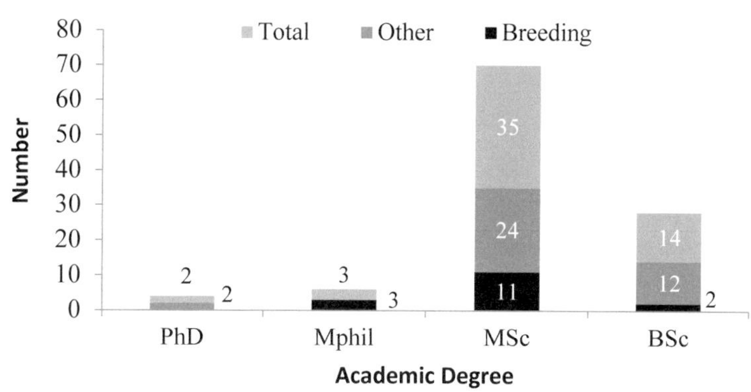

Fig. 6.3 Numbers of scientists working at Horticultural Crops Research and Development Institute (HORDI) system and their academic qualifications

qualifications in plant breeding and genetics. Even among the HORDI scientists, only 22% are having postgraduate qualifications in plant breeding (Fig. 6.3).

6.5.3 Genetic Resources

Germplasm for the vegetable breeding programmes was derived from foreign (De Silva 1890) and local sources since pre-independence. At the beginning, responsibility of the collection of local varieties was given to experimental station established in different districts of the country (Administrative Reports 1887–1889, Anon 1890a, b). The performance of the germplasm collection was shown to the public as agricultural exhibitions. Collections of yams, okra, brinjal, manioc and many other vegetables were the popular exhibits in these exhibitions (Mac Millan Mac Millan 1905a, b). Exotic vegetables such as cauliflower and Chinese cabbage were grown in Nuwara Eliya gardens at an experimental level using imported germplasm (Administrative reports 1909 and 1910, Pieris 1945). Germplasm collection and characterization was handled by systematic botanist at Peradeniya after the formation of the DOA in 1912. Varieties were collected from central, north-western division, etc., and evaluated at different seed stations. These *improved strains* of vegetables were later released for general cultivation (Table 6.4). After decentralization of research activities, regional stations too had their own collection of local and foreign germplasm (derived from international institutes) and were maintained by the respective breeders. After the establishment of the Plant Genetic Resource Center (PGRC) in 1988, most of the germplasm collected at regional centres were handed over to PGRC, and the centre itself started the collection of local germplasm through expeditions and foreign germplasm through official channels. The germplasm stored at PGRC of some selected vegetable crops is given in Table 6.11.

Most of the local accessions collected by PGRC are landraces and traditional varieties grown by the farmers and possess heterogeneity to a certain degree. Therefore, breeders quite often find that duplicates and low genetic purity in these accessions.

Instances of receiving vegetable germplasm from international institute (World Vegetable Centre formally AVRDC) were very few. Such an occasion is South Asia Vegetable Research Network (SAVERNET) initiated in the year 1997 where some tomato lines of AVRDC were tested for local adaptability and later released a variety known as T 245 (AVRDC 2001).

Table 6.11 Germplasm availability of selected vegetable crops at PGRC in 2018

Plant category	Number of samples
Bean	532
Eggplant	253
Tomato	283
Pumpkin	82
Cucumber	104
Kekiri	150
Cabbage	15
Onion	19
Okra	483
Leafy vegetables	209
Other vegetables	2877
Total	5007

Source: Progress Report of PGRC- 2018

6.5.4 Breeding Strategies

The major goal of variety development in vegetables is to increase productivity through the development of high-yielding varieties, with resistance to major biotic (pest and diseases) stresses while maintaining high quality. With the threat imposed by global climate change, inclusion of strategies to address tolerance or resistance to abiotic stress such as heat, drought and flood became imperative. However, until recent times, vegetable breeding for abiotic stresses is confined only to heat tolerance in tomato, which seems to achieve mainly through introduction and selection of heat-tolerant varieties from international institutes like the World Vegetable Center (formerly Asian Vegetable Research and Development Center, AVRDC, in Taiwan). As a result, variety KC1 was introduced during 2008.

Since 1970, introduction and selection seemed to be the cheap and popular strategy of the development of varieties for other economic traits. Table 6.10 shows the number of vegetable varieties developed by the DOA and breeding strategy used since 1940. Majority of varieties released (or recommended) so far were the result of introductions (in case of exotic vegetables) or selections (through evaluation of local germplasm or landraces). In addition to the availability of genetic variation among local germplasm collections, breeders seem to prefer this method since knowledge, skills, effort, funds and time needed are less compared to the other methods of breeding. Breeding strategies adopted in developing the varieties released and recommended calculated on the basis of information given during the variety release procedure of DOA are given in Table 6.12.

Except for a very few studies such as Fonseka et al. (2002) on floral biology and other basic research related to pollination and reproduction of vegetables are very scanty. Due to various reasons, development of pure-line varieties through hybridization and selection seems very limited (Regunathan 1980) in the DOA. Therefore, genetic studies and use of molecular breeding techniques in the future may suffer greatly as in the past due to non-availability of segregating populations and mapping populations.

Development of F1 hybrid vegetables has recently attracted a lot of attention (Giritharan and Arulandhy 1994–1995; Alwis et al. 2005; Millawithanachchi et al. 2006; Ilangakoon et al. 2004; Ratnapala et al. 2010; Hitinayake et al. 2017). The F1 hybrid development programmes initiated for tomato and eggplant were culminated

Table 6.12 Breeding methods used and number of varieties released from 1940 to 2017 by DOA (Source: Unpublished data of DOA)

Breeding method	Number of varieties released
Exotic varieties introduced and recommended (almost all are upcountry vegetables)	20
Varieties selected and released from local/ indigenous/foreign germplasm	58
Varieties hybridized and selected by DOA breeders	10[a]
Total	88

[a]Number of varieties with pedigree

in 2005 by releasing two eggplant and one tomato varieties. At present, experiments are underway for F1 hybrid development of other vegetables. Once a compatible parental line is identified prior to hybrid development, it is a matter of crossing those selected parents to produce F1 hybrid.

As F1 hybrid development requires less time, money, space and attention compared to pure-line varieties, lots of breeders are now embarked on it. Germplasm availability for the development of parental lines for F1 hybrids may not be a serious problem currently but could be a major problem in the near future for the crops having less germplasm availability with low genetic diversity (capsicum, bitter gourd, luffa, etc.). The only solution is to identify promising inbred lines from segregating populations.

However, promising varieties produced ("pipeline varieties") by any means have to be tested for their adaptability under researcher-managed conditions at different agro-ecological regions of the country in the National Coordinated Varietal Trial (NCVT) as well as under farmer-managed conditions in Varietal Adaptability Trial (VAT). To have a continuous flow of pipeline varieties into NCVT and VAT, there should be a continuous breeding programme starting from germplasm evaluation to Major Yield Trial (MYT) passing all the other essential steps in the process of variety development.

6.5.4.1 Resistant Breeding

In Sri Lanka, very few research stations undertake screening of vegetable germplasm for major pest and diseases as a routine task. Even in those stations, there is no systematic screening of available germplasm for major pests of vegetables. Objectively developed breeding programmes for pest and disease resistance are scarce (Abeygunawardena and Siriwardana 1963; Arasakesary et al. 2013; Rajapakse et al. 2005; Samarajeewa et al. 2005). Back crossbreeding has not been practised by the breeders. Instead, cross-breeding progenies of resistant varieties are evaluated mostly for field resistance. Under these circumstances, luck and chance will decide success.

Instances of use of wide hybridization are confined to a few vegetable crops, i.e. okra (Samarajeewa et al. 1998) and brinjal (Perera and Wrashamana 1987); however, the country has not been able to release varieties based on such techniques to date. Use of wild relatives of crop plants to incorporate valuable traits has not been systematically tried in any vegetable breeding programmes even though tissue culture and other facilities are available to undertake such research programmes. A tri-nation research project on pre-breeding of eggplant using wild relatives emanating from different gene pools is now underway, of which Sri Lanka is a partner country with Spain and Ivory Coast (Maria et al. 2012; Plazas et al. 2015).

6.5.4.2 Molecular Breeding

Molecular breeding and its applications have resulted in a breakthrough in plant breeding from its introduction to agriculture in the late 1980s. Mapping of genomes and detection of genetic markers of qualitative and resistance traits of cultivated and wild types of crop varieties in the USA, China and various other countries have attained their ultimate target of having a series of transgenic crop varieties.

Tomato, sweet pepper (*C. annuum*), carrot, cabbage and lettuce are among the vegetable crops subjected to extensive improvements through molecular breeding. The result of this effort has benefitted farmers through improved resistance to insect pests, diseases and abiotic stresses and also the consumers through improved nutritional value (Ortiz and Dias 2011). For example, ten transgenic tomato varieties have been approved for commercialization in the USA and China by the year 2000. The number of successful genetic engineering efforts on fruit quality and abiotic and biotic stress tolerance in tomato germplasm in the world during 2000–2013 was nearly 85 (Gerszberg et al. 2015). Halterman et al. (2016) reported that a record of 18 million farmers have planted 447 million acres (180.9 million ha) of GM food crops in 28 countries.

Despite the fact that several molecular genetic laboratories have been established few years ago their programmes are restricted to basic molecular genetic studies (Prematilake et al. 2002) and training of technicians in the field. No vegetable variety has been developed in Sri Lanka using molecular techniques. Tissue culture techniques so far have not been used as a tool to facilitate breeding programmes in vegetables. Even isozyme markers have rarely (Pathmaraja et al. 2005) been used in vegetable breeding. Use of molecular techniques in vegetable breeding is confined to characterization of germplasm using RAPDs and some other markers (Samarajeewa et al. 2007; Millawithanachchi et al. 2006; Weerasinghe et al. 2004; Samarasinghe et al. 2009; Welegama et al. 2015). Due to financial constraints and lack of facilities and trained human resources, use of specific genetic markers such as RFLP, AFLP, SSR and SNP in vegetable breeding is very limited. Molecular accelerated breeding (MAB) techniques (i.e. tagging of desirable commercial traits using molecular markers, marker-assisted selection and incorporation of resistances) have not been attempted at all on vegetable crop species to date.

Despite a considerable number of varieties that have been released in Sri Lanka recently, the productivity of vegetable crops has not improved significantly. During both pre- and post-independence era, the vegetable crop improvement has not received due attention and recognition compared to rice and other revenue-generating crops. There was no concerted national effort to develop vegetable breeding programmes during the past 40–50 years. Lack of prioritized national breeding objectives, deficiency in objectively developed breeding programmes, insufficient financial allocations and lack of trained personnel are some of the major constraints in local vegetable breeding efforts.

6.5.5 Breeder Seed Production

Breeder seed production of varieties released by DOA is entrusted to respective breeders of relevant crops at various research stations. Breeder seeds of more than 35 vegetable varieties are produced at various research institutes located at different agro-ecological zones. The service of breeders of the DOA to maintain trueness to the type of some vegetable varieties released before independence until to date

Table 6.13 Summary of the present crop improvement strategies adopted for certain vegetable crops by plant breeders of DOA

Crop	Improvement strategy
Beans	Introduction, selection and evaluation (ISE)/hybridization and selection (HS)/ maintenance breeding (MB)[a]/adaptability testing (AT)
Bitter gourd	ISE/hybrid breeding (HB)/purification (PF)/AT
Brinjal	ISE/HS/HB/MB/AT
Capsicum	ISE/MB/HB/PF/participatory breeding/AT
Carrot	ISE/HS/AT
Chilli	ISE/HS/MB/HB/AT
Cucumber	MB/HB/AT
Exotic vegetables	AT
Exotic vegetables	AT
Luffa	ISE/HS/MB/HB/AT
Mae	ISE/MB/AT
Moringa	ISE/CP
Okra	ISE/HS/MB/HB/AT
Pumpkin	ISE/MB/AT
Radish	MB
Root and tubers	ISE/CP
Tibbatu	ISE/MB
Tomato	ISE/HS/MB/HB/AT/mutation breeding
Thumba karawila	MB/CP
Vegetable cowpea	MB
Winged bean	MB

[a]MB includes breeder seed production

should be highly commended. Breeders who engaged in the breeder seed production of "old varieties" are confronted with the challenge to maintain original characters of some of those varieties since characterization details are not available at present. Summary of crop improvement strategies presently adopted by vegetable breeders of the DOA is presented in Table 6.13.

Most of the seeds of upcountry vegetables are imported by the private sector. Figure 6.4 presents the trends in quantity and number of vegetable varieties imported to Sri Lanka during the last 4 years. It indicates almost similar quantities of vegetable seeds were imported during 2015 to 2018 while, total number of vegetable varieties imported was increased tremendously.

Though the recommended varieties of almost all low-country vegetable seeds are produced under the government seed production programme, seeds of some of the low-country vegetable varieties are among the top 10 vegetable seeds imported to Sri Lanka in the recent years (Fig. 6.5).

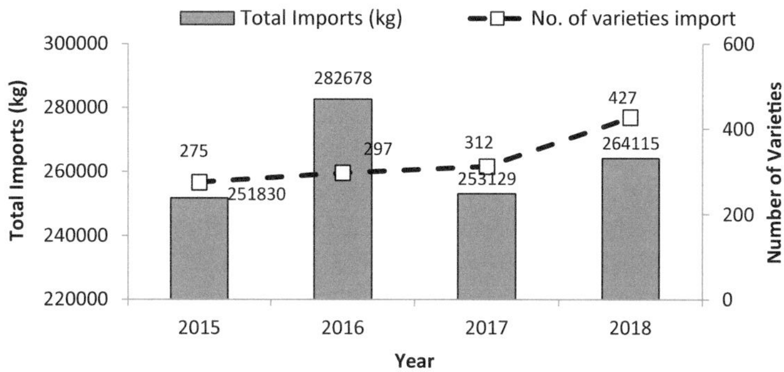

Fig. 6.4 Number of varieties and quantities of vegetable seed imports (kg) to Sri Lanka during 2015–2018 (based on unpublished data of Seed Certification and Plant Protection Centre (SCPPC) of DOA, Sri Lanka)

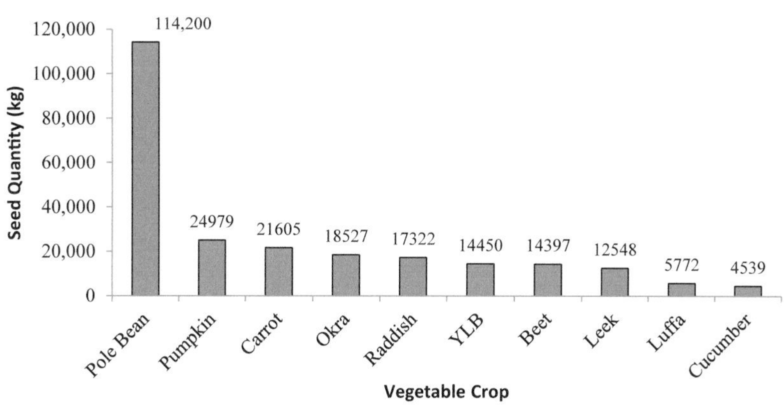

Fig. 6.5 Top 10 crops of vegetable seed imports from 2018 to Sri Lanka (Source: Unpublished data of Seed Certification and Plant Protection Centre [SCPPC] of DOA, Sri Lanka)

To improve the quality and to reduce the cost involved in seed production, very few studies such as extended pollination duration for seed production (Fonseka et al. 2008, 2010) and fruit thinning and canopy management to improve seed quality (Fonseka et al. 2013; Sooriyapathirana et al. 2013) have been reported.

The quality-assured vegetable seed supply, formally shared by the public and private sectors, is around 25–30%, while the informal supply sector (without quality assurance) (i.e. farmers' saved seeds, seeds exchanged between farmers) fulfils the rest of the requirement. Price of the seeds supplied by the formal sector is expensive than the informal sector as the quality assurance process by the former is costly. However, compared to imported "low-country vegetable" seeds, locally produced seeds within the formal sector hold a lower share. However, locally produced F1 hybrid vegetable seeds are cheaper than their imported counterparts. Though seeds produced locally are cheap, the quantities available in the market during the growing season cannot fulfil the demand in certain crops. The insufficient quantities produced may be attributed to the following: (1) unavailability of basic seed of some vegetable crops in required quantities, making it impossible to multiply them into certified seeds, (2) size of the seed market being not profitable for a given vegetable crop as result of cultivation of a large number of vegetable crops varieties, (3) requirement of a large isolation facility due to cross-pollinated nature of vegetable crop specics, (4) restrictions in production areas as seed production is normally confined to certain agroclimatic zones, (5) high production costs mainly due to labour shortage and higher wages, (6) fresh market vegetable being more profitable sometimes than the seed production (i.e. hot pepper/chilli) resulting in the reluctance of contract growers to engage in seed production, (7) time-consuming seed certification process resulting in delays in getting payments and (8) release of low-quality seeds (e.g. less viable) due to lapses in the seed certification process.

6.5.6 Dissemination of New Improved Technologies

The importance of technology transfer from a development perspective is nothing new. More than three decades back, Mansfield (1975) pointed out that "One of the fundamental processes that influence the economic performance of nations and firms is technology transfer". It is well recognized that the transfer of technology is the key to the process of economic growth in both developed and developing countries. Transfer of technology generated by the researchers to farmers is considered to be the responsibility of the agriculture extension system. Despite large investments made on technology transfer by both developed and developing countries over the last three decades, anticipated agricultural production is yet to achieve. It is optional that either the technology is to be developed locally or be borrowed from other places/countries with similar environmental conditions and socio-economic situations. The burden on the local extension system is less in introducing packages of borrowed technologies such as imported seeds compared to the technologies developed locally. For instance, a new F1 hybrid vegetable variety imported by the private sector after obtaining the recommendation of DOA could become popular among farmers than a locally developed F1 hybrid variety of the same crop category released by the DOA. Differences and efficiency of the agricultural technology transfer system and subsequent adoption rate depend firstly on the macro factors such as agro-ecological, political-economic, socio-cultural, policy, market intervention, infrastructure, transportation and communication, while secondly on the institutional factors including research, education and training, input supply, credit and farmer organizations.

Although not comparable with regional countries, the recent developments in vegetable breeding in Sri Lanka, as evidenced by the release of new varieties, are commendable compared to other sub-sectors of agriculture. However, the genetic base of vegetable crops is yet to be strengthened through intensive exploration and conservation. Further, lack of some useful conventional breeding strategies (i.e. resistance breeding, pure-line development and male sterility-based heterosis breeding) and use of molecular genetics and other non-conventional genetic tools (i.e. marker-aided selection and gene pyramiding) are considerable deficiencies in the local vegetable breeding programmes. Meanwhile, the effect of breeding programmes is never reflected in the statistics on crop productivity. This phenomenon is justifiable by the fact of less popularity of locally recommended vegetable crop varieties, whereas imported vegetable hybrids are receiving higher attention from farmers owing to their yield and quality advantages as well as their improved seed viability and longevity. Hence, the introduction of exotic and wild germplasm from other countries, safeguarding the IPR issues, together with a technological boost in breeding programmes and seed technology would be a timely need. When the strength of the regional gene banks and the progress in hybrid seed production are considered, collaboration with South Asian countries would immensely benefit Sri Lanka. Meanwhile, the breeding objectives need to be redefined by developing varieties that suit different cropping systems and technological intensities: new product qualities and prolonged shelf-life of vegetables.

6.6 Critical Gaps

Despite the considerable number of varieties developed, the productivity of vegetable crops has not improved. During both pre- and post-independence era, vegetable has not received due attention and recognition compared to rice and other revenue-generating crops by the policymakers, even by the hierarchy of the DOA.

There was no concentrated national effort to develop vegetable breeding programme during the last 40–50 years since scarcity or less availability of vegetables, not as in the case of rice, neither poses serious threat to politicians nor to the consumer as diversity of the group permit to gather wide variety of them from every nook and corner in this country.

National crop improvement programmes serve to bring coordination, communication and empowerment and are focused on the relevance, impact and quality of research. Role of national institutes should be to established breeding priorities, provide programme review and evaluation and recommend appropriate redirection of research. Though the breeding programmes of commodity research institutes of DOA are national programmes run by the public funds, the above characteristics are lacking except rice breeding programme of the country.

Lack of prioritized national breeding objectives, deficiency in objectively developed breeding programmes, insufficient financial allocations and lack of trained persons are some of the major constraints in vegetable breeding. Some salient critical gaps identified are given below:

1. Lack of germplasm of certain vegetables and non-availability of resistant sources for some pest and diseases
2. Lack of basic research programmes to strengthen applied research activities
3. Dependence on data of foreign origin on modes of reproduction, floral biology, frequency of out-crossing, etc.

4. Arbitrary use of available MAB techniques
5. Lack of central coordination and therefore no proper follow-up procedures
6. Lack of infrastructure and funds available for research on vegetables
7. Frequent change in research priorities and crop priorities
8. Lack of appreciation/incentives for breeders

It is high time that Sri Lanka critically reviews its vegetable breeding programmes and develops national goals and lays down strategies to achieve them.

6.7 Recommendations for the Future

Sri Lanka has pledged to pursue sustainable development goals in meeting the global agenda of UNDP by 2030. Some of the 17 global goals are directly or indirectly related to agriculture. No poverty, zero hunger and good health and well-being can be achieved through increased availability of fruits and vegetables for consumption. High-yielding vegetable varieties with enhanced nutraceutical properties (Fonseka et al. 2007) and eating quality will reduce the incidence of malnutrition and related disorders among children and improve quality of life.

Varietal development programmes provides the nucleus to all other agricultural activities. Without a variety, pest and disease management, agronomic and cultural management and extension and training programmes are redundant. Therefore, it is essential to have sustainable breeding programmes to keep live the other disciplines and activities related to the agricultural production process.

1. All island coordinated vegetable improvement programmes should be initiated to provide a national grid for newly developed varieties and technologies by various research institutes and agricultural universities through their inter-disciplinary multi-location research approach.
2. Farmer involvement during the selection of elite materials developed at research stations should be encouraged through participatory breeding programmes.
3. Breeding for resistance/multiple resistance to major biotic and abiotic stresses through well-formulated breeding techniques in collaboration with entomologists/pathologists has to be initiated.
4. Logical use of available molecular accelerated breeding (MAB) techniques to enhance the efficiency of breeding methodologies.
5. Emphasis should be given to improve nutraceutical properties and processing qualities of vegetables.
6. Facilities should be developed for the use of molecular techniques for fingerprinting, tagging of desirable commercial traits using molecular markers and incorporation of resistances.
7. Intensification of research on export-oriented vegetables and indigenous/underexploited vegetables.
8. Establishment of appropriate plant variety protection and intellectual property management laws.

9. Public plant breeding must adjust to meet the demands of private plant breeding since the major opportunities to interact with the private sector are in the discipline of plant breeding than any other discipline in agriculture.

References

Abeygunawardena DVW, Siriwardana AAP (1963) Studies on resistance in tomato to Bacterial wilt. Trop Agri 119(2):55–66

Acquaah G (2007) Principles of plant genetics and breeding. Blackwell Publications, Oxford

Administration Reports of Ceylon 1874–1965 (2013) Agricultural statistics (AgSTAT). Department of Agriculture, Peradeniya

Alwis LMHR, Perera ALT, Fonseka HH, Peiris BCN (2005) Breeding and selection of tomato F1 hybrids for yield and fruit quality characters. Trop Agri Res 17:31–38

Anon (1890a) A retrospect. Mag School Agric Colombo 1(12):113–114

Anon (1890b) An account on the Matara market garden. Mag School Agric Colombo 2(1):4–5

Anon (1908) Plant breeding in Ceylon and its future. Trop Agric Mag Ceylon Agric Soc 30:6

Anon (1940a) Objectives in vegetable breeding: answer to a questionnaire issued by the botanist. Departmental Notes. Trop Agri 95:344–349

Anon (1940b) Vegetable variety trials at Matale, Tabbowa and Karadiyanaru, 1937-1940. Trop Agri 95:149–164

Arasakesary S, Bavaleeswaram B, Athputhachandran P, Vijayarathnam S (2013) Agronomic evaluation and characterization of local landraces, resistant to shoot and fruit borer in brinjal (*Solanum melongena*). Ann Sri Lanka Depart Agri 15:357–360

AVRDC (2001) AVRDC Report 2000. Asian vegetable research and Development Center, Shanhua, Tainan, Taiwan. vii + 152

De Silva (1890) Cultivation of tomato. Mag School Agric Colombo 2(5):38–39

FAOSTAT (2017) FAO statistics on regional agricultural development. Food and Agriculture Organization of the United Nations, Rome

Fonseka H, Chandrasekara A, Fonseka M, Wickramasinghe IP, Kumara KPDRSP, Wickramarchchi WCN (2007) Determination of anti-amylase and anti-glucosidase activity of different bitter gourd (*Momordica charantia* L.) genotypes. Proceedings of the First International Conference on Indigenous Vegetables and Legumes; Prospects for Fighting Poverty, Hunger and Malnutrition. Acta Hortic 752:131–137

Fonseka H, Malkanthi R, Herath AHMAJ, Rathnayake RMC (2002) Floral biology and fruit formation in eggplant (*Solanum melongena* L.) genotypes. Ann Sri Lanka Depart Agric 4:119–128

Fonseka H, Herath K, Devasinghe D, Izuka K (2013) Effect of fruit thinning, canopy management and a number of harvests on seed quality of eggplant (*S. melongena*). Ann Sri Lanka Depart Agric 15:217–226

Fonseka H, Warnakulasooriya K, Senanayake G, Fonseka R (2010) Studies on the effect of extended pollination duration on fruit set and seed development and storage temperature on viability and storability of pollen of eggplant (Solanum melongena L.). Proceedings of Advances in Genetics and breeding of Capsicum and eggplant, pp 473–482

Fonseka HHD, Weerasinghe WAYSP, Bandara DC, Fonseka RM (2008) Effect of temperature on yield, viability and pollen of brinjal (*Solanum melongena* L.) and Capsicum (*Capsicum annum* L.). *Sri Lankan*. J Agric Sci 45:11–20

Gerszberg A, Hnatuszko-Konka K, Kowalczyk T, Kononowicz AK (2015) Tomato (*Solanum lycopersicum* L.) in the service of biotechnology. Plant Cell Tissue Org Cult 120:881–902

Giritharan D, Arulandhy V (1994–1995) Studies on hybrid vigour in the crosses of okra (*Abelmoschus esculentus*) genotypes of diverse origin. Trop Agri 150:65–69

Halterman D, Guenthner J, Collinge S, Butler N, Douches D (2016) Biotech potatoes in the 21st century: 20 years since the first biotech potato. Am J Potato Res 93(1):1–20

Hancock JF (2004) Plant evolution and the origin of crop species, 2nd edn. CAB International, Wallingford, Oxon

Hitinayake H, Sumanarathne J, Abeysekara W, Madushika K, Danushka W, Swarnalatha (2017) Yield improvement of spin gourd through gynomonoecious hybrid. Ann Sri Lanka Depart Agric 19(2):71–78

IASRI (2018) Vegetable research and development in India: Role of AICRP on vegetable crops. http://iasri.res.in/aicrpvc. Accessed 18 Jun 2018

Ilangakoon TK, Bandara DC, Fonseka H (2004) Evaluation of physio- agronomic and chemical traits in relation to the productivity of eggplant (*S. melongena* L.). Trop Agri Res 16:14–24

INFORM (Information for agricultural Research Managers) (2005) Sri Lanka council for agricultural research and policy. SLCARP, Colombo, p 182

Kumaravelu V (1889) Brinjal cultivation. Mag School Agric Colombo 1(12):115

Lock RH (1908) Plant breeding and tropical agriculture. Trop Agric Mag Ceylon Agric Soc 30:466–468

Lock RH (1904) The experimental station Peradeniya. Trop Agric Mag Ceylon Agric Soc 32(4):313–314

Mac Millan HF (1905a) Notes on Diascoreas. Circ Agric J R Bot Garden Ceylon 3(1):1–19

Mac Millan HF (1905b) Fruit cultivation in Ceylon. Circ Agric J R Bot Garden Ceylon 3(14):203–222

Mansfield E (1975) International technology transfer: forms, resource requirements, and policies. Am Econ Rev 65(2):372–376

Maria H, Vilanova S, Plazas M, Gramazio P, Fonseka HH, Fonseka R et al (2012) Diversity and relationships of eggplants from three geographically distant secondary centers of diversity. PLoS One 7(7):1–14

Millawithanachchi MC, Perera ALT, Peiris BCN, Fonseka HH (2006) Development of new high yielding chili hybrids (*Capsicum annum* L.) based on heterobeltiosis and characterization of parental germplasm for DNA *polymorphisms*. Trop Agric Res 18:209–216

Ortiz R, Dias JS (2011) Transgenic vegetables for 21st century horticulture. Proc Symp Genet Modif Organ Hort 974:15–30

Pathmaraja KAB, Eeswara JP, Fonseka H (2005) Isozymic and morphological characterization of brinjal (*Solanum melongena* L.) accessions and cultivars available in Sri Lanka. *Sri Lanka*. J Agric Sci 42:20–33

Perera KDA, Wrashamana (1987) Graft compatibility, wilt resistance and graft induced changes on *S. melongena* by two wild species, *S. torvum* and *S. indicum*. *Trop Agric* 143:27–35

Pieris HA (1945) Progress of cultivation of exotic vegetables in Udukinda Division, Province of Uva. Trop Agri 101:156–157

Plazas M, Vilanova S, Gramazio P, Rodríguez-Burruezo A, Fita A, Herraiz FJ et al (2015) Interspecific hybridization between eggplant and wild relatives from different gene pools. J Am Soc Hort Sci 141:34–44

Prematilake DP, Power JB, Davey MR (2002) Genetic transformation of cultivated tomato (*Lycopersicon esculentum*) with *agrobacterium*. Ann Sri Lanka Depart Agric 4:207–214

Progress Report (2017) Plant genetic resources center. Department of Agriculture, Colombo

Rajapakse R, Fonseka H, Karunarathne R, Wickramasinghe I (2005) Evaluation of brinjal (*Solanum melongena*) germplasm for foot rot disease. Ann Sri Lanka Depart Agric 7:369–374

Ratnapala DGS, Jayalatha WD, Dharmasena MK (2010) A Study on Combining Ability and Heterosis in Okra (*Abelmoschus esculentus* (L.) Moench). Ann Sri Lanka Depart Agric 12:167–174

Regunathan R (1980) A note on the improvement of okra (*Hibiscus esculentus* L). Trop Agric 136:149–151

SAARC Report (2017) Vegetable production in the South Asia. SAARC Secretariat, Kathmandu

Samarajeewa P, Meegahakumbura M, Rajapaksha R, Gammulla C, Sumanasinghe V (2005) Molecular and morphological identification of Tomato Yellow Leaf Curl Virus in Tomato. Ann Sri Lanka Depart Agric 7:233–244

Samarajeewa PK, Attanayake P, Gamage NST (1998) Interspecific crosses between *A. esculentus* L. and *A. angulosus* L. Trop Agric 152:45–51

Samarajeewa PK, Liyange HM, Gimhani DR, Meegahakubura MGMK (2007) Molecular characterization of tomato (*Lycopersicon esculentum*) cultivars using RAPDS and microsatellite markers. Ann Sri Lanka Depart Agric 9:149–160

Samarasinghe WLG, Samarajeeva PK, Liyanage HM, Gimhani DR (2009) A PCR technique to detect yellow vein mosaic virus (YVMV) infection in okra breeding progenies. Ann Sri Lanka Depart Agric 11:131–140

Schokman D (1981) Regional Research; A new concept of agricultural development. Trop Agri 137:29–39

Sooriyapathirana C, Sajeevika C, Fonseka R, Fonseka H (2013) Effects of canopy management and fruit thinning on seed quality of tomato (Variety: *Thilina*—(*Solanum lycopersicum* L.)). Trop Agri Res 25(2):171–179

Stockdale FA (1921) Scientific research and investigation in regard to the Agricultural industries of Ceylon. Trop Agri 58(3):153–165

Weerasinghe OR, Perera ALT, De Costa WAJM, Jinadasa GA, Vishnukanthasingha AC (2004) Production of tomato hybrids for dry zone conditions of Sri Lanka using combining ability analysis, heterosis and DNA testing procedures. Trop Agri Res 16:79–90

Welegama HMVT, Jayarathne RGYH, Alwis LMHR, Fonseka H (2015) Estimation of combining ability, heritability and heterosis in Tomato (*Solanum lycopersicum* L.) using full diallel mating design. Ann Sri Lanka Depart Agric 17:242–259

Whitaker TW (1979) The breeding of vegetable crops: highlights of the past seventy five years- A review. Hortic Sci 14(3):359–363

World Food Programme (2018) Fill the nutrient gap Sri Lanka, Summary Report. www.ips.lk. Accessed 21 Aug 2019

Genetic Improvement for Sustainability of Coconut Production: The Sri Lankan Experience

7

S. A. C. N. Perera

Abstract

Coconut (*Cocos nucifera* L.) is a perennial tree crop offering a multitude of uses and a major component in the daily diet of Sri Lankans. The crop is economically and socially blended into the lifestyle of people. Both inherent and external factors affect the productivity of the coconut palm and the sustainability of coconut industry. Comparatively long juvenile phase, long economically productive life span (generally exceeding 50 years), varying genetic potential of different cultivars and levels of germplasm diversity are a few of the main inherent factors, while the environmental factors, biotic and abiotic stresses, represent the main challenges for the sustainability of the coconut cultivation. In addition to the above factors, the general factors such as low soil fertility, old age or senility of palms and the recent undesirable trends in global climate change are important factors that need to be addressed in ensuring the sustainability of coconut cultivation. The sustainability of the food systems in which coconut is a main constituent depends on the productivity of cultivations and the viability of the coconut industry. This chapter elaborates the contribution of genetic improvement programmes for the productivity enhancement, industry development and sustainability of the food systems related to coconut in Sri Lanka with reference to global scenario as appropriate.

Keywords

Coconut palm · Genetic improvement · Sustainable production · Value addition · Income generation

S. A. C. N. Perera (✉)
Department of Agricultural Biology, Faculty of Agriculture, University of Peradeniya, Peradeniya, Sri Lanka

B. Marambe et al. (eds.), *Agricultural Research for Sustainable Food Systems in Sri Lanka*, https://doi.org/10.1007/978-981-15-2152-2_7

7.1 Introduction

7.1.1 Coconut as a Global Crop

The coconut (*Cocos nucifera* L.) belongs to the family Arecaceae and is the only member of the genus *Cocos*. The crop is globally important with over 82 countries in the tropical region growing coconut and millions of people across the world utilizing various products offered by the coconut palm as food or as multitude of other uses. Coconut thrives in the coastal regions of the tropical zone in areas between latitudes 20° north and south of the equator and up to an altitude of about 1200 m above mean sea level. Conditions of high humidity, temperatures of 27–30 °C and deep soils with moderate to higher aeration are the best suited environments for coconut.

Coconut is cultivated world over, in an estimated extent of 11,988,000 ha producing a total of 67.04 billion nuts (Anon 2015). Coconut cultivation is mainly centred in Asia with Africa and South America also growing the crop. Indonesia, the Philippines and India contribute over 75% followed by Brazil and Sri Lanka in terms of the cultivated extent. In these countries, the coconut cultivation is practised both in small scale as a smallholder crop and a major plantation crop.

Coconut is predominantly an oil crop with the preferred oil characteristics of relatively high melting point and narrow melting range, comparative resistance to oxidation and rancidity and lack of unpleasant odour (Anon 1979). Coconut oil is rich in short-chain fatty acids and displays high digestibility (Banzon 1977) making it a fat source of high quality in infant milk. Coconut oil is used as an important ingredient in various food products including ice cream and filling cream and as confectionery oil and hard butter and in a variety of confectionery formulations (Marcus and Puri 1978). It is also widely used for domestic consumption as cooking and frying oil in many of the coconut-growing countries including Sri Lanka. The novel form of coconut oil is virgin coconut oil (Bawalan and Chapman 2006) which is expelled from fresh coconut meat under low heat preserving its natural vitamins and enzymes (Marikkar et al. 2007). Virgin coconut oil has gained the attention of many even outside the producing countries, for its merits, and has secured a significant place in international markets.

In addition to its food uses, coconut oil is also used as an industrial vegetable oil as a raw material in soap, glycerine, detergents, cosmetics and pharmaceutical industries. Apart from oil, the other significant kernel products are desiccated coconut, coconut cream, coconut milk powder, defatted coconut, etc. Toddy, arrack, vinegar, sugar and jaggery from the tapped inflorescence, fibre and fibre products, shell charcoal and timber are other main products of various parts of the coconut palm. In the recent times, coconut water both from tender and mature nuts has gained special attention around the world as a healthy, natural, energy and sports drink. In addition, coconut milk is gaining significance as a substitute for cows' milk among the community having lactose intolerance. Due to these multitude of uses, coconut palm has been termed as 'one of nature's gift to mankind', 'tree of life', 'tree of thousand uses' and the 'milk bottle on the doorstep of mankind' (Bourdeix et al. 2005).

In view of the vast array of raw material that the coconut palms offer for various cottage industries, the palm has been identified as having a huge potential for poverty alleviation in rural communities in the coconut-growing countries.

7.1.2 Coconut Palm: The Sri Lankan Context

Coconut is an introduced crop to Sri Lanka, but unlike other major plantation crops, the history of coconut cultivation in the island dates back to the era of kings' rule. South East Asia and the Pacific islands are considered to be the centres of origin of coconut (Harries 1978). The domestication processes have also taken place in the same parts of the word, while a separate domestication process has been recorded in South Asia, including Sri Lanka, India and East Africa (Harries 1978). There are two hypotheses on the dissemination of coconuts from its centres of origin to other parts of the world, i.e. (1) coconuts being carried by early sea voyagers as food and drink and dumping the remaining in the shores resulting in cultivations along the coastal belt and (2) distribution of coconuts by regular ocean currents and also by occurrences such as tsunami at a faster rate. Both the hypotheses may hold true in Sri Lanka because the highest diversity of coconuts is found along the southern coastal belt of the country and the adjacent home gardens (Ekanayake et al. 2010).

In Sri Lanka, coconut cultivation is spanned over an area of 440,000 ha (Anon 2017), occupying the second largest agricultural extent in the country being second only to the staple crop, rice. The annual national requirement of coconuts is to satisfy the need for domestic consumption and to sustain the processing industries. The food chains have been calculated to be 3000 million nuts, yet there are yearly variations in production, and a yield of 2450 million mature coconuts was recorded in 2017 (Anon 2017). At present coconut is socially and economically blended with the lives of Sri Lankans. It is an important constituent in the food basket and provides direct and indirect employment to about half a million Sri Lankans and is an important export commodity of Sri Lanka while gaining the major portion of foreign exchange through non-traditional coconut products such as fibre, virgin oil and beverage coconuts (Anon 2017).

7.2 Genetics and Breeding of the Coconut Palm

Coconut is a monocot perennial with a comparatively long vegetative phase of about 6–7 years and an economical life span of over 50 years. It possesses 16 pairs ($2n = 2x = 32$) of chromosomes in a diploid genome. Breeding the coconut palm for desirable traits records a history nearing its centenary, with attempts being led by several Asian countries, i.e. Sri Lanka, India, Malaysia and Indonesia, and Ivory Coast in Africa.

7.2.1 Breeding Systems of Coconut

Coconut bears a monoecious inflorescence, called a spadix, having both male and female flowers enclosed in a spathe which splits open to expose the mature flowers of both sexes into the environment. The inflorescences originate at each leaf axil and are emitted one at a time at approximately monthly intervals producing about 12–14 inflorescences a year. The male and female phases are temporally separated in tall coconut inflorescences, thereby naturally preventing self-pollination, while the reverse takes place in dwarf coconuts facilitating self-pollination. Both types of coconuts are amenable to self- and cross-pollination artificially, which is vital in genetic improvement of the palm.

7.2.2 Objectives of Genetic Improvement of Coconut

Sri Lanka is among the pioneering nations to adopt scientific research for the development of coconut cultivation. Studies on genetics of the coconut palm leading to genetic improvement programmes were initiated with the establishment of the Coconut Research Institute (CRI) in Sri Lanka in 1929. Similar to global situation, increased yields was the primary objective in the early days of coconut breeding. Coconut oil occupied a prominent place in the world trade of edible oil until the latter part of the twentieth century and, thus, was the main yield component of coconut. However, with other uses of the kernel products in Sri Lanka such as desiccated coconuts and as fresh kernel for culinary purposes, the focus was more on the increase in nut yield.

The long vegetative phase of coconuts is instrumental in delaying returns to investment in the commercial coconut holdings, and thus, shortening of the vegetative phase or introducing early bearing coconuts was also included as primary objective of coconut breeding in the early days (Liyanage 1955). These initial objectives have now been expanded further in view of the emerging stressful environments created by biotic and abiotic stresses. Accordingly, during the last two decades, the national coconut breeding programme in Sri Lanka has been geared to encompass the novel objectives: (a) breeding coconuts for biotic stresses; (b) *Aceria* mite infestation and Weligama coconut leaf wilt disease (WCLWD); (c) abiotic stresses, especially drought and heat stress, enhanced by the changing global climates and (d) diverse uses of the palm, especially considering the variety of processed products that the coconut offers.

7.2.3 Classification of Coconut

Two groups of coconuts are mainly identified in the global coconut classification, namely the tall (*Typica*) and dwarf (*Nana*) types, differing in their stature and breeding behaviour. The *Typica*-type coconuts are tall in habit, producing a prominent root bole (a swollen base of the stem), and display relatively larger crowns and bigger nuts. These coconuts are naturally out-crossing, resulting in heterozygous individuals and heterogeneous populations. Vegetative phase of tall coconuts is longer, taking about 6–7 years for the flower initiation; however, the life span exceeds 60 years under favourable conditions. At the reproductive stage, the tall coconuts continuously produce inflorescences at regular intervals resulting in non-seasonality in nut production.

The *Nana*-type coconuts are short in stature, and they lack prominent root boles displaying uniformly slender trunks. Smaller crowns and smaller nuts are characteristics of dwarf coconuts, and they display naturally self-pollinating breeding behaviour due to overlapping of the male and female phases of a single inflorescence. As a result of this naturally inbreeding nature, dwarf coconut individuals are homozygous, and the populations are pure lines. In comparison to tall coconuts, dwarf coconuts are early flowering, taking about 3–4 years for flower initiation. Yet, the dwarf coconuts record shorter life spans and seasonality in nut production due to delayed inflorescence emission under unfavourable environments (Liyanage et al. 1988).

The classification of coconuts in Sri Lanka includes a third coconut type/variety named intermediate (*Aurantiaca*) in addition to *Typica* and *Nana*. The variety *Aurantiaca* displays mixed morphologies of *Typica* and *Nana*. Members of the variety *Aurantiaca* grow tall in stature, produce root boles and develop smaller crowns and medium-sized nuts. They are early flowering and display naturally self-pollinating breeding behaviour. Different forms of king coconuts and Sri Lanka yellow semi-tall coconut form are included in the variety *Aurantiaca*. Sri Lankan coconut classification further includes several different morphotypes within each coconut variety. A complete description of the forms within coconut varieties are given in Table 7.1.

7.2.4 Genetic Resources of Coconut

Availability of genetic diversity, which is defined as the amount of genetic variability among individuals of a variety, or population of a species (Brown 1983) is a prerequisite for breeding plants for desirable traits. Diverse germplasm of a crop plant is represented by the wild relatives, landraces, different varieties and breeding lines, etc. Wild relatives of coconuts are extinct from the world, and coconut, being a mono-species genus, does not have any close relatives for gene introgression in genetic improvement. Accordingly, the best option available to coconut breeders is the collections of farmers' varieties, indigenous forms and exotic varieties in addition to the main coconut varieties and forms available in the coconut classification.

Despite the drawback of not having the wild or close relatives, coconut is benefitted by global coconut germplasm conservation programmes, which was mediated by the International Coconut Genetic Resources Network (COGENT) of the Bioversity International. Collected global coconut germplasm has been conserved ex situ in five international field gene banks in India, Papua New Guinea, Ivory Coast, Brazil and Indonesia. In addition, the national repositories are maintained by individual countries including Sri Lanka, for characterization, evaluation and

Table 7.1 Varieties and forms of coconut in Sri Lanka and their key morphological and reproductive phases (Sources: Liyanage 1958; Wickremaratne 1984; Perera et al. 1997; Ekanayake et al. 2010; Kamaral et al. 2016)

Common name (variety/form)	Specific morphological and reproductive features
Tall (*Typica*/*Typica*)	Tall stature, allogamous, heterogeneous, flower in 6–7 years, medium-sized nuts, 20–25 nuts per bunch, 60–80 nuts per palm per year
Gon thembili (*Typica*/*Gon thembili*)	Similar to tall. Exhibits ivory colour in nuts, leaves, petioles and inflorescences
Nawasi (*Typica*/*Nawasi*)	Similar to tall. Contains soft mesocarp which is edible at the immature stage producing soft fibre in the mature nut
Pora pol (*Typica*/*Pora pol*)	Similar to tall. Contains remarkably thick-shelled nuts
Ran thembili (*Typica*/*Ran thembili*)	Similar to tall except for the pink-coloured mesocarp in immature fruit and the pink whorl under the perianth. Large nuts
Kamandala (*Typica*/*Kamandala*)	Similar to tall except for the large-sized nuts and low number per bunch (2–5 nuts per bunch). This is regarded as the largest local nut
Bodiri (*Typica*/*Bodiri*)	Similar to talls. But produces small-sized nuts and high number per bunch (30–100 nuts per bunch). Display seasonality in nut production
Dikiri (*Typica*/*Dikiri*)	Similar to tall. Some nuts contain a jelly-like albumen/endosperm
Ran pol (*Typica/Ran pol*)	Similar to tall. Stout fast-growing stems. Very large round coconuts of mostly brown or less commonly green fruits. Plicata leaves (attached leaflets, especially in young fronds)
Juwan pol (*Typica/Juwan*)	Similar to tall. Produces round small nuts. A high number (10–20) nuts per bunch
Rath Gon Thembili (*Typica*/*Rath Gon thembili*)	Similar to tall. Bear orange-coloured coconuts. The petiole fronds and the inflorescences also are of the same colour
King coconut (*Aurantiaca*/*King coconut*)	Intermediate stature, autogamous, homogeneous, fruits in 6–7 years, seasonal flower production, slightly small-sized nuts with orange epicarp and sweet nut water, 25–50 nuts per bunch
Nawasi thembili (*Aurantiaca*/*Nawasi thembili*)	Similar to king coconut. Has a soft and thus edible mesocarp like nawasi
Rathran thembili (*Aurantiaca*/*Rathran thembili*)	Similar to king coconut. Contains pink-coloured mesocarp and a pink whorl under the perianth
Bothal thembili (*Aurantiaca/Bothal thembili*)	Similar to king coconut but with elongated nuts of unique shape
Sri Lanka yellow semi tall (*Aurantiaca*)	Semi-tall in stature. Autogamous. Produce yellow-coloured nuts
Green dwarf (*Nana*/*Green dwarf* or *Pumila*)	Dwarf stature. Autogamous. Homogeneous populations. Fruits in 3–4 years, small-sized nuts with green epicarp, low copra content, 80–150 nuts per palm per year
Yellow dwarf (*Nana*/*Yellow dwarf* or *eburnea*)	Similar to green dwarf except that nuts have a yellow epicarp

(continued)

Table 7.1 (continued)

Common name (variety/form)	Specific morphological and reproductive features
Red dwarf (*Nana*/*Red dwarf* or *regia*)	Similar to green dwarf except for that nuts have a red epicarp
Brown dwarf (*Nana*/*Brown dwarf* or *braune*)	Similar to green dwarf except for that nuts have a brown epicarp

utilization in breeding programmes. The characterization and evaluation programmes, which are being carried out in Sri Lanka, are shedding light on the diversity of conserved material and assist in the selection of parental material for the hybridization programme (Ekanayake et al. 2010; Perera et al. 2015a).

A systematic programme for collecting and conservation of Sri Lankan coconut germplasm has been in progress since 1984 (Wickremaratne 1984). This programme had the dual objectives of conserving the existing biodiversity in all its features selected through random sampling and identifying material for conservation through biased sampling for phenotypic variants. Currently, the total number of accessions conserved is over 150 including both the local collections and 24 exotic accessions as per the records at the CRI of Sri Lanka. Field gene banks are the only viable option for ex situ conservation of coconut largely due to the recalcitrant nature of the coconut seed. Currently the CRI of Sri Lanka manages 11 ex situ field gene banks of coconut preserving its biodiversity available in Sri Lanka. A detailed description of these accessions covering passport data and some characterization data pertaining to stem, leaf inflorescence and fruit morphology of the conserved accessions is available in the international database on Coconut Genetic Resources (CGRD), which can be accessed only by the member countries of COGENT at present.

7.2.5 Coconut Breeding is a Challenging Task

Lengthy vegetative phase and generation interval and naturally cross-pollinating behaviour of tall coconuts resulting in heterozygous individuals and heterogeneous populations leading to F_1 populations that are not true to type pose inherent genetic limitations for coconut breeding. Lack of a viable vegetative propagation method, low seed output per palm, recalcitrant nature of seeds and the massive stature of the palm make coconut a difficult crop to breed. In addition, the limited availability of genetic stocks and genotypes imposes hindrances for pre-breeding activities. These inherent characteristics have limited the genetic improvement of coconut to mass selection and hybridization mainly between tall and dwarf coconuts (Perera 2010a).

7.2.6 Conventional Techniques of Coconut Breeding

Sri Lanka and India are the pioneering countries to set up research institutes devoted to coconut and thus are the forerunners in coconut breeding. Accordingly, scientific attempts of genetic improvement of coconuts began in the 1930s following the conventional plant breeding techniques. Such conventional coconut breeding methods have been limited to intra-species level due to *Cocos* being a mono-species genus. The most fundamental coconut breeding method that has been used by the plant breeders is mass selection (Liyanage 1955). Consequently, the majority of the coconut plantations in Sri Lanka and around the world have been derived from mass selection. Nut and/or copra yields were the main criteria used in the middle of the twentieth century for mass selection. Later, mass selection was supplemented by selfing or intercrossing between selected parents to produce genetically improved cultivars with desirable traits (Liyanage et al. 1988).

The techniques developed in Sri Lanka for controlled cross-pollination in coconut facilitated hybridization between different varieties to extract the hybrid vigour (Liyanage 1972). Hybridization of coconut became the most favoured choice of coconut breeders, and to date the majority of genetically improved cultivars of coconut in Sri Lanka and other coconut-producing countries are hybrids. Hybrid coconuts between tall and dwarf coconut varieties have successfully been developed to extract the hybrid vigour for economically important traits (Wright 1980). The CRIC65, CRISL2004 (Kapruwana), CRISL2012 (Kapsuwaya) and CRISL2014 are examples for the Dwarf × Tall coconut hybrids developed and recommended in Sri Lanka (Fig. 7.1), and these hybrids are improved basically for precocity, nut yield and fruit components (Table 7.2).

Fig. 7.1 A 5-year-old dwarf × tall coconut hybrid developed in Sri Lanka

Table 7.2 Characteristics of genetically improved and recommended coconut cultivars in Sri Lanka

Cultivar (breeding principle)	Parents (female × male)	Improved characteristics
CRIC60 (selection)	SLT × SLT	Nut yield, husked nut weight, agronomic characteristics
CRIC65 (hybridization)	SLGD × SLT SLYD × SLT	Precocity, nut yield, higher productivity
CRISL98 (hybridization)	SLT × SR	Higher weights for all fruit components including fibre and good-quality copra (280–300 g)
CRISL2004—Kapruwana (hybridization)	SLGD × SR	Precocity, nut yield, higher weights of fruit components, good-quality copra (250–260 g), higher productivity
CRISL2013—Kapsuwaya (hybridization)	SLBD × SLT	Precocity, nut yield, higher productivity
CRISL2014—Kapsetha (hybridization)	SLT × SLBD	Precocity, higher weights for all fruit components

SLT Sri Lanka tall, *SLGD* Sri Lanka green dwarf, *SLYD* Sri Lanka yellow dwarf, *SR* San Ramon, *SLBD* Sri Lanka brown dwarf

The second type of hybrid coconuts was the hybrids between two different tall coconut populations. This method has also been widely used in Sri Lanka, the Philippines and Indonesia, with the hybrid between 'Sri Lanka tall' and 'San Ramon' tall being the best example in Sri Lanka. The third type of coconut hybridization was between two different dwarf coconut forms. Though Malaysia has recommended a few of such hybrids, there is no such recommendation made as yet in Sri Lanka. In coconut hybridization, the two parents are selected to combine the desirable features of one parent and to contradict the weaknesses of each parent with the second parent. Despite the widespread usage of the term 'hybrid coconuts', apart from the hybrids between two dwarf coconut forms, the other two hybrids (Dwarf × Tall and Tall × Tall) do not qualify to be true hybrids as per the genetic definition, which specifies a hybrid as a cross between two pure lines. Yet, the recommended 'hybrid coconuts' have proved their practical advantage over unimproved cultivars and thus have been widely accepted by the Sri Lankan coconut growers as well as the global coconut community.

In most of the coconut-growing countries including Sri Lanka, the promising hybrids are identified at the F_1 stage without proceeding to advanced generations due to longer periods involved in crossing for several generations. However, the termination at F_1 stage limits the complete exploration of possible desirable recombination of a cross. Nevertheless, countries such as Thailand and Ivory Coast have tested multiple crosses to develop varieties with important multiple traits. Moreover, the production of synthetic varieties has been reported in the Philippines. Sri Lanka still lags behind the application of both multiple crosses and the synthetic cultivars although efforts in developing synthetic varieties are underway for breeding for resistance to Weligama coconut leaf wilt disease (Perera 2015).

7.2.7 Application of Novel Techniques in Coconut Breeding

Although considerable progress has been made through the conventional coconut breeding techniques, they have suffered from a few major drawbacks, i.e. long time involved to develop a cultivar, the difficulty of distinguishing the actual genetic effect from the environmental effect and tagging genes linked to desirable traits being the most critical issues. The modern molecular marker technology has been applied in different crops in effectively reducing the time period of a breeding programme, in addition to targeting the direct gene/QTL effect independent from the environmental effect. Over the last two decades, molecular techniques have been used in several countries to enhance the breeding programmes. It is noteworthy that Sri Lanka has played a pioneering role in application of molecular markers for coconut breeding. Application of molecular marker technology in coconut was initially aimed at assessing the genetic diversity of germplasm. The pioneering work with universal marker techniques such as RAPD (Everard 1996; Ashburner et al. 1997; Duran et al. 1997), RFLP (Lebrun et al. 1999), AFLP (Perera et al. 1998; Teulat et al. 2000) and ISTR (Duran et al. 1997; Rohde et al. 2000) included the evaluation of Sri Lankan coconut germplasm (Everard 1996; Perera et al. 1998). Sri Lanka was also the first among the research groups in developing and utilizing the coconut specific microsatellite (SSR) markers sharing the credit with the Philippines, with the technical assistance from the United Kingdom (Perera et al. 1999, 2003; Rivera et al. 1999). Microsatellites being the co-dominant markers have been highly useful in analysing the genetic diversity of naturally heterozygous coconut populations (Perera et al. 2000; Teulat et al. 2000; Dassanayaka et al. 2003; Meerow et al. 2003) and DNA fingerprinting of coconut varieties (Kamaral et al. 2017). A SSR marker kit comprising 14 pairs of primers with an associated software for data analysis has been developed (Baudouin and Lebrun 2002) with the objective of standardizing the protocols across laboratories for the comparison of results for effective and efficient detection of diversity. Microarray-based diversity array technology (DArT) has been validated for coconut using Sri Lankan coconut germplasm (Perera and Kilian 2008) opening up avenues for high-throughput genotyping in coconut breeding research.

Coconut breeding has further been benefited by molecular marker technology with the use of SSR markers for the construction of genetic linkage and QTL maps (Rohde et al. 1999; Herran et al. 2000; Lebrun et al. 2001; Baudouin et al. 2006) in the Ivory Coast. Baudouin et al. (2006) has successfully mapped the QTL for traits including fruit characters. Coconut genome mapping research is in progress in Sri Lanka with a F_2 population (Perera et al. 2016), yet with the limitation of a low number of individuals in the mapping population. Research is in progress in Sri Lanka for the utilization of DArT markers to search for marker-trait associations to facilitate marker-assisted selection, which will be the direct and practical application of molecular marker technology in coconut breeding.

Lack of a commercial scale and viable vegetative propagation technique hinders both the true-to-type multiplication of superior genotypes and the application of certain novel in vitro breeding techniques such as gene transformation. Sri Lanka has made some successful attempts for in vitro culture of coconut and has achieved considerable success in tissue culture via the use of immature inflorescence as the explant (Perera et al. 2007, 2010; Bandupriya et al. 2016).

7.3 Genetic Improvement of Coconut for Sustainable Production

7.3.1 Factors Affecting Sustainability in Coconut Production

Sustainability of coconut production depends on inherent factors in addition to the environmental factors. Comparatively long juvenile phase, lengthy economically productive life span, the genetic potential of different cultivars, diversity in germplasm and the multitude uses offered by the palm are listed as the universal inherent factors affecting the sustainability. Furthermore, the environmental factors, i.e. biotic and abiotic stresses, are a serious threat to the sustainability of coconut productions globally, and their effect in Sri Lanka has also been strongly felt. Drought is the most important abiotic stress of coconut in Sri Lanka posing serious threats for coconut production, which has also limited the cultivable area in many coconut-growing countries.

Other abiotic stresses such as lowlands with high water table and cyclones cause mortality especially in seedling coconuts and occasional uprooting of large number of palms, respectively, in Sri Lanka. However, the damages due to strong winds may be more common occurrences in certain other parts of the world. Depletion of soil fertility, especially in coconut lands with long-standing coconut plantations, is a serious threat to sustainability of the plantations in Sri Lanka. Moreover, the coconut productivity has been affected negatively due to senility, which has been common in many coconut-growing countries. Since recently, the effects of global climate change has become a universal problem threatening the sustainability of coconut plantations with elevated temperature and heat levels resulting in yield reduction and even the death of palms. Cultivation of coconuts in home gardens, smallholder and plantation scales and the diverse uses of the palm result in both positive and negative effects on the sustainability. The smaller scale of cultivation in home gardens fulfils at least a part of domestic needs of coconuts leaving the harvest from large-scale plantations to sustain the coconut industry. However, the smaller-scale farm yards may not have sufficient levels of maintenance resulting in low yields. Furthermore, diverse uses of the palm may encourage many farmers to grow coconut, but, if the proportioning of the produce among different uses is not balanced, certain coconut industries may have to run under-capacity owing to shortage of coconuts during the lean yield periods.

7.3.2 Coconut Breeding for Biotic Stresses: Pests

Red weevil (*Rhynchophorus ferrugineus*) and black beetle (*Oryctes rhinoceros*) causing lethal damages to adult and seedling coconuts, respectively, have been the traditional coconut pests in Sri Lanka. However, the eryophyid mite (*Aceria guerreronis*), which started spreading in the dry and intermediate zones in Sri Lanka during the late 1990s, was the first coconut pest for which the solutions were sought through plant breeding. *Aceria* mite causes high economic damages to developing coconut palms resulting in premature nut fall and smaller-sized nuts. With positive indications on genetic governance for the tolerance of *Aceria* mite infestation, research have been initiated in Sri Lanka to screen coconut varieties/hybrids for tolerance to *Aceria* mite infestation, followed by hybridization among potential parents to deliver tolerance to their offspring (Perera 2005, 2011; Perera et al. 2013).

Aceria mite infestation is a common problem for coconut especially in the drier areas in India, Oman and several other coconut-growing countries. Research teams in India and Sri Lanka are searching for sustainable solutions for this pest using plant breeding strategies in combination with molecular biological approaches. Shalini et al. (2006) revealed molecular markers associated with *Aceria* mite damage using a palm pool selected on a standard scale varying from resistance to susceptibility and genotyping with RAPD and SSR markers. Yet another study conducted in Sri Lanka has revealed the potential for marker-assisted selection of coconuts for tolerance to *Aceria* mite infestation (Perera et al. 2015b) indicating the possibility of sustainable options to this pest via genetic improvement.

Fig. 7.2 A coconut palm infected with Weligama coconut leaf wilt disease in Southern Sri Lanka

7.3.3 Coconut Breeding for Biotic Stresses: Diseases

Coconut plantations in Sri Lanka have been plagued with several diseases, i.e. stem bleeding, ganoderma and bud rot caused by fungi and collar rot caused by soil-borne bacteria. However, the efforts towards sustainable solutions for coconut diseases were initiated only when WCLWD was identified from Southern Sri Lanka in 2006 (Fig. 7.2). The WCLWD is a debilitating, incurable disease caused by an intracellular pathogen phytoplasma, which spreads via insect vectors. The coconut variety Sri Lanka green dwarf was reported to be highly resistant to WCLWD indicating genetic governance for the trait (Perera et al. 2014). Consequently, a coconut breeding programme has been initiated and in progress to transfer the resistant genes to cultivated coconut varieties (Perera and Dissanayake 2013; Perera et al. 2014).

Phytoplasma diseases are causing substantial damages to coconuts in many of the coconut-growing countries, with lethal yellowing being the most devastative (Dollet et al. 2009). Trees infected with lethal yellowing disease display symptoms of nut fall, yellowing of fronds, blackening of inflorescences, rotting and the ultimate falling off of the crown resulting in death of the palm within a period of 3 months (Dollet et al. 2009). This lethal disease first appeared in the Caribbean and by now has caused severe damages to coconuts in many countries in Central America and the African continent. Kerala (root) wilt has been plaguing coconuts in Southern India for over 150 years. It is also a debilitating phytoplasma disease, which is symptomatically similar to WCLWD. Genetic improvement of the coconut palm for resistance to phytoplasma diseases has been identified as the most viable and sustainable option, and accordingly, genetic improvement of coconut against phytoplasma diseases is being carried out in Jamaica (Baudouin et al. 2009; Omamor et al. 2011; Zizumbo-Villarreal et al. 2006), Ghana (Nkansah-Poku et al. 2009; Quaicoe et al. 2009) and India (Nair et al. 1996, 2006). In these studies, the coconut varieties 'Sri Lanka green dwarf', 'Chowghat green dwarf' and 'Malayan green dwarf' have been identified as the main varieties possessing resistant alleles for phytoplasma diseases (Nair et al. 2000). Furthermore, advances in molecular screening and resistant genotypes among root wilt disease-affected palms have been reported in India (Devakumar et al. 2011).

7.3.4 Coconut Breeding for Abiotic Stresses: Drought

Drought is a multidimensional factor of stress which affects the plants at various levels of growth, development and reproduction (Lambers et al. 1998). It is a natural climatic phenomenon when less than expected precipitation occurs within a specified period of time (Feng and Zhang 2005). Drought causes significant yield reduction in coconut and, in prolonged droughts, even death of adult palms (Fig. 7.3) in certain drier areas in Sri Lanka.

Coconut grows well in areas with a well-distributed mean annual rainfall of 1500 mm. Both the amount and the distribution of rain are especially important for coconut fruit set and growth, with a monthly rainfall of 150 mm being the ideal

Fig. 7.3 Adult coconut palms succumbed to severe drought in the dry zone of Sri Lanka in 2016

(Rajagopal et al. 1996). Yet, being a perennial palm with a long productive life span, coconut is frequently exposed to soil and atmospheric water deficit (Rajagopal and Kasturi Bai 2002). Owing to its perennial nature, coconut is not capable of escaping drought with strategies such as adjustments made to time of planting as can be done in annual crops. Breeding coconuts to incorporate an acceptable level of tolerance to moisture deficit is considered the most viable option to mitigate the ill-effects and ensure the sustainability of coconut production. Despite its importance, the efforts towards developing drought-tolerant coconut varieties still remain in its infancy in Sri Lanka and in other coconut-growing countries. Agronomic performance of coconut trees grown under water stress can largely be explained by physiological and morphological traits (Gomes and Prado 2007) indicating the possibility of breeding for moisture-tolerant varieties. Currently, physiological studies on screening and genetic improvement of coconuts to drought stress are in progress; however, the actual gene identification and introgressing research remains scarce.

7.3.5 Coconut Breeding for Abiotic Stresses: Changing Climates

The climate of the earth is known to be constantly changing due to natural causes such as volcanic activity. Certain rapid changes have been observed during the last few decades resulting in an increase of global temperature levels, negatively affecting the production and sustainability of perennials such as coconut. An analysis conducted in Sri Lanka reported the reduction of coconut yields with the global climate change (Waidyarathne et al. 2017), and the disease incidence is expected to increase due to faster evolution of pathogens and the emergence of new pathogenic life forms, increase in pest populations, etc. Harnessing the genetic potential of coconuts, by developing cultivars with the ability to perform well under varying environments, is envisaged as the best option in mitigating the damages due to global climate change. Conventional coconut breeding research have addressed this issue to develop cultivars, which are adapted to varying agro-ecological regions in Sri Lanka (Dissanayaka et al. 2008), but research integrating modern plant breeding techniques directly related to global climate change is still at its initial stages, not only in Sri Lanka but in other coconut-growing countries as well.

7.3.6 Other Factors Affecting the Sustainability of Coconut

In addition to the aforementioned limiting factors, the sustainability of coconut also depends on less apparent factors such as the levels of input for higher yields, urban agriculture requiring short stature plants, need for dual purpose varieties, etc. Breeding programmes should also be geared towards catering to these needs ensuring sustainability by adopting genetic improvement measures for the development of varieties suitable for low input conditions, for urban agriculture and homestead farming, and dual or multi-purpose cultivars fulfilling the demands for tender nut and copra production, aromatic endosperm, sweet kernel products, etc.

7.4 Opportunities, Constraints and Strategies for the Development and Sustainability of Coconut Industry

Increased production and productivity is the major consideration in increasing profitability and sustainability of the coconut industry. Conventional breeding approaches, though being hindered with typical constraints, have addressed this issue to considerable extents by developing coconut cultivars, which are early bearing and highly productive.

Availability of diverse germplasm is a key for the development and genetic improvement of any crop. Global coconut genetic resources are both a limitation and an opportunity in the sustainable development of coconut by breeding. The major limitation with coconut germplasm is the extinction of wild relatives of coconut and the lack of genetically closer relatives for techniques such as alien gene transfer via wide hybridization or introgression of desirable characters by crossing. The opportunity rendered by the genetic resources of coconut is the availability of a vast majority of unidentified or uncharacterized germplasm either conserved in regional or national germplasm repositories or in the field as farmers' varieties. Farmer participatory selection approaches will thus be highly useful to identify promising genotypes to be incorporated into the genetic improvement programmes as and when required.

Advances in molecular breeding techniques provide effective and efficient methods to complement and enhance the conventional breeding techniques and overcome certain limitations of the conventional techniques. Molecular breeding will pave way for identification of elite genotypes possessing desirable traits for tolerance to stresses through QTL mapping and marker-assisted selection (MAS). The

MAS will be effective only with fine genome maps with higher marker density. Hence, molecular breeding techniques on coconut should be directed towards high-throughput molecular marker systems for the effective use of novel technology (Perera 2010b).

The application of linkage disequilibrium or association mapping will provide greater opportunities in overcoming the limitations of linkage mapping for gene discovery, although the literature is scare on such approaches. Novel technologies such as transcriptome profiling for stress-tolerant genes can be used for candidate gene identification in coconut. Research can also harness the benefits of advanced high-throughput genotyping facilities and more genomic resources, which are emerging (Varshney et al. 2010) even for an 'orphan crop' such as coconut. Physical map is yet another novel technology that can be used for genome-wide gene discovery, EST mapping (functional genomics) or comparative genomics (synteny studies) in coconut. The recently published coconut genome sequence (Xiao et al. 2017) will be a huge boost in the efforts of finding sustainable solutions through novel molecular marker technology.

Being a perennial, coconut faces several difficulties in agronomy and plant breeding. In addition to these common difficulties, coconut development suffers greatly due to the lack of a viable vegetative propagation method resulting in the inability of obtaining superior genotypes fixed by vegetative propagation. In this respect, Sri Lanka has been leading with reports that have shed green light on the feasibility of coconut tissue culture using various explants (Perera et al. 2007), with unfertilized ovary being the most successful explant so far. Vegetative propagation via tissue culture would be a major step towards greater achievements in the development of coconut cultivation. The attempts of in vitro culture research need to be further strengthened to make anther/haploid culture for deriving homozygous lines, in vitro screening for stress tolerance, cryopreservation of coconut embryos/pollen for germplasm conservation and also for multiplying rare genotypes cloning and characterization of genes induced during somatic embryogenesis. Rapid multiplication of elite varieties for scaling up of the planting material production to cater to the demand for true-to-type quality planting material will be a practical reality.

The genetic and morphological uniformity of plantations gained much attention of the coconut researchers over the years. Yet, the inclusion of genetic buffering capacity in plantations, especially against newly emerging biotic and abiotic stresses, will be an important phenomenon in the context of sustainability of the palm. Consequently, establishing coconut seed gardens to produce synthetic varieties of coconut will be highly advantageous even at the expense of sacrificing certain levels of genetic uniformity. Furthermore, the use of supplementary tools such as in vitro multiplication of elite genotypes using appropriate explants with high regeneration ability to produce parents or true-to-type progeny for seed gardens would be highly attractive in the future.

Coconut is versatile in its offer of diverse uses to varying groups of consumers not only in coconut-growing countries but world over. This provides diverse markets for coconut products creating a viable opportunity to justify the efforts to development and sustainability of the coconut industry. The genetically improved coconut varieties in Sri Lanka differ in their comparative proportions of fruit components. There are better suited cultivars for different uses, viz. culinary, beverage, oil and DC production, fibre, etc. Yet, directing the produce of different cultivars towards proper market channels is important for the optimal use of the produce.

Improvements to post-harvest handling of coconut products in Sri Lanka are envisaged as an immediate measure for the sustainability of the food chains and the coconut industry. Seventy per cent of the coconut kernel produce in the country is consumed within Sri Lanka, but with a high percentage of wastage in the households. The improvements to processing and post-harvest technologies should focus on processed products of coconut milk for culinary purposes, because a wastage of 40% is reported in the traditional domestic coconut milk extraction methods. Furthermore, despite a well-developed local and international market for shell charcoal, the major portion of coconut shells used in households in Sri Lanka are still wasted. The wastage is common in the industry scale as well, with respect to coconut water. Millions of litres of coconut water that can be processed to be exported are dumped freely, resulting in environmental pollution as well. Thus, while focusing on development of coconut varieties for sustainability of the industry through genetic improvement, it is imperative to focus on the optimum usage of available produce and making the diverse uses to the advantage of the coconut producers and for the sustainability of food chains and market channels.

Sustainable coconut cultivation is a balance between economic, social and environmental priorities. The genetic improvement programmes can deliver highly sustainable and attractive solutions for factors which are negatively affecting the sustainability of the coconut industry. Efforts to enrich the coconut gene pool with trait-specific germplasm and utilizing them for the development of varieties suitable for different end uses and specific industrial uses will be a key factor leading to the sustainability of cultivations, assuring the sustainability of food systems as well as the industry. Promotion of value addition and enhancing income generation of coconut farmers can be targeted and achieved through selection of specific genotypes suitable for the production of novel products such as coconut chips, recovery of inflorescence sap and preparation of shell products, etc., thereby improving the socio-economic status of the coconut farmers.

Acknowledgements Mr. Roshan Silva and Mrs. Nilmini Jayalath of the Coconut Research Institute (CRI) of Sri Lanka are acknowledged for providing Figs. 7.2 and 7.3, respectively. The untiring efforts of all the past and present coconut breeders at the CRI of Sri Lanka are greatly appreciated.

References

Anonymous (1979) Technical data handbook on the coconut, its products and by products, a compilation. Philippines Coconut Authority, Quezon City, p 123

Anonymous (2015) Asia Pacific statistical year book. Asia Pacific Coconut Community, Jakartha

Anonymous (2017) Annual report of the Central Bank of Sri Lanka. Colombo: Central Bank of Sri Lanka. https://www.cbsl.gov.lk/en/publications/economic-and-financial-reports/annual-reports/annual-report-2017. Accessed 24 Jun 2019

Ashburner GR, Thompson WK, Halloran GM (1997) RAPD analysis of South Pacific coconut palm populations. Crop Sci 37:992–997

Bandupriya HDD, Fernando SC, Vidhanarachchi VRM (2016) Micropropagation and androgenesis in coconut: an assessment of Sri Lankan implication. Cocos 22:31–47

Banzon JA (1977) Industrial coconut research and development. Philip J Coconut Stud 2:13–17

Baudouin L, Lebrun P (2002) The development of a microsatellite kit and dedicated software use with coconuts. Burot Bull 17:16–20

Baudouin L, Lebrun P, Konan JL, Ritter E, Berger A, Billottee N (2006) QTL analysis of fruit components in the progeny of a Rennell Island tall coconut (*Cocos nucifera* L.) individual. Theor Appl Genet 112:258–268

Baudouin L, Philippe R, Quaicoe R, Dery S, Dollet M (2009) General overview of genetic research and experimentation on coconut varieties tolerant/resistant to Lethal Yellowing. OCL 16(2):127–131

Bawalan DD, Chapman KR (2006) Virgin coconut oil; production manual for macro- and village-scale processing. FAO, Thammada Press, Bangkok

Bourdeix R, Santos G, Labouisse JP, Baudouin L (2005) Useful definition of terms and nomenclature. In: Batugal P, Ramanatha Rao V, Oliver J (eds) Coconut genetic resources. IPGRI, Rome, pp 9–10

Brown WL (1983) Genetic diversity and genetic vulnerability: an appraisal. Econ Bot 37(1):4–12

Dassanayaka PN, Everard JMDT, Karunanayake EH, Nandadasa HG (2003) Characterization of coconut germplasm by microsatellite markers. Trop Agric Res 15:51–61

Devakumar K, Thomas RJ, Nair RV, Jerard BA, Rajesh MK, Jacob PM et al (2011) Analysis of population structure and genetic relatedness among root (wilt) disease-resistant and susceptible west coast tall coconut palms (*Cocos nucifera* L.) using microsatellite markers. Indian J Agric Sci 81:487–493

Dissanayaka HDMAC, Perera SACN, Fernando WBS, Attanayake RB, Meegahakumbura MGMK, Perera L (2008) Evaluation of the comparative performance of five commercial coconut cultivars under two different agro-ecological zones in Sri Lanka. In: Ninanayake A, Jayamanne E (eds) Proceedings of the second plantation crop research symposium. Samayawardena Printers, Colombo, pp 126–138

Dollet M, Quaicoe R, Pilet F (2009) Review of coconut lethal yellowing diseases. Diversity, variability and diagnosis. OCL 16(2):1–5

Duran Y, Rohde W, Kullaya A, Goikoetxea P, Ritter E (1997) Molecular analysis of east African tall coconut genotypes by DNA marker technology. J Genet Breed 51:279–288

Ekanayake GK, Perera SACN, Dassanayaka PN, Everard JMDT (2010) Varietal classification of new coconut (*Cocos nucifera* L.) forms identified from southern Sri Lanka. Cocos 19:41–50

Everard JMDT (1996) Use of molecular markers for breeding of the coconut palm (*Cocos nucifera* L.), MSc thesis, University of New England, Armidale, Australia

Feng LH, Zhang XC (2005) Quantitative expression on drought magnitude and disaster intensity. Nat Haz Earth Syst Sci 5:495–498

Gomes F, Prado CHBA (2007) Ecophysiology of coconut palm under water stress. Braz J Plant Physiol 19(4):377–391

Harries HC (1978) The evolution, dissemination and classification of *Cocos nucifera* L. Bot Rev 44:205–317

Herran A, Estioko L, Becker D, Rodriguez MJB, Rohde W, Ritter E (2000) Linkage mapping and QTL analysis in coconut (*Cocos nucifera* L.). Theor Appl Genet 101:292–300

Kamaral LCJ, Dassanayaka PN, Perera KLNS, Perera SACN (2017) Characterization of Sri Lanka yellow dwarf coconut (*Cocos nucifera* L.) by DNA fingerprinting with SSR markers. J Natl Sci Found 45(4):405–412. https://doi.org/10.4038/jnsfsr.v45i4.8234

Kamaral LCJ, Perera SACN, Dassanayaka PN (2016) Sri Lanka yellow semi tall; a new addition to the coconut (*Cocos nucifera* L.). Cocos 22:49–55

Lambers H, Chapin IFS, Pons TL (1998) Plant physiological ecology. Springer-Verlag, New York

Lebrun P, Baudouin L, Bourdeix R, Louis Konan J, Barker JHA, Aldam C, Herran A, Ritter E (2001) Construction of a linkage map of the Rennell Island tall coconut type (*Cocos nucifera* L.) and QTL analysis for yield characters. Genome 44:962–970
Lebrun P, N'Cho YP, Bourdeix R, Baudouin L (1999) Le cocotier. In: Hamon P, Seguin M, Perrier X, Glaszmann JC (eds) Diversité génétique des plantes cultivées. CIRAD, Montpellier, pp 219–240
Liyanage DV (1955) Planting materials for coconut. Ceylon Cocon Quart 6:75–80
Liyanage DV (1958) Varieties and forms of coconut palms grown in Ceylon. Ceylon Cocon Quart 9(1):1–10
Liyanage DV (1972) Production of improved coconut seeds by hybridization. Oleagineux 27(12):597–599
Liyanage DV, Wickramaratne MRT, Jayasekera C (1988) Coconut breeding in Sri Lanka. Cocos 6:1–26
Marcus JG, Puri PS (1978) Manufactured hard butters. J Am Oil Chem Soc 55:610A–612A
Marikkar JMN, Jayasundara JMMA, Prasadika SAH, Jayasingha CVL, Premakumara GAS (2007) Assessment of stability of virgin coconut oil during deep frying. CORD 23(1):62–70
Meerow AW, Wisser RJ, Brown JS, Kuhn DN, Schnell RJ, Broschat TK (2003) Analysis of genetic diversity and population structure within Florida coconut (*Cocos nucifera* L.) germplasm using microsatelite DNA, with special emphasis on the Fiji dwarf cultivar. Theor Appl Genet 106:715–726
Nair MK, Koshy PK, Nair RV, Rao EVVB, Nampoothiri KUK, Iyer RD (1996) A root (wilt) disease resistant coconut hybrid and strategy for resistance breeding. Ind Cocon J 27(1):2–5
Nair RV, Jacob PM, Nair SN (2006) Breeding for resistance to coconut root (wilt) disease. In: Nair RV, Krishnakumar V, Thomas RJ, Muralidharan K, Thomas GV (eds) Breeding for resistance to diseases and insect pests in plantation crops. CPCRI, Kasaragod, pp 63–69
Nair RV, Rajesh MG, Jacob PM (2000) Chowghat green dwarf—a suitable variety for the coconut root (wilt) diseased tract. Ind Cocon J 31(3):55
Nkansah-Poku J, Philippe R, Quacoe N, Dery S, Ransford A (2009) Cape Saint Paul Wilt Disease of coconut in Ghana: surveillance and management of disease spread. OCL 16(2):111–115
Omamor IB, Eziashi EI, Asemota OA, Aisagbonhi CI, Ogunkanmi LA (2011) DNA molecular analysis between lethal yellowing disease and non-lethal yellowing disease of coconut palms (*Cocos nucifera* L.) in Nigeria. Trends Mol Sci 3:25–31
Perera KNS, Herath HMNB, Attanayaka DPSTG, Perera SACN (2015a) Assessment of the diversity in fruit yield and fruit components among Sri Lanka tall coconut accessions conserved *ex-situ*. CORD 31(2):33–41
Perera L, Padmasiri MHL, Peries RRA (1997) Brown dwarf: a recent addition to the varieties and forms of coconut palm grown in Sri Lanka. Cocos 12:82–84
Perera L, Russell JR, Provan J, McNicol JW, Powell W (1998) Evaluating genetic relationships between indigenous coconut (*Cocos nucifera* L.) accessions from Sri Lanka by means of AFLP profiling. Theor Appl Genet 96:545–550
Perera L, Russell JR, Provan J, Powell W (1999) Identification and characterization of microsatellites in coconut (*Cocos nucifera* L.) and the analysis of coconut populations in Sri Lanka. Mol Ecol 8:344–346
Perera L, Russell JR, Provan J, Powell W (2000) Use of microsatellite DNA markers to investigate the level of genetic diversity and population genetic structure of coconut (*Cocos nucifera* L.). Genome 43:15–21
Perera L, Russell JR, Provan J, Powell W (2003) Studying genetic relationships among coconut varieties/populations using microsatellite markers. Euphytica 132:121–126
Perera L, Sarathchandra SR, Wickramananda IR (2013) Screening coconut cultivars for tolerance to infestation by the coconut mite, *Aceria guerreronis* (Keifer) in Sri Lanka. CORD 29(1):46–51
Perera PIP, Hocher V, Verdeil J, Doulbeau S, Yakandawala DMD, Weerakoon LK (2007) Unfertilized ovary: a novel explant for coconut (*Cocos nucifera* L.) somatic embryogenesis. Plant Cell Rep 26:1–8

Perera PIP, Yakandawala DMD, Verdeil JL, Hocher V, Weerakoon LK (2010) Morphological aspects of coconut anther culture derived structures. J Nat Sci Found 38(1):69–74
Perera SACN (2005) Report of the genetics and plant breeding division, Annual Report of the Coconut Research Institute of Sri Lanka. Coconut Research Institute of Sri Lanka, Lunuwila
Perera SACN (2010a) In: Gupta SH (ed) Technological innovations in major world oil crops. Springer, New York, pp 201–218
Perera SACN (2010b) QTL analysis in coconut via genome mapping; principles, requirements and prospects. Cocos 20:1–9
Perera SACN (2011) Development of coconut hybrids for tolerance to Aceria mite infestation. Annual Report of the Coconut Research Institute of Sri Lanka. Coconut Research Institute of Sri Lanka, Lunuwila, pp 36–37
Perera SACN (2015) Breeding coconuts for resistance to Weligama Coconut Leaf Wilt Disease. In: Annual report of the Coconut Research Institute of Sri Lanka. Coconut Research Institute of Sri Lanka, Lunuwila
Perera SACN, Dissanayake HDMAC (2013) Management of the Weligama Coconut Leaf Wilt Disease: screening and breeding coconuts for resistance/tolerance to WCLWD. Weligama Coconut Leaf Wilt Disease. Coconut Research Institute Sri Lanka, Lunuwila, pp 96–106
Perera SACN, Herath HMNB, Perera KNS, Fernando WBS (2016) Detection of Marker Trait Associations: a step towards marker assisted selection in coconut (*Cocos nucifera* L.). In: Vidhanaarachchi VRM, Herath HMIK, Meegahakumbura MK, Kumara ADNT, Nadheesha MKF (eds) Proceedings of the sixth symposium on Plantation Crops Research. Coconut Research Institute of Sri Lanka, Lunuwila, pp 229–237. isbn:978-955-9013-19-8
Perera SACN, Herath HMNB, Wijesekera HTR, Subhathma WGR, Weerakkody WATL (2014) Evaluation of coconut germplasm in Weligama and Matara area of the Southern Province of Sri Lanka for resistance to Weligama coconut leaf wilt disease. Cocos 21:15–20
Perera SACN, Kamaral LCJ, Fernando WBS (2015b) Molecular assessment of Cocos nucifera L. Var. Sri Lanka yellow dwarf for genetic purity and aceria mite tolerance. Intl J Mol Evol Biodivers 5(1):1–5. https://doi.org/10.5376/ijmeb.2015.05.0001
Perera SACN, Kilian A (2008) Diversity arrays technology: a high throughput molecular marker system for coconut. J Inst Fund Stud xix(1 Special issue):60–64
Quaicoe R, Dery S, Philippe R, Baudouin L, Nipah J, Nkansah Poku J et al (2009) Resistance screening trials on coconut varieties to Cape Saint Paul Wilt Disease in Ghana. OCL 16(2):132–136
Rajagopal V, Kasturi Bai KV (2002) Drought tolerance mechanism in coconut. Burot Bull 17:21–22
Rajagopal V, Shivashankar S, Mathew J (1996) Impact of dry spells on the ontogeny of coconut fruits and its relation to yield. Plant Res Dev 3:251–255
Rivera R, Edwards KJ, Barker JHA, Arnold GM, Ayad G, Hodgkin T et al (1999) Isolation and characterization of polymorphic microsatellites in *Cocos nucifera* L. Genome 42:668–675
Rohde W, Becker D, Kullaya A, Rodriguez MJB, Herran A, Ritter E (1999) Analysis of coconut germplasm biodiversity by DNA marker technologies and construction of a first genetic linkage map. In: Oropeza C, Verdeil JL, Ashburner GR, Cardega R, Santamaria JM (eds) Current advances in coconut biotechnology. Kluwer Academic Publishers, Dordrecht, pp 99–120
Rohde W, Herran A, Estioko L, Sinje S, Becker D, Kullaya A, et al (2000) Mapping of DNA markers, homeotic genes and QTLs in coconut (*Cocos nucifera* L.) and synteny studies with oil palm, International Symposium on Oil Palm genetic resources and Utilization, AC1–AC2
Shalini KV, Manjunatha S, Lebrun P, Berger L, Baudouin N, Pirany N et al (2006) Identification of molecular markers associated with mite resistance in coconut, *Cocos nucifera* L. Genome 50:35–42
Teulat B, Aldam C, Trehin R, Lebrun L, Barker GM, Karp A et al (2000) An analysis of genetic diversity in coconut (*Cocos nucifera*) populations from across the geographic range using sequence–tagged microsatellites (SSRs) and AFLPs. Theor Appl Genet 100:764–771
Varshney RK, Glaszmann J, Leung H, Ribaut J (2010) More genomic resources for less-studied crops. Trends Biotechnol 28:452–460

Waidyarathne KP, Dissanayake HDMAC, Perera SACN, Chandrathilake TH (2017) Temporal yield stability of coconuts to extreme weather events—a case study. Proceedings of the international statistics conference. Colombo, IASSL, p 90

Wickremaratne MRT (1984) Report of the genetics and plant breeding division. Coconut Research Institute of Sri Lanka, Lunuwila, p 45

Wright H (1980) Commercial hybrid production. In: Fehr WR, Hadley HH (eds) Hybridization of crop plants. The American Society of Agronomy, Wisconsin, p 162

Xiao Y, Xu P, Fan H, Baudouin L, Xia W, Bocs S et al (2017) The genome draft of coconut (*Cocos nucifera*). Gigascience 6(11):1–11. https://doi.org/10.1093/gigascience/gix095

Zizumbo-Villarreal D, Ruiz-Rodriguez M, Harries H, Colunga-Garcia Martin C (2006) Population genetics, lethal yellowing disease, and relationships among Mexican and imported coconut ecotypes. Crop Sci 46:2509–2516

Breaking the Mold: Pave the Way for Future Cereals

8

Dimanthi Jayatilake and Venura Herath

Abstract

Cereals have been the staple food of Sri Lankans even before the dawn of its agricultural sector. Over the years, Sri Lankan agriculture has grown from subsistence-level farming to a profit-oriented industry that supplies for both local and international markets. Sri Lankan cereal sector adopted innovative techniques such as green revolution and hybrid seed technology to meet production targets; however, the adoption of technology came with much caution and at a slower pace. Today the booming population and the changing climate that leads to unfavorable growth conditions have left the country's production at a shortage to an ever-increasing demand. To bridge the gap between the current status and future expectations, modern approaches such as molecular breeding and genetic engineering need to be adopted. To break the yield barriers and take cereal production to the next level, change needs to be positively engulfed across all dimensions of the sector. In this chapter, we look at yesterday's events that mold the path to reach the current standing of the cereal sector and discuss how the timely adoption of modern technology will pave the way to feed tomorrow's world.

Keywords

Cereals · Technology adoption · Molecular breeding · Genetic engineering

D. Jayatilake (✉) · V. Herath
Department of Agricultural Biology, Faculty of Agriculture, University of Peradeniya, Peradeniya, Sri Lanka

B. Marambe et al. (eds.), *Agricultural Research for Sustainable Food Systems in Sri Lanka*, https://doi.org/10.1007/978-981-15-2152-2_8

8.1 Cereal Production in Sri Lanka: A Glimpse on Yesterday and a Purview on Tomorrow's Expectations

While the fine details of when and where the early settlers began agriculture in Sri Lanka are not scripted, the earliest records come from a *pali* script named *Mahavamsa* dated sixth century B.C. to fourth century A.D. In the *Mahavamsa*, there are written accounts of rituals and celebrations that took place in the grandest of royalty involving produce, irrigation systems, and payments made to the peasants. All of this reflects a well-established agricultural system that thrived in ancient Sri Lanka (Smith 2006). The written history backed by the still standing irrigation and cultivation systems shows off how deep the roots of agriculture have penetrated into the civil society. In early days, the crops were selected based on the needs of the rulers. As a result, cereals especially rice dominated the agricultural landscape, in which it was the major crop grown under the traditional tank-based villages of ancient Sri Lanka.

Over the years, Sri Lankan agriculture has evolved from a subsistence-level venture to a profit-earning industry. Going along, the cultivation of cereals has transformed from subsistence farming into a capitalistic trade controlling a significant portion of today's economy. At present, rice is the leading cereal produced in the country, followed by maize, millet, and sorghum (Fig. 8.1; FAOSTAT 2018). While the production of rice paddy has witnessed nearly a 4-fold increase, and maize a whopping 26-fold increase, between the years 1961 and 2014, the production of sorghum and millet has gone down in 5- and 9-folds, respectively (Fig. 8.1). This transition from subsistence- to commercial-level cereal production was facilitated by the adoption of technology that was introduced as a part of the green revolution and hybrid seed production.

Green revolution refers to a newline of thinking and research that took the world by a storm in the 1950s and 1960s. This movement and the out-of-the-box thinking that came along with it lead to the development of the modern or high-yielding varieties (HYVs). These high yielders changed the course of global production statistics of cereals. The thinking of pioneer researchers such as Nobel Peace Prize winner Dr. Norman Borlaug, and the research conducted in the early 1950s, led to the development of dwarf varieties with higher grain yield, better fertilizer response, and low leaf biomass (Borlaug 2002). This approach to crop improvement directed global agriculture, especially the production of cereals into a leap of success. The revolution that salvaged billions of people worldwide from hunger found its way into Sri Lanka in the mid-1960s. A cabinet sub-committee was appointed in the year 1966 to look into the possibilities of adopting these newer technologies, and the path to facilitate the needed research was paved.

The green drive led to the development of rice HYVs in Sri Lanka. However, as noted by Herath (1981), the significant returns of the green revolution started to die down within a decade or so but managed to achieve a rice production growth rate of 4.9% from 1955 to 1978. The adoption came slow to the farmers' fields, where by the year 1978, only 50% of the paddy land was cultivated with HYVs (Chandrawathi et al. 2013). The main reasons for the low returns in adopting HYV under the

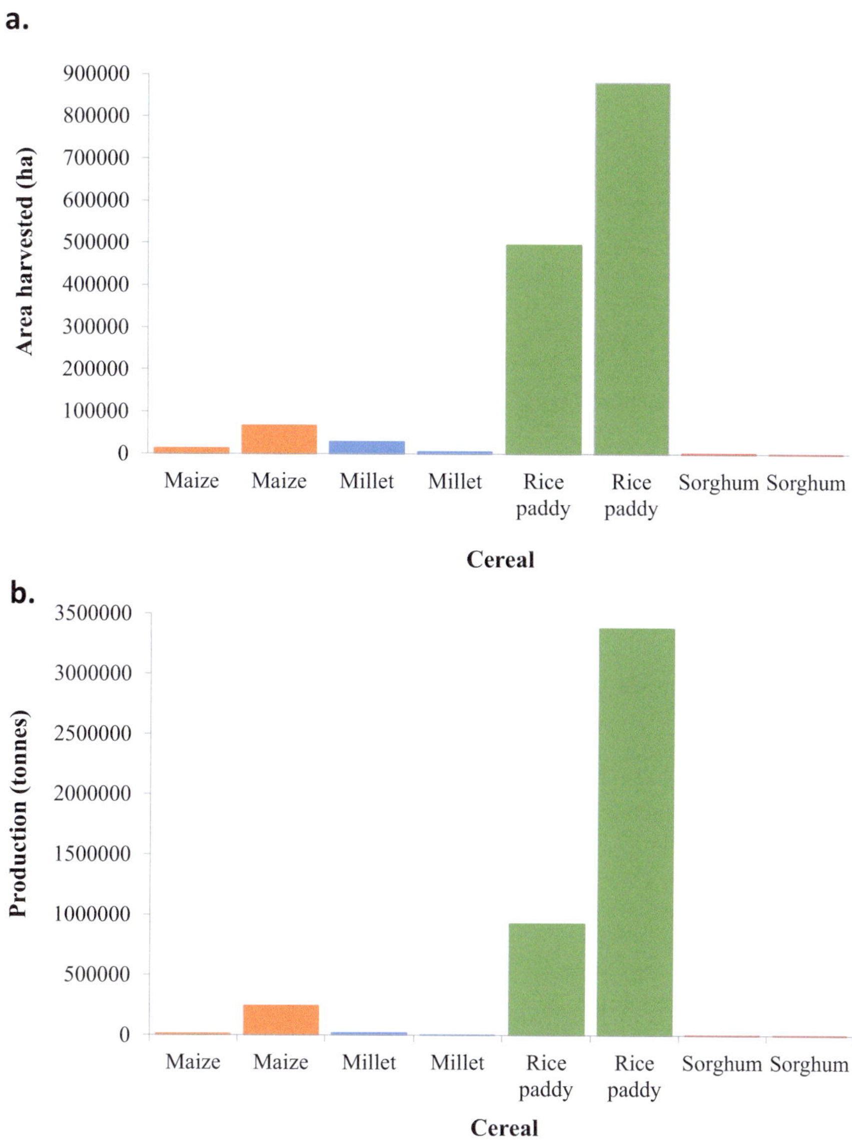

Fig. 8.1 Comparison of (**a**) area harvested and (**b**) production of cereals in Sri Lanka between 1961 and 2014 (Source: FAOSTAT 2018)

Sri Lankan conditions were identified as the high dependency on rain-fed water, uncertainty of crop success, and low technology adoption tendency (Herath 1981). As a result, the replacement period of an HYV was estimated to be over a decade in Sri Lanka (Chandrawathi et al. 2013). However, over the last decade, almost every year, the Department of Agriculture has released a new improved rice variety.

In 1994, the rice breeding program in Sri Lanka turned a corner when the Rice Research Institute at Bathalagoda partnered with the International Rice Research Institute in the Philippines, to initiate the hybrid rice program. The hybrids BgHR1, BgHR6, and BgHR12 were developed through this program and had higher yields than the standard improved varieties such as Bg 357 and Bg 403, yielding best at the time (Abeysekera et al. 2003). Since its inception, the hybrid rice production was carried out dominantly by the government sector. While increasing the yield parameters being the main objective, later on producing rice with the desired eating and cooking quality characters was also part of the breeding objectives. However, the hybrid rice production faced the challenges of limited heterosis occurring from the local germplasm in terms of rice yield, widespread practice of direct seeding, tendency of farmers to use saved seeds, inadequate resources to produce hybrid seeds to meet the demand, and the lack of uniformity in hybrid seeds due to genetic impurities in parental lines (Abeysekera et al. 2003).

The story of maize takes a different direction than rice in the breeding programs conducted at the Field Crops Research and Development Institute (FCRDI) in Mahailluppallama in the North Central Province of Sri Lanka. The crop improvement research boom in Sri Lanka that happened along with the green revolution led to the release of the first open-pollinated maize variety T-48 in the early 1960s (Ranaweera et al. 1988). With close links getting established in 1968 with the Inter-Asian Corn Improvement Centre in Thailand and with the International Maize and Wheat Improvement Center (CIMMYT) in Mexico, improved maize germplasm was introduced to uplift the local breeding programs. As a result, in 1977, the variety *Bhadra 1* was released, and it became a promising and popular maize variety. Maize breeding program in Sri Lanka has been focusing on producing dwarf, early maturing varieties with quality protein profiles and high yields (Ranaweera et al. 1988). Since the initial attempt on germplasm improvement, many high-yielding open-pollinated varieties have been developed and released by the Department of Agriculture.

With the introduction of hybrid technology, attempts were made both to produce hybrids locally and to introduce hybrids from the USA. However, they came with no yield advantage compared to the open-pollinated varieties, and they were not suitable for the local conditions. The Indian hybrid *Ganga-3* was tested for local conditions in 1968/1969 and showed promising yield improvements; however, given the considerable amount of money needed for seed purchase, the researchers looked for a local option (Ranaweera et al. 1988). Some of the more recent hybrids released by the FCRDI at Mahailluppallama includes MI Maize Hybrid 01 and MI Maize Hybrid 02 (Bentota 2013; Weerasinghe et al. 2016). Today, maize production is a commercial venture with privately owned companies such as CIC Agri-Business and Ceylon Agro Industries entering into the industrial production through farms and farmer-by-back systems using high-yielding hybrids.

The crop improvement programs targeting other minor cereals are carried out at the FCRDI at Mahailluppallama. Through these breeding programs, new varieties with improved characteristics are being developed for sorghum and millet; however, no varieties were released recently. However, the high dependency on imported hybrid seeds could pose a threat in the future.

According to the predictions of the Food and Agriculture Organization of the United Nations (FAO), the world population is expected to increase by 2.3 billion, between the years 2009 and 2050, reaching 9.5 billion heads in 2050 (FAO 2012). This population boom will mostly happen in the developing countries, and for the majority of the population, cereals will be the staple food. To feed the growing population, the production needs to be intensified to reach a target that is nearly a double of the current global cereal production (Ray et al. 2013). According to the projections made by Ray et al. (2013), under the current circumstances, the rice production in Sri Lanka is not going to have a significant increase to meet the expected leap in production by 2050. Sri Lanka has a current population of 21.4 million (CBSL 2018), and according to the 2016 estimates, the population is increasing at a rate of 1.1% (CBSL 2017) and will be confined to a land area of 62,705 km^2 excluding the inland waterbodies (CBSL 2018). With the increasing population, urbanization, and alternative land use, Sri Lanka will soon face the limitation in agricultural land. While some agricultural land involved in cereal cultivation gets protected by laws (i.e., the Paddy Lands Act No. 1 of 1958), the other agricultural lands are vulnerable to become victims of urbanization and industrialization. As a result, soon we will face the challenge of increasing the per unit production to feed the demand of the increasing heads.

Globally we are experiencing a shift in the climate, and its impact on Sri Lanka is profoundly felt. In one angle, there is a frequent occurrence of dry spells in the drier areas of the country, and from another perspective, the intense heavy rainfall leads to inundation of the crop lands in the low-lying areas of the country. As a result, the cereal belt of Sri Lanka which is mainly falling over the drier agro-ecological zones is under the risk of drought, and the areas that can act as the buffer production zones get frequently affected by flooding in the terrestrial rains. Under the persisting drought-/flood-prone climatic conditions, the cereal production is expected to fall. As an example, the *Maha* rice harvest of 2017 experienced a 45% decline in the production, a 20-year low due to the prolonged drought that swept the country (FAO 2017). A similar scenario is experienced in 2018.

Although most felt and talked about reason for crop failure is lack/excess of precipitation, adverse climate effect comes in many forms. One such is the rising atmospheric CO_2 levels. Crop plants with C_3 photosynthesis pathway such as rice are at a disadvantage compared to crops with a C_4 pathway such as maize with concerning efficient CO_2 assimilation at lower and higher atmospheric CO_2 levels (Von Caemmerer et al. 2012). Rice scientists consider genetically engineering a C_4 rice as the "grand challenge" in creating the next phase in green revolution (Von Caemmerer et al. 2012). With the C_4 Rice Project, researches are underway to biochemically and anatomically equip rice to meet the expected transition (C4 Rice Project 2018). Meanwhile, the atmospheric temperature is on the rise, and all living beings including the crop plants feel the change undoubtedly. While the C_4 crops are more resilient to the increasing temperatures compared to C_3 crops, C_4 are vulnerable as well. In maize, it was reported that the net photosynthesis starts to get inhibited when leaf temperatures reach 38 °C, due to the decreased activation state of *Rubisco*, a typical scenario for both C_3 and C_4 plants (Crafts-Brandner 2002).

With the changing climate, another main challenge for future cereal production looms around the corner; soil toxicities and deficiencies, such as salinity and iron toxicity, and Zn deficiency, especially in the wet lowland production areas of the world, are an upcoming challenge that needs immediate attention (Shahbaz and Ashraf 2013; Neue et al. 1998). These are two of the major reasons for the reduction in land available for cultivation and lowered yield potential; however, these are not simple enough issues that can be rectified or reversed with a simple solution. Hence, it is an urgent necessity to develop cereals that are tolerant to abiotic stresses to ensure that the cereals grown are climate-ready to face the irregular weather patterns we are currently experiencing. Resistance to pests and diseases is no exception; the future cereals need to be equipped to produce, despite the hikes in the pests and disease occurrence frequencies and widespread damage. While the complex relationships of the changing weather with pests and disease are not understood correctly, it is expected to be more frequent and severe causing significant yield and quality losses (Chakraborty and Newton 2011). This will make the farmers ready for the natural adversities they face such as the salinity experienced after the 2004 tsunami and the drought spells that the country experienced in 2015/2016 and 2017/2018.

The future cereal consumers are expected to be demanding for quality over quantity. The consumer preference seems to be changing due to the increasing purchasing power of some fractions of the society and the public awareness of eating healthy for a better life. This trend has led to the emergence of niche markets, and the public awareness of eating healthy to lead a better life. To cater to this demand, the breeding strategies need to be focused on improving physicochemical and nutritional qualities of cereals. For example, future rice needs to be looked upon with improved quality attributes such as eating, cooking, milling, nutritional quality, and desirable appearance (Birla et al. 2017; Kumar et al. 2015). The market attributes such as purity, white, and slender grains, with high amylose content that produces a fluffy, grain-separated cooked rice, will be key targets (Walisinghe and Gunaratne 2012). With the booming tourist industry, "Basmati"-type rice will have a greater demand, and the rice breeding programs must have a directive to cater to the growing demand than importing. In the future, breeding cereals targeting fortification of cereals to alleviate "hidden hunger" by enriching nutrients such as vitamins and minerals and reducing postharvest losses of nutrients during processing and preparation will standout (Birla et al. 2017; Strobbe and Van Der Straeten 2018). Some cereals such as sorghum are naturally deficient in essential amino acids such as lysine, threonine, and tryptophan (Salunkhe et al. 1977). Through biofortification, the nutrient value of such cereals could be enhanced. The presence of anti-nutritional factors such as phytic acid in sorghum can reduce the bioavailability of trace elements (Salunkhe et al. 1977). Through breeding programs, the levels of anti-nutritional properties need to be reduced in due course.

The recent consumer drift toward nutritious eating has created a trend for consumption of cereals that were on the sidebar all this time. As an example, it is trending to consume red and traditional rice varieties to seek health benefits, and more attention has been received by minor cereals such as millet, due to their low glycemic index leading to benefits in formulating diet plans for people with diabetics. While only few of the available accessions are suitable for wider cultivation, a collection of cereal accessions (i.e., rice, maize, sorghum, and millet) are conserved at the Plant Genetic Resource Centre, Gannoruwa. These accessions could be effectively used in the breeding programs to improve the attributes of commonly cultivated varieties. The tendency toward consuming fragrant rice has also swept its way into the local markets. These trends indicate that the future cereal breeding strategies must not be focused solely on the quantity but also must invest in the improvement of the physicochemical and nutraceutical properties to find better market potentials. Another important aspect in the future cereal breeding programs would be the diversification of the food options. For that cereal such as maize and millet must be fortified and bred according to consumer preference to supplement the now rice-dominant diets of Sri Lankans.

With the challenges we have for tomorrow's cereal production, Sri Lanka will not be able to seal the yield gap and produce good-quality cereals, if only a classical breeding approach is implemented. While the world is advancing in science and technology, and globally the cereal breeders are adopting the most novel techniques available out there for crop improvement, it is hard to say that we as a nation are going alongside. We are going forward into the next decade with limited technology resources and a slow pace and a reluctance to embrace technology. To close the yield gap and enhance the quality attributes of cereal production, there is no doubt that the genetic gain must be improved by adopting new interventions to go hand in hand with the rest of the world.

8.2 Bridging the Gap with Modern Technology

Over the last two decades, new bio-engineering technologies have revolutionized the field of crop breeding. Among these, marker-assisted selection (MAS) and genetic engineering are the forefronts. Scientists have been able to manipulate the germplasm resources to convey desirable traits such as higher yields, biotic resistance, and abiotic stress tolerance in major crops while significantly reducing the input requirements (Dively et al. 2018). Emerging novel technologies like bio-engineering have paved its way into the crop improvement programs. The national research priorities involving biotechnology has been identified by the national committee on agricultural biotechnology representing the Sri Lanka Council for Agricultural Research Policy of the Ministry of Agriculture for the period of 2017–2021. For the food crop sector, six main priorities were identified (Perera et al. 2017), namely:

1. Molecular characterization of genetic resources of important food crops,
2. Gene/QTL (quantitative trait loci) mapping and MAS for important yield, nutritional quality, and biotic and abiotic stress tolerance traits,
3. Use of novel molecular marker techniques,
4. Development and application of molecular diagnostic tools for diseases diagnosis,
5. Production of quality, disease-free planting material, and
6. In vitro conservation of crop germplasm.

To achieve the goals set under the main biotechnology priority areas, we need to embrace and use the latest technology available.

Crop improvement is mainly focused on exploring the genetic diversity to bring out the desired phenotypic traits. A diversity assessment allows identification of the population structure and allele richness, reconstructs phylogenetic relationships, and permits breeders to select suitable germplasm for crop improvement (Govindaraj et al. 2015). The diversity is naturally found among the crop varieties and in their close/wild relatives. In the modern day, accurate assessment of diversity is made easier with the availability of rapid high-throughput molecular marker technologies and sequencing approaches. Among the crop species, cereals are a group of crops that have received the privilege of getting their genome assembled ahead of other groups. With a genome size of 389 Mb, rice is one of the first crops to get a completely assembled physical map (International Rice Genome Sequencing Project 2005), followed by barley (Beier et al. 2017), maize (Jiao et al. 2017), and a partially completed genome of wheat given that it is the most complex of all cereals with a 15.3 Gbp hexaploid genome (Zimin et al. 2017). For rice, the genomic sequences of nearly 3000 accessions sequenced through next-generation sequencing (NGS) platforms are available through the 3K Rice Genomes Project (Li et al. 2014). With this project, whole genome sequence of 47 Sri Lankan rice varieties has been published at Rice SNP-Seek Database of IRRI (http://snp-seek.irri.org/) providing excellent opportunity for rice breeders of the country.

The next-generation sequencing platforms together with in silico or computational methods allow identification of single nucleotide polymorphisms (SNP) and insertion-deletion length polymorphisms (InDels) present in intergenic and intragenic positions of the genome, thereby giving insights into the identification of agronomically important genes and their *cis*-acting regulatory mechanisms. Recently, the whole genome of *Godawee*, a traditional Sri Lankan rice variety, was sequenced using Illumina paired-end NGS technology, revealing 2,231,717 SNPs and 480,460 InDels comparatively to *Oryza sativa* cv. Nipponbare, the *japonica* reference genome (Wijesinghe 2017). The availability of these genomic sequence resources has had a huge impact on the assessment of existing diversity in the major cereals and development of molecular markers for MAS in breeding programs.

Diversity assessments of cereals have been reported for many important genes, i.e., submergence tolerance (Iftekharuddaula et al. 2016; Septiningsih et al. 2009), grain traits (Edzesi et al. 2016), flowering time (Navarro et al. 2017; Naranjo et al. 2014), eating and cooking quality traits (Tian et al. 2009), and pest and disease resistance (Zhao et al. 2016) are a handful to state. When a collection of germplasm is screened for allelic diversity, it facilitates the assessment of the germplasm based on known alleles and also sets the platform to identify novel alleles. Once the alleles are identified and their functionality is validated, they can be effectively used in improving the effectiveness and efficiency of the ongoing breeding programs. The diversity assessment along with phenotypic trials will enable the breeders to select the most suitable parents to perform crosses. These selections will be the base for development of biparental mapping populations for QTL mapping or for genome-wide association mapping to enable detection of loci that are highly associated with the trait. This facilitates the confirmation of existing QTL and/or identification of novel QTL in local germplasm. Given the island geography and the restricted movement of genetic material due to strict quarantine regulations, Sri Lanka is home to an exotic germplasm collection, especially with rice, and hence needs to be utilized sustainably.

Molecular markers can be developed by differentiating the identified haplotypes to track the desired alleles starting from a very early crop growth stage. This enables the accurate genetic selection of desired alleles and facilitates selections/pyramiding alleles of interest during breeding programs. Marker-assisted selection leads to an accurate section in a time-cost-effective way, as it offers selection under various conditions that otherwise would create delays. Marker-assisted selection cuts down the time from initial parental crosses to the varietal release and enables the breeders to focus on the genetics of the traits, without getting masked by the environment. Further, it is a cost-space-efficient screen, as MAS can be performed at seed/seedling stages. The molecular tools used in MAS minimize linkage drag, enable selection of even low heredity traits and allow to conduct selection for traits that cannot be field tested well in advance due to the risk of accidental releases. Most importantly MAS enables pyramiding genes responsible for delivering elite varieties with multiple desired traits for the future (Collard et al. 2005). Introgressions through marker-assisted backcross breeding (MABB) is currently done to produce varieties that can produce to meet tomorrow's needs, i.e., some examples from rice include introgression of *Sub1* submergence tolerance alleles (Neeraja et al. 2007); pest and disease resistance genes such as brown plant hopper resistance alleles (*Bph14* and *Bph15*), rice stripe resistance alleles (*Stv-bi*), and bacterial leaf blight resistance alleles of *Xa21*, *Xa13*, and *xa5* (Singh et al. 2001); abiotic stress tolerance alleles such as *Pup1* for phosphorus tolerance (Chin et al. 2011) and *Saltol* for salinity tolerance (Huyen et al. 2013); eating and cooking quality alleles such as *waxy* (Huyen et al. 2013); and other important genes such as thermo-sensitive genetic male sterility alleles *tms2*, *tgms*, and *tms5* (Huyen et al. 2013).

In early 2005, through the Rice Annotation Project, 18,828 Class 1 simple sequence repeats (SSRs) with hypervariable loci were identified (International Rice Genome Sequencing Project 2005). However, decades later even SSRs have not been adequately utilized in rice MAS in Sri Lanka. A similar scenario has been experienced with other cereals as well. To facilitate genetics and breeding, the world is currently moving fast toward high-throughput marker technologies such as high-throughput SNP genotyping (Chen et al. 2014), diversity array technology (DArT; Yong et al. 2006), Kompetitive allele specific PCR (KASP), and genotyping by sequencing (GBS; Furuta et al. 2017). These are scored based on high-throughput detection assays such as capillary gel electrophoresis, high-resolution melting technology (HRM), endpoint genotyping, and next-generation sequence analysis pipelines. In these large data resources are a wealth of information that can be tapped into advancing our way forward. For a MAS program to be successful, the markers used must be either intragenic, flanking, or be at least tightly linked to the gene (<5 cM) to achieve reliable selection. The new-generation technologies facilitate the development of such intragenic diagnostic markers. Sri Lankan cereal breeding programs are now in the trend of using MAS, especially with simple forms such as SSR and allele-specific markers, to track desired traits at the molecular level with high precision in a fraction of time. However, the use is limited. Limited adoption of MAS in breeding programs could be due to poor availability of facilities to the researchers and the lack of policies that support the adoption of advancements in science and technology. These reasons have hindered the progression of pre-breeding research in Sri Lanka.

The diversity of major crops was significantly narrowed during the process of crop domestication where only a few traits were looked upon. As an example, in maize domestication, the focus was mainly on five genes, including *tga1* and *tb1* (Studer et al. 2017; Wang et al. 1999). However, diversity is an essential component in crop improvement, which was foreseen when yield-related traits were made the focus at early days leading to the crisis we face today. In the absence of natural diversity, genetic variation can be created through conventional approaches and with modern technology. The conventional approaches include hybridization and introgression to acquire desired alleles from wild/close relatives and mutation breeding which creates variations. While inter- and intra-species hybridization and introgression can occur naturally, the reproductive barriers limit its occurrence (Twyford and Ennos 2012). However, these two scenarios are often used to create diversity in its absence in the natural form through conventional and modern breeding techniques. In cereal improvement programs, introgression of desired alleles from wild relatives/landraces is commonly practiced in the absence of the ideal phenotypes, both at inter- and intra-species level. An example of inter-species introgression was the successful introgression of bacterial leaf blight resistance into susceptible rice varieties by using the resistance allele of the *Xa21* gene from *Oryza longistaminata* (Song et al. 1995). Further, an example of an intra-species introgression in rice is the introgression of the *Sub1* submergence tolerance allele to elite Asian lines from the variety FR13A (Neeraja et al. 2007). Approaches as such have made huge changes in securing food safety and livelihood of rice growers, and there are many other success stories yet to be discovered and put into work.

In the absence of genetic variability, induced mutagenesis is a key technique that can be used to create the needed diversity. According to the Mutant Variety Database of the International Atomic Energy Agency, currently there are 821 rice varieties, 96 maize varieties, 15 sorghum varieties, and 258 wheat varieties developed through direct mutation breeding or by crossing mutated lines (https://mvd.iaea.org). As reviewed by Tai (2007), the age-old technique of mutation breeding has resulted in breakthroughs since early as the green revolution days where two Chinese varieties were released as products of mutation breeding. Since then, many important breakthroughs in creating needed diversity have occurred. Two of the success stories that were developed through spontaneous mutagenesis are semi-dwarf rice developed through X-irradiation (Hu 1973) and monogenic male sterility in rice with chemical mutagens (Singh and Ikehashi 1981). In more recent years, site-directed mutagenesis has become a hot topic in science. Through the site-directed mutagenesis followed by homologous recombination, some crops have been improved successfully; two success stories includes the development of herbicide tolerance rice (Endo et al. 2007) and high tryptophan-containing rice (Saika et al. 2011). In recent years, the inducing of site-specific mutations through a novel tool called CRISPR/Cas9 has gained the attention of plant breeders in targeted genome editing. CRISPR/Cas9 tool has been developed adapting the principles of natural genome editing found in bacteria, and the tool introduces precise and targeted changes in genomes of the target organisms. Given that the tool is more efficient than conventional genome editing, the approach can be considered as a promising way forward in crop improvement.

In crop genomes, diversity is mostly a result of the variations in their non-coding (promoter) regions than that of the coding regions (Tatarinova et al. 2016). Genome editing can be effectively used to generate promoter variations without altering protein coding regions (Rodríguez-Leal et al. 2017). During the domestication and crop improvement processes, the major focus was given to the variations located in protein coding regions compared to non-coding regions. This provides scientists with an unexplored resource that can be used to introduce diversity back to crop genomes. The *cis*-regulatory elements (CREs) located on gene promoters are essential for the modulation of gene expression. Their effects are governed by their occurrence and vicinity compared to the gene of interest. These CREs can be precisely introduced, modified, or removed using genome editing tools to create a series of plants with diverse phenotypic outcomes. For the first time, the CRISPER/Cas9-based targeted editing was induced in the promoter of tomato *WUS* (*SlWUS*) and *CLV3* (*SlCLV3*) genes to generate novel mutant CRE alleles with increased fruit size, enhanced inflorescence architecture, and improved plant architecture. This displays the potential use of genome editing approaches to engineer genes rapidly and efficiently compared to time-consuming and less effective conventional strategies (Li et al. 2017; Rodríguez-Leal et al. 2017). However, this novel approach is yet to be used in cereal crops. The major advantage of this strategy is that resulting crops are transgene-free crops. As a result, these crops will be more appealing to the end consumers. The availability of comprehensive sequence information on major cereal genomes enables scientists to utilize genome editing both easily and effectively. In the context of local scenario, the task is made easier for rice with the availability of genomic sequences of more than 40 Sri Lankan rice varieties in public databases (Mansueto et al. 2016; Tello-Ruiz et al. 2018). Therefore, there is a huge potential for local crop improvement programs to incorporate genome editing tools to improve the production of superior cultivars to withstand unfavorable conditions while maximizing their yield. Hence, it is timely to start focused research programs toward developing advanced varieties with enhanced attributes to address the ever-increasing calorie requirement.

While conventional plant breeding approaches and molecular breeding tools can be effectively used for the improvement of cereal crops, to seal the gap, a more far-reaching tactic is needed. The stepping stone is genetic engineering (GE), the oldest and the most widely used bio-engineering technology in crop improvement. The GE technology came with ground-breaking inventions that changed the course of agriculture, and the farmers around the globe embraced the use of GE crops given their effectiveness. In 2016, more than 185.1 million ha of crop land in 26 countries, including 19 developing and 7 industrial countries, is cultivated with GE crops, with soybean, maize, cotton, and canola representing the biggest share (ISAAA 2016).

Due to the huge potential of genetic engineering, a multitude of traits can be easily incorporated to cereal genomes including yield, nutritional quality, disease resistance, and stress tolerance as reviewed by Kamthan et al. (2016). As an example, pesticide misuse is a huge concern in agriculture, and GE can bring a sustainable solution to mitigate the issue. One such GE product is the Bt crops that carry Bt Cry protein-encoding gene obtained from soil bacterium *Bacillus thuringiensis*. The Bt crops have reduced the pesticide usage by almost 50%, by providing a safe environment for many beneficial insects living in farmlands. The same Bt protein is approved in organic agriculture as a safe compound to control insect pests (Dively et al. 2018; Pellegrino et al. 2018). The only available commercialized rice variety carrying Bt protein (cry1Ab) for lepidopteran insect resistance has been developed by the Agricultural Biotech Research Institute, Iran. The use of these varieties could pave for the establishment of an ecologically safer agricultural system. In the meantime, there are 15 commercial bioengineered varieties of maize in the world. These maize varieties carry traits such as insect resistance (Lepidoptera and Coleoptera), herbicide tolerance (glufosinate and glyphosate), and drought stress tolerance, and also some are optimized for bioethanol production (thermostable alpha-amylase). These enhanced traits enable farmers to use significantly fewer quantities of pesticide and herbicide and decrease the environmental footprint during the cultivation (Brookes and Barfoot 2016; Klümper and Qaim 2014). Also farmers can obtain higher yields under the changing climate conditions. The success stories of GE maize hint about the existing gap in rice GE work. Currently, there are 22 GE rice varieties in the production pipeline at pre-commercial, regulatory, and advanced developmental stages (Parisi et al. 2016).

None of the traditional or conventional crops have undergone rigorous testing compared to bioengineered crops both considering the human health and environmental safety (Nicolia et al. 2014). Still, the majority of the consumers look at these technologies with skepticism due to religious beliefs and misguided and fabricated information. As a result, these modern crops need to undergo unnecessary extensive regulatory barriers that sometimes take up to 7–8 additional years before their commercial release (Prado et al. 2014). Golden rice is one of the most promising results of bioengineering with provitamin A biofortification. It was ready in 2000 commercial cultivation. However, golden rice is yet to reach the market due to unfound fears caused by internationally financed anti-GMO lobby. Confined field trials were conducted in the Philippines and Bangladesh and Taiwan (Potrykus 2017; Stone and Glover 2017). The International Rice Research Institute filed applications for the approval of golden rice for commercial cultivation in Bangladesh, the Philippines, and Bangladesh in 2016 (Owens 2018).

To make the matter worse, currently there is no act for biosafety available in Sri Lanka, though the Ministry of Mahaweli Development and Environment, the national focal point for the formulation of the biosafety act was assigned to work on it a decade ago. This is a barrier to researchers both at universities and research institutes, to develop much needed future crops. Developing bioengineered crop from gene/trait identification to commercial release is a long process, i.e., ~17 years. These delays detrimentally affect the contribution of bioengineered crops to national food security (Prado et al. 2014). To remedy the situation, it is important to have a trait-based approval process preceding a risk assessment than a blanket regulatory process for each trait. It will make bioengineering an affordable approach for researchers coming from public universities and research institutes. None of these approaches will be tangible unless the end users are made aware of the benefits of bioengineered crops. To achieve that, there should be a multifaceted approach to disseminate the real science of bioengineering to the public. Given the future needs, it is timely that we invest in the potential of modern technologies.

8.3 An Integrated Platform to Keep Out Pseudoscience and Engulf Change

If science and technology are out there, what is keeping us from moving forward?—a question that all stakeholders involved in cereal production need to think deeply. Whether you are a policymaker, a pre-breeder, a breeder, an extension officer, or a farmer, there is a role for each one of us to play. The main causes that limit us from adopting these novel techniques are the reluctance to change, unawareness of the science behind, and the limitations of the available facilities. If an integrated platform could be created to address these three main limitations, Sri Lanka could go into the year 2050 with a target of self-sufficiency in cereal production.

In this digital era, access to information is not farfetched. With a click of an icon, one is connected to a wealth of information via the Internet. Blogs and popular social media platforms such as Facebook™ and Twitter™ are at one's doorstep returning information on demand. Social media can be effectively used for science communication to change attitudes by reaching out to all demographics. However, the truth may not be said as it is at all times in these forums. Often it is the case of fake facts, altered interpretations, and pseudoscience taking over the evidence-based science. The use of social media as a tool to get news is rapidly increasing across the globe. In the USA, 67% of the population receives news via social media, making social media one of the top most influential factors that affect the modern human being (Shearer and Gottfried 2017). Even though there are no proper surveys conducted, given its popularity, a similar trend could be expected in Sri Lanka as well. This is providing a platform for ill-informed celebrities, influencers, organizations, and companies to spread misleading pseudoscience. Last 5 years, anti-science, anti-technology, and anti-commercial agriculture content was used to create fear mongering among both farmers and consumers. This is mostly done with the intention of creating a public trend which will lead to the feeding of expensive niche agriculture markets that flourish on the public fear. This problem is further aggravated by the poor awareness among general public on how to evaluate the credibility of news sources and to fish out the real from fake under the current scenario. Unfortunately, that group includes farmers, consumers, scientists, and policymakers and thus boils down to all stakeholders of today's agriculture.

To address the issue, the scientists need to play the role of social influencers. Unfortunately, this is becoming a herculean task for the scientists over celebrities, since social media reach is based on likes, shares, and retweets (Galetti and Costa-Pereira 2017; Mojarad 2017; Rodger-Withers 2017). To combat the issue, there should be a platform where scientific information can be effectively disseminated covering all printed, digital, and social media to the general public in a context where the reports are written by trained journalists in consultation with the scientists.

A nationwide awareness program that targets all layers of stakeholders is a timely need. The unawareness of the underlying science, the deep-rooted myths that have over the time become norms, and the reluctance to engulf change can only be addressed through creating awareness through public campaigns led based on scientific conversations. These awareness programs must target the general public to the policymakers. Creating awareness on the most current science and technology advancements must be done at a very young stage and should be a priority that is identified during the development of education and higher education policies and curricula in Sri Lanka. Science education must be reformed in such a way that our young generation gains the ability to investigate and synthesize their own opinion on important scientific issues. Through scientific forums, conversations must be built to educate our youth on the pressing issues that define us as a nation from moving forward. Not stalling there, our youth needs to voice their opinion so that the next generation becomes more vocal and demanding to perceive an education that will make them open-minded and challenge-ready.

For the scientists involved in the agriculture sector, it is important that timely training is given to update their knowledge and skills on the advancements in their respective fields. The scientific community must organize focus group discussions, and the decisions made must be convened to the policymakers to influence their thinking to lead the nation toward a science-friendly future that embraces the positive notes on science for the betterment of the country.

It is important to develop a research culture that embraces both local and international collaborations. Often in the Sri Lankan context, the collaborations between private sector industries, research stations, and universities come rarely. Close collaborations between these institutes will ensure that the research funds are put to the best use and the research outcomes will reach the end consumer catering to their needs. The local and international collaborations will promote efficient resource utilization and sharing of knowledge.

Awareness alone, however, is not going to give solutions to the said problem. The nation needs to go into a revolution in adopting science and technology to intensify the agricultural production. For that the research infrastructure needs to undergo a drastic improvement. Currently, it is the research stations of the Department of Agriculture, the universities, and few private sector institutes such as the CIC Agri-Business that are involved in agricultural research in Sri Lanka. Of these, the government institutes are the major stakeholders of agricultural research that is conducted in Sri Lanka. Hence, it should be made a national priority to improve the research facilities at these institutes and make them centers of excellence with state-of-the-art facilities suited to conduct research that matters. Science is a field that changes constantly. To keep up investment in the latest technologies is a must. However, investments do not come cheap; therefore, there should be national policies set up to boost the funding for scientific research. At the grassroots level, it is the farmer that finally keeps the food chain intact. The government needs to make investments on research and development to ensure that the famers get the best technology out there to face the challenges they encounter.

Almost four decades of research has proved how safely and effectively these modern technologies can be used in crop improvement while minimizing the environmental impact of agriculture on our ecosystem. In the end, every stakeholder from researchers, farmer, merchants, regulators, policymakers, nongovernmental organizations, and journalists to consumer has a role to play to come up with safe and sustainable food production systems in Sri Lanka. We are only able to pave the way for future cereals if our approaches are not blindfolded by conventional thinking and our actions no less than done with the noble aim of bringing out the best in science to feed a nation. For that we need to break the mold and pave the way for future cereal crops through an integrated platform combining the best of both worlds, conventional and modern.

References

Abeysekera SW, Jayawardena SN, Kiriwaththuduwage KDS, Abeysiriwardena DS d Z (2003) Hybrid rice research and development in Sri Lanka. In: Virmani SS, Chang XM, Hardy B (eds) Hybrid rice for food security, poverty alleviation, and environmental protection. International Rice Research Institute, Manila, pp 341–355

CBSL (2017) Annual report 2016. Central Bank of Sri Lanka, Colombo. https://www.cbsl.gov.lk/en/publications/economic-and-financial-reports/annual-reports/annual-report-2016

CBSL (2018) Annual report 2017. Central Bank of Sri Lanka, Colombo. https://www.cbsl.gov.lk/en/publications/economic-and-financial-reports/annual-reports/annual-report-2017

Beier S, Himmelbach A, Colmsee C, Zhang X-Q, Barrero RA, Zhang Q et al (2017) Construction of a map-based reference genome sequence for barley, *Hordeum vulgare* L. Sci Data 4:170044. https://doi.org/10.1038/sdata.2017.44

Bentota AP (ed) (2013) Released and recommended new crop varieties by the varietal release committee of the Department of Agriculture Sri Lanka. Department of Agriculture, Peradeniya

Birla DS, Malik K, Sainger M, Chaudhary D, Jaiwal R, Jaiwal PK (2017) Progress and challenges in improving the nutritional quality of rice (*Oryza sativa* L.). Crit Rev Food Sci Nutr 57(11):2455–2481. https://doi.org/10.1080/10408398.2015.1084992

Borlaug NE (2002) The green revolution revisited and the road ahead. Nobelprize.org, Stockholm

Brookes G, Barfoot P (2016) Environmental impacts of genetically modified (GM) crop use 1996–2014: impacts on pesticide use and carbon emissions. GM Crops Food 7(2):84–116. https://doi.org/10.1080/21645698.2016.1192754

Rice Project (2018). https://c4rice.com/. Accessed 6 Mar 2018

Chakraborty S, Newton AC (2011) Climate change, plant diseases and food security: an overview. Plant Pathol 60(1):2–14. https://doi.org/10.1111/j.1365-3059.2010.02411.x

Chandrawathi LM, Muthunayaka MMP, Herath RM (2013) Adoption of new rice varieties in Sri Lanka. Sri Lanka J Agrar Stud 16(1 & 2):27–40

Chen H, Xie W, He H, Yu H, Chen W, Li J et al (2014) A high-density SNP genotyping array for rice biology and molecular breeding. Mol Plant 7(3):541–553. https://doi.org/10.1093/mp/sst135

Chin JH, Gamuyao RL, Dalid C, Bustamam M, Prasetiyono J, Moeljopawiro S et al (2011) Developing rice with high yield under phosphorus deficiency: *Pup1* sequence to application. Plant Physiol 156:1202–1216

Collard BCY, Jahufer MZZ, Brouwer JB, Pang ECK (2005) An introduction to markers, quantitative trait loci (QTL) mapping and marker-assisted selection for crop improvement: the basic concepts. Euphytica 142(1–2):169–196

Crafts-Brandner SJ (2002) Sensitivity of photosynthesis in a C4 plant, maize, to heat stress. Plant Physiol 129(4):1773–1780. https://doi.org/10.1104/pp.002170

Dively GP, Venugopal PD, Bean D, Whalen J, Holmstrom K, Kuhar TP et al (2018) Regional pest suppression associated with widespread Bt maize adoption benefits vegetable growers. Proc Natl Acad Sci U S A 115:201720692. https://doi.org/10.1073/pnas.1720692115

Edzesi WM, Dang X, Liang L, Liu E, Zaid IU, Hong D (2016) Genetic diversity and elite allele mining for grain traits in rice (*Oryza sativa* L.) by association mapping. Front Plant Sci 7:787

Endo M, Osakabe K, Ono K, Handa H, Shimizu T, Toki S (2007) Molecular breeding of a novel herbicide-tolerant rice by gene targeting. Plant J 52(1):157–166

FAO (2012) World agriculture towards 2030/2050: the 2012 revision. FAO, Rome. http://www.fao.org/economic/esa/esag/en/

FAO (2017) Rice market monitor. FAO, Rome. http://www.fao.org/economic/est/publications/rice-publications/rice-market-monitor-rmm/en/

FAOSTAT (2018). http://www.fao.org/faostat/en/#data. Accessed 29 Jul 2018

Furuta T, Ashikari M, Jena KK, Doi K, Reuscher S (2017) Adapting genotyping-by-sequencing for rice F_2 populations. G3 (Bethesda) 7(3):881–893. https://doi.org/10.1534/g3.116.038190

Galetti M, Costa-Pereira R (2017) Scientists need social media influencers. Science 357(6354):880–881. https://doi.org/10.1126/science.aao1990

Govindaraj M, Vetriventhan M, Srinivasan M (2015) Importance of genetic diversity assessment in crop plants and its recent advances: an overview of its analytical perspectives. Genet Res Int 431487:14

Herath HG (1981) The Green Revolution in rice: the role of the risk factor with special reference to Sri Lanka. Asian Surv 21(6):664–675

Hu C-H (1973) Evaluation of breeding semidwarf rice by induced mutation and hybridization. Euphytica 22(3):562–574. https://doi.org/10.1007/BF00036657

Huyen LTN, Cuc LM, Ham LH, Khanh TD (2013) Introgression the SALTOL QTL into Q5DB, the elite variety of Vietnam using marker assisted selection (MAS). Am J Biol Sci 1(4):80À84

Iftekharuddaula KM, Ghosal S, Gonzaga ZJ, Amin A, Barman HN, Yasmeen R et al (2016) Allelic diversity of newly characterized submergence-tolerant rice (*Oryza sativa* L.) germplasm from Bangladesh. Genet Resour Crop Evol 63(5):859–867

International Rice Genome Sequencing Project (2005) The map-based sequence of the rice genome. Nature 436:793. https://doi.org/10.1038/nature03895

ISAAA (2016) Global status of commercialized biotech/GM crops: 2016. ISAAA Brief No. (vol. 52). ISAAA, Ithaca, NY. http://www.journals.cambridge.org/abstract_S0014479706343797

Jiao Y, Peluso P, Shi J, Liang T, Stitzer MC, Wang B et al (2017) Improved maize reference genome with single-molecule technologies. Nature 546:524. https://doi.org/10.1038/nature22971

Kamthan A, Chaudhuri A, Kamthan M, Datta A (2016) Genetically modified (GM) crops: milestones and new advances in crop improvement. Theor Appl Genet 129(9):1639–1655. https://doi.org/10.1007/s00122-016-2747-6

Klümper W, Qaim M (2014) A meta-analysis of the impacts of genetically modified crops. PLoS One 9(11):e111629. https://doi.org/10.1371/journal.pone.0111629

Kumar S, Kulkarni KS, Thirunavukkarasu N, Singh G, Sharma R (2015) Biofortification for selecting and developing crop cultivars denser in iron and zinc. In: Rakshit A, Singh HB, Sen A (eds) Nutrient use efficiency: from basics to advance. Springer, New Delhi, pp 237–253. https://doi.org/10.1007/978-81-322-2169-2_17

Li J-Y, Wang J, Zeigler RS (2014) The 3,000 rice genomes project: new opportunities and challenges for future rice research. Gigascience 3(1):8. https://doi.org/10.1186/2047-217X-3-8

Li X, Xie Y, Zhu Q, Liu Y-G (2017) Targeted genome editing in genes and cis-regulatory regions improves qualitative and quantitative traits in crops. Mol Plant 10(11):1368–1370. https://doi.org/10.1016/j.molp.2017.10.009

Mansueto L, Fuentes RR, Chebotarov D, Borja FN, Detras J, Abriol-Santos JM et al (2016) SNP-Seek II: a resource for allele mining and analysis of big genomic data in *Oryza sativa*. Curr Plant Biol 7–8:16. https://doi.org/10.1016/j.cpb.2016.12.003

Mojarad S (2017) Social media: more scientists needed. Science 357(6358):1362–1363. https://doi.org/10.1126/science.aap9655

Naranjo L, Talón M, Domingo C (2014) Diversity of floral regulatory genes of *japonica* rice cultivated at northern latitudes. BMC Genomics 15:101. https://doi.org/10.1186/1471-2164-15-101

Navarro JAR, Willcox M, Burgueño J, Romay C, Swarts K, Trachsel S et al (2017) A study of allelic diversity underlying flowering-time adaptation in maize landraces. Nat Genet 49(3):476

Neeraja CN, Maghirang-Rodriguez R, Pamplona A, Heuer S, Collard BCY, Septiningsih EM et al (2007) A marker-assisted backcross approach for developing submergence-tolerant rice cultivars. Theor Appl Genet 115(6):767–776

Neue HU, Quijano C, Senadhira D, Setter T (1998) Strategies for dealing with micronutrient disorders and salinity in lowland rice systems. Field Crop Res 56(1–2):139–155. https://doi.org/10.1016/S0378-4290(97)00125-1

Nicolia A, Manzo A, Veronesi F, Rosellini D (2014) An overview of the last 10 years of genetically engineered crop safety research. Crit Rev Biotechnol 34(1):77–88. https://doi.org/10.3109/07388551.2013.823595

Owens B (2018) Golden Rice is safe to eat, says FDA. Nat Biotechnol 36:559. https://doi.org/10.1038/nbt0718-559a

Parisi C, Tillie P, Rodríguez-Cerezo E (2016) The global pipeline of GM crops out to 2020. Nat Biotechnol 34(1):31–36. https://doi.org/10.1038/nbt.3449

Pellegrino E, Bedini S, Nuti M, Ercoli L (2018) Impact of genetically engineered maize on agronomic, environmental and toxicological traits: a meta-analysis of 21 years of field data. Sci Rep 8(1):1–12. https://doi.org/10.1038/s41598-018-21284-2

Perera SACN, Samarasinghe WLG, Girihagama PC (2017) National research priorities in agricultural biotechnology 2017–2021. Sri Lanka Council for Agricultural Research Policy, Colombo, Sri Lanka

Potrykus I (2017) The GMO-crop potential for more, and more nutritious food is blocked by unjustified regulation. J Innov Knowl 2(2):90–96. https://doi.org/10.1016/j.jik.2017.03.003

Prado JR, Segers G, Voelker T, Carson D, Dobert R, Phillips J et al (2014) Genetically engineered crops: from idea to product. Annu Rev Plant Biol 65(1):769–790. https://doi.org/10.1146/annurev-arplant-050213-040039

Ranaweera, N. F. C., de Silva, G. A. C. de, Fernando, M. H. J. P., & Hindagala, H. B. (1988). Maize production in Sri Lanka. Regional Co-Ordination Centre for Research and Development of Coarse Grains, Pulses, Roots and Tuber Crops in the Humid Tropics of Asia and the Pacific Bogor

Ray DK, Mueller ND, West PC, Foley JA (2013) Yield trends are insufficient to double global crop production by 2050. PLoS One 8(6):e66428. https://doi.org/10.1371/journal.pone.0066428

Rodger-Withers SM (2017) Social media: actions speak louder than likes. Science 357(6358):1363–1363. https://doi.org/10.1126/science.aap8595

Rodríguez-Leal D, Lemmon ZH, Man J, Bartlett ME, Lippman ZB (2017) Engineering quantitative trait variation for crop improvement by genome editing. Cell 171(2):470–480.e8. https://doi.org/10.1016/j.cell.2017.08.030

Saika H, Oikawa A, Matsuda F, Onodera H, Saito K, Toki S (2011) Application of gene targeting to designed mutation breeding of high-tryptophan rice. Plant Physiol 156(3):1269–1277. https://doi.org/10.1104/pp.111.175778

Salunkhe DK, Kadam SS, Chavan JK (1977) Nutritional quality of proteins in grain sorghum. Qual Plant 27(2):187–205. https://doi.org/10.1007/BF01092359

Septiningsih EM, Pamplona AM, Sanchez DL, Neeraja CN, Vergara GV, Heuer S et al (2009) Development of submergence-tolerant rice cultivars: the *Sub1* locus and beyond. Ann Bot 103(2):151–160. https://doi.org/10.1093/aob/mcn206

Shahbaz M, Ashraf M (2013) Improving salinity tolerance in cereals. Crit Rev Plant Sci 32:237–249. https://doi.org/10.1080/07352689.2013.758544

Shearer E, Gottfried BYJ (2017) News use across social media platforms 2017, vol 17. Pew Research Center, Washington, DC. http://www.journalism.org/2017/09/07/news-use-across-social-media-platforms-2017/

Singh RJ, Ikehashi H (1981) Monogenic male-sterility in rice: induction, identification and inheritance 1. Crop Sci 21(2):286–289

Singh S, Sidhu JS, Huang N, Vikal Y, Li Z, Brar DS et al (2001) Pyramiding three bacterial blight resistance genes (*xa5*, *xa13* and *Xa21*) using marker-assisted selection into indica rice cultivar PR106. Theor Appl Genet 102:1011–1015. https://doi.org/10.1007/s001220000495

Smith ML (2006) The archaeology of food preference. Am Anthropol 108(3):480–493. https://doi.org/10.1525/aa.2006.108.3.480

Song W-Y, Wang G-L, Chen L-L, Kim H-S, Pi L-Y, Holsten T et al (1995) A receptor kinase-like protein encoded by the rice disease resistance gene, *Xa21*. Science 270(5243):1804–1806

Stone GD, Glover D (2017) Disembedding grain: golden rice, the green revolution, and heirloom seeds in the Philippines. Agric Hum Values 34(1):87–102. https://doi.org/10.1007/s10460-016-9696-1

Strobbe S, Van Der Straeten D (2018) Toward eradication of B-vitamin deficiencies: considerations for crop biofortification. Front Plant Sci 9:443. https://doi.org/10.3389/fpls.2018.00443

Studer AJ, Wang H, Doebley JF (2017) Selection during maize domestication targeted a gene network controlling plant and inflorescence architecture. Genetics 207:755–765. https://doi.org/10.1534/genetics.117.300071

Tai TH (2007) Induced mutations in rice (*Oryza sativa* L.). Israel J Plant Sci 55(2):137–145

Tatarinova T, Chekalin E, Nikolsky Y, Chebotarov D, McNally KL, Alexandrov N (2016) Nucleotide diversity analysis highlights functionally important genomic regions. Sci Rep 6, 35730. https://doi.org/10.1038/srep35730

Tello-Ruiz MK, Naithani S, Stein JC, Gupta P, Campbell M, Olson A et al (2018) Gramene 2018: unifying comparative genomics and pathway resources for plant research. Nucleic Acids Res 46(D1):D1181–D1189. https://doi.org/10.1093/nar/gkx1111

Tian Z, Qian Q, Liu Q, Yan M, Liu X, Yan C et al (2009) Allelic diversities in rice starch biosynthesis lead to a diverse array of rice eating and cooking qualities. Proc Natl Acad Sci U S A 106(51):21760–21765. http://www.pnas.org/content/106/51/21760.abstract

Twyford AD, Ennos RA (2012) Next-generation hybridization and introgression. Heredity 108(3):179–189. https://doi.org/10.1038/hdy.2011.68

Von Caemmerer S, Quick WP, Furbank RT (2012) The development of C4 rice: current progress and future challenges. Science. https://doi.org/10.1126/science.1220177

Walisinghe BR, Gunaratne LHP (2012) Consumer preferences for quality attributes of rice: a conjoint analysis. Sri Lankan J Agric Econom 10:1–13. https://doi.org/10.4038/sjae.v10i0.4589

Wang RL, Stec A, Hey J, Lukens L, Doebley J (1999) The limits of selection during maize domestication. Nature 398(6724):236–239. https://doi.org/10.1038/18435

Weerasinghe P, Hettiarachchi K, Keerthisena RSK, Munasinghe MAK, Peiris PC, Ranatunga KNS (2016) Performance report 2016. Department of Agriculture, Peradeniya

Wijesinghe C (2017) Whole genome sequencing and analysis of *Godawee*, a salt tolerant *indica* rice variety. J Rice Res 5:177. https://doi.org/10.4172/2375-4338.1000177

Yong X, Kenneth M, Cheng Yun L, Hei L, You Yong Z (2006) A high-throughput genomic tool: diversity array technology complementary for rice genotyping. J Integr Plant Biol 48(9):1069–1076. https://doi.org/10.1111/j.1744-7909.2006.00298.x

Zhao Y, Huang J, Wang Z, Jing S, Wang Y, Ouyang Y et al (2016) Allelic diversity in an NLR gene *BPH9* enables rice to combat planthopper variation. Proc Natl Acad Sci U S A 113(45):12850–12855. https://doi.org/10.1073/pnas.1614862113

Zimin AV, Puiu D, Hall R, Kingan S, Clavijo BJ, Salzberg SL (2017) The first near-complete assembly of the hexaploid bread wheat genome, *Triticum aestivum*. Gigascience 6(11):1–7. https://doi.org/10.1093/gigascience/gix097

Recent Developments in Vegetable Production Technologies in Sri Lanka

9

W. A. P. Weerakkody and S. M. M. R. Mawalagedera

Abstract

The yields of major vegetables produced in Sri Lanka as at present are 10–70% below the potential yields in countries like Japan, USA, and India. Further, the per capita vegetable consumption in Sri Lanka is 50% lower than the levels recommended by the World Health Organization (WHO). The vegetable production in Sri Lanka is burdened by decreasing arable lands due to rapid urbanization and ever-increasing demand for food by an exponentially growing population. All the major commercial vegetable-based cropping systems in Sri Lanka predominantly follow the conventional production technologies using agrochemicals. However, there is a growing trend of sustainable vegetable production in organic farms, homegardens, and peri-urban systems. Thus, it is clear that technological innovations are vital in local vegetable cropping to increase yield productivity, production efficiency, food quality, and food safety. Technological trends that would benefit the modern vegetable production in Sri Lanka include rapid multiplication and production of propagules, development of modern nursery techniques, micro irrigation along with fertigation and greenhouse crop production, proper pruning, training and pollination strategies, Good Agricultural Practices (GAP) and Integrated Crop Protection Technologies (ICPT), traceability initiatives, and methods to ensure quality and safety of food. Meanwhile, a comparative assessment of the national vegetable subsector with the countries in the region would elucidate the need of immediate attention for further improvements. This emphasizes the need of capitalizing on intensification of cropping systems, viable seed production and quality assurance,

W. A. P. Weerakkody (✉)
Department of Crop Science, Faculty of Agriculture, University of Peradeniya, Peradeniya, Sri Lanka
e-mail: palithaw@pdn.ac.lk

S. M. M. R. Mawalagedera
School of Education, Faculty of Arts, Deakin University, Geelong, VIC, Australia

B. Marambe et al. (eds.), *Agricultural Research for Sustainable Food Systems in Sri Lanka*, https://doi.org/10.1007/978-981-15-2152-2_9

restructuring and revitalizing of the research and extension system, etc., with a view to ensure the sustainable production of high-quality vegetables.

Keywords
Vegetables · Present status · Conventional · Sustainable · Cropping systems

9.1 Introduction to Vegetable Crops and Cropping Systems

Vegetable crops provide many vital food constituents and meet the essential nutritional requirements of the global community. Vegetables are available in fresh as well as value-added forms and consumed in raw, cooked, or partially processed forms. Together with fruits, vegetables can be considered as the main source of vitamins A and C, potassium, folic acid, and dietary fiber. Hence, they can be considered as a cheap and readily available healthy food for the much health-conscious consumer markets as well as communities that are vulnerable to malnourishment in the world today. This situation is much prominent in Asia as vegetables become a vital side dish in their predominantly rice-based staple diets.

Steady supply of vegetables through continuous production to meet the ever-changing market demand is a huge task for the responsible regulatory bodies and the market-oriented production sectors. The production sector and value-chain management need continuous technological improvements that are sensitive to volatile market forces (Johnson 2008). Hence, bridging the gap between the demand and supply must involve novel technologies that would resolve the current constraints in production, postharvest, and marketing of vegetables.

Vegetable subsector plays an important role in the Sri Lankan economy. It provides livelihood to a large proportion of poor people, contributes to increase the level of national income and export revenue, generates new employment opportunities, increase farm income, and indirectly enhances the nutrition and health of the nation. There is a need to increase the vegetable production by several folds with the gradual increase in population, increase in per capita income, and expansion of exports and tourism industry in the country. Consumers tend to have health-safe and quality foods with the increase in income, thus enhancing the consumption levels of vegetables. This increasing demand should be met with the increased supply of quality vegetables. Even though the prevailing diverse agroclimatic conditions in Sri Lanka make it possible to grow a wide variety of vegetable crops (about 40 species) throughout the year in different parts of the country, the per capita consumption of vegetables is only about 112 g/day (SAARC Report 2017), which is far below the recommended dietary requirement (200 g/day). Thus, it is essential to increase the vegetable production at least by twofold to meet the growing demand and attain the recommended per capita vegetable intake.

The position of Sri Lanka with regard to area of cultivation, production, and productivity of vegetables in comparison with its neighbors and rest of the world is shown in Table 9.1. The average productivity of main vegetable crops in comparison

Table 9.1 The mean extent of vegetable cultivation, production, and productivity in Sri Lanka and selected countries in comparison to global average (based on statistics from 2012 to 2016) (FAOSTAT 2017)

Country	Area (ha × 10^6)	% Area (share of global)	Production (million t)	% Production (share of global)	Productivity (t/ha)
Sri Lanka	**0.081**	**0.15**	**0.92**	**0.1**	**11.4**
India	8.26	14.9	119.0	11.6	14.4
Bangladesh	5.33	0.96	4.5	0.4	8.4
Japan	0.38	0.68	9.9	0.96	26.4
USA	10.2	1.8	34.4	3.35	33.6
China	22.7	41	513.7	0.1	23.0
World	55.4	100	1026.7	100	18.6

Table 9.2 Comparison of average yield of selected vegetable crops grown in Sri Lanka (Source: IASRI 2018)

Crops	Average yield (t/ha)			Potential productivity (t/ha)	Max. productivity reported (t/ha)
	Sri Lanka	India	World		
Tomato	**18.9**	20.7	33.8	60–80	70.5 (USA)
Eggplant	**12.8**	18.6	26.5	40–50	34.7 (Japan)
Okra	**11.1**	12.0	7.8	15–20	17.8 (Jordan)
Cabbage	**25.0**	**22.9**	29.4	30–40	42.6 (Japan)

with India and the leading producers of the world is given in Table 9.2. These statistics clearly indicate the need for further improvements in vegetable production to win the market competition at least within the south-Asian neighborhood.

There are many reasons for the average yield differences among countries. Some of these reasons are agronomic, while others are socioeconomic. Irrigation is important in the achievement of high yields in several countries. Further, agroecological and demand factors influence the mix of varieties of the same crop grown in each country, i.e., traditional vegetables and commercial vegetable varieties. Therefore, in the final analysis of intercountry yield gaps, it is necessary to separate out the part that is caused by agroecological diversity from that caused by other factors.

In the last several decades, the carrying capacity of our land resources has surpassed by the constant decline in cultivable land due to other development requirements coupled with a booming population with ever-increasing demand for food. In Sri Lanka, a large variety of vegetable crops are grown. To feed the increasing population, it is necessary to utilize agricultural lands to the best of its capacities. Cropping systems, a blend of crops with a package of crop management within a set of environmental boundaries, is the basis of intensification of crop production. This approach improves the effective use of inputs such as sunlight, soil, water, labor, and other external inputs. The environmental requirements of different vegetable crops and crop groups make them best fit into each of the cropping system. Hence, either output per unit area or the quality of harvest increases with manifold returns to the growers.

Meanwhile, the Research Division of the Department of Agriculture (DOA) is the main institution responsible for vegetable crops research in Sri Lanka. Universities, especially Faculties of Agriculture, private sector organizations, etc., also contribute to the vegetable crops developments in Sri Lanka by means of research and extension. Prior to 1994, vegetable research in DOA was carried out at almost all the research stations located in different agroclimatic zones (Stockdale 1921; Senanayake 2000). With the introduction of commodity-based research institutes in 1994, vegetable research in DOA was restricted to the stations coming under the purview of the Horticultural Crop Research and Development Institute (HORDI) in Gannoruwa, while rest of the stations were given the research mandate for other commodity crops such as field crops and fruit crops. This impeded the testing of technologies developed in vegetable research under various agroecological regions. In addition, allocation of village-level agricultural extension officers to other institutions, severing of linkages of the researchers' with the other stakeholders, absence of a streamlined mechanism to obtain stakeholder feedback on the production and marketing issues in agriculture, lack of well-planned long-, medium-, and short-term research and development programs, frequent changes in the research thrust, heavy workloads on the research staff due to the involvements in non-research activities (i.e., exhibitions, and production programs), disorganized human resource development process, and lack of qualified human resources are some of the other constrains and issues in mainstreaming vegetable research and developments in the government set up in Sri Lanka which are impediments for reaching the national goals in vegetable crop production.

In this backdrop, analyzing the progress and assessing the quality and impact of recent research and developmental programs within the vegetable subsector would be helpful to identify the technological gaps that need to be bridged through future improvement programs. Thus, it will pave the way to an array of future technologies for upgrading the market supply of vegetable-based foods and open up new avenues for all stakeholder groups within the vegetable subsector.

9.2 Vegetable-Based Cropping Systems

In ancient Sri Lanka, vegetables were produced in mixed cropping systems, predominantly in homegardens, "*Koratuwa*" gardens[1] and under *Chena* (shifting) cultivation (traditional systems) for direct use or for exchange purposes. With the diversification of livelihoods during the colonial era, these systems have gradually invaded by market-oriented production systems, which were fundamentally high input-based monocropping systems. These commercial vegetable cropping systems have been expanding during the past 100 years, leaving the traditional systems to be shrunk over the time (Ranaweera and de Silva 1994). A substantial account of the technological background of these systems is discussed below.

[1] *Koratuwa is a small enclosure maintained by the farmer for home vegetable needs at the boundary of the upland fields.*

The fairly large-scale commercial cropping systems, (i) continuous commercial cultivation in the up-country and (ii) rice-based seasonal commercial cultivations in the low country, introduced nearly a century ago, are exhausted at present due to the overuse of natural resources and high external input use. Although the traditional cropping systems were sustainable, the commercial or "conventional" systems are comparatively unsustainable, despite their yield and quality advantages (Weerakkody et al. 2000; Suriyagoda et al. 2012).

As a solution, organic farming and/or ecological farming have gained momentum. However, their popularity and rate of adoption have been limited and slow, even decades after the introduction. Meanwhile more intensive (high input use) forms of commercial-scale vegetable-based cropping systems in rain shelters and film-plastic greenhouses are emerging due to upmarket value of their productions (Kumara et al. 2015).

9.2.1 Intensive Commercial Gardening in the Open Field

The vegetable production system in Sri Lanka has two main divisions, i.e., (i) continuous vegetable cultivation in the sloppy lands in high elevations (Nuwara Eliya and Badulla districts) and (ii) rice-based (seasonal) vegetable cultivation in the mid and low elevations (Perera 1990). These two systems are basically found in the Central Province (CP) and Uva Province (UP) of Sri Lanka, and represents 45% of the total vegetable production area of the country. The largest extents of up-country vegetables are cultivated in the Nuwara Eliya, Kandy, and Matale districts in the Central Province.

In the continuous vegetable cultivation system in the up-country, mostly cabbage (*Brassica oleracea*), carrot (*Daucus carota*), leeks (*Allium ampeloprasum*), beet (*Beta vulgaris*), radish (*Raphanus raphanistrum* subsp. *sativus*), knoll khol (*Brassica oleracea var. gongylodes*), lettuce (*Lactuca sativa*), etc., are cultivated. These crops have been introduced to Sri Lanka from the temperate regions in the world. This system uses imported hybrid seeds and synthetic chemical fertilizers and pesticides. As a result of high environmental risk on production, farmers apply excessive dosages of agrochemicals to protect the crop and increase the marketable yields. Farming is practiced in small land holdings (0.2–0.4 ha) in hilly land terrains under continuous cultivation, thus exposing soil to rain and wind erosion, further deteriorating the natural fertility of the farm lands and causing threats to water resources downstream. However, the economic sustainability of this system is the highest concern among the vegetable-based cropping systems in Sri Lanka (Suriyagoda et al. 2012).

In the rice-based vegetable cropping system, relatively large extents (0.2–0.4 ha) of lowland paddy (rice) (*Oriza sativa*) fields in the dry and intermediate zones are cultivated with vegetables during the dry season (*Yala*—March to September), provided supplementary irrigation is available. The popular crops in this system are capsicum (*Capsicum annuum*), eggplant (*Solanum melongena*), chili (*C. annuum*), onion (*Allium cepa*), and some trellising-type cucurbits. Meanwhile, in the up- and mid-country regions, located in elevations 900 mamsl, terraced paddy fields are used to cultivate potato (*Solanum tuberosum*), beans (*Phaseolus vulgaris*), tomato (*Solanum lycopersicum*), and cabbage during the *Yala* season (Ranaweera and de Silva 1994; Weerakkody et al. 2000). The system falls second to the continuous vegetable cultivation in the up-country, in terms of input use as well as the income generated. Meanwhile the characteristic vegetable cultivation system at Kalpitiya area in the North Western Province resembles the intensive commercial gardening found in the up country due to its overdependence on external inputs. In this system, year-round production is practiced owing to assured irrigation from shallow wells with the use of sprinklers. Fairly large numbers of shallow wells have been established in the farming area to support this. However, precautions should be taken to avoid any environmental problems that may arise due to overexploitation of ground water (Aheeyar et al. 2016).

9.2.2 Homegardening

There are about 1.42 million homegardens in Sri Lanka, accounting for about 76,483 ha. These make a substantial contribution to agricultural production in the country, and play a perceptible role in maintaining high species and genetic diversity of fruits, vegetables, and spices, canopy cover in the island ameliorating the microclimate (Kottawa-arachchi and Wijeratne 2017). This is the best system to meet the fresh vegetable requirement at the household level. The vegetables produced under homegardens are more or less eco-friendly with respect to input use and thus safe and healthy.

These systems also have permanent vegetable trees, such as jackfruit (*Artocarpus heterophyllus*) and breadfruit (*Artocarpus altilis*) in the Wet Zone (WZ) and Intermediate Zone (IZ) and drumstick (*Moringa olifera*) in the Dry Zone (DZ). Ash plantain (*Musa* sp.) and *kathuru-murunga (Sesbania glandiflora*) are also common in the homegardens of wet and intermediate zones. Many green leafy vegetables, having high fiber, vitamins, and mineral contents, namely *Gotukola* (*Centella asiatica*), *Mukunuwenna* (*Alternanthera sessilis*), *Amaranth* (*Amaranthus* spp.), *Kankung (Ipomoea aquatica)*, and Ceylon Spinach (*Basella alba syn B. cordifolia*), are easily grown in homegardens in any part of the island (Ranaweera and de Silva 1994).

Traditional vegetables seem to be the major contributor in the location-specific food production systems, i.e., homegardening systems, as they are well-adapted to adverse environmental conditions over the years, and generally resistant to pests and diseases. Even though the vegetable crops fit well into homegarden systems, particularly under less intensive management conditions, they can easily be made available year-around assuring less seasonality in production. However, promotion of the traditional vegetables in homegardens is limited. Precise statistics on extents and production of vegetables in homegardens is not available. However, homegardens alone play a significant role as a contributing factor not only to national production but also to daily consumption of vegetables at household level in most parts of the country.

9.2.3 Chena Cultivation

Historical shifting cultivation (*Chena* cultivation) is still practiced in the low country DZ, especially in the North Central and Eastern Provinces of Sri Lanka. Many warm-season vegetables, tomato, eggplant, cucumber (*Cucumis sativus*), pumpkins (*Cucurbita maxima*), luffa (*Luffa acutangula*), bitter gourd (*Momordica charantia*), capsicum, and okra (*Abelmoschus esculentus*), together with some cereals, pulse crops, and some fruit trees are cultivated in *Chena* farms in a mixed manner. The bulk of shifting cultivation is practiced during rainy season (*Maha*; from October to February) with little or no supplementary irrigation (Weerakkody 2004). Even though the original form of *Chena* utilizes the inherent fertility and the slash and burn type protection for growing crops, the present-day *Chena* faming systems show a blend of modern agrotechnology such as use of improved varieties and agrochemicals to a certain extent. The population pressure and scarcity of natural vegetation have forced the *Chena* farmers to return to the same farm land, leaving little or no room for fallowing or returning the soil and environment to normal. This could be one of the reasons for applying agrochemicals by modern-day *Chena* farmers.

9.2.4 Peri-urban System

Urban agriculture refers to small areas (e.g., vacant plots, gardens, balconies, and containers) within the city for growing crops and raising some livestock. In urban areas vegetable cultivation is practiced by poor and landless city dwellers. Intensive horticulture can be practiced on small plots, making efficient use of limited resources (water and land), whereas peri-urban refers to farm units closer to the town, which operates as semi or fully commercial farms to grow vegetables and other horticultural crops, raise chickens, and other livestock. Intensive cultivations of leafy vegetables (*keerakotu*) are concentrated near Colombo and suburbs as peri-urban production systems to cater to the needs of the urban and semi-urban populations. Leafy vegetables are particularly perishable, and postharvest losses can be reduced significantly when production is located closer to consumer markets (Ranaweera and de Silva 1994).

9.2.5 Organic and Eco-Friendly Systems

The driving force behind the fast-growing demand for organic food is the consumer preference for chemical residue-free foodstuff. According to latest reports, the number of registered organic farms in Sri Lanka is 3301 by 2006. This covers 15,215 ha, comprising nearly 4.6% of Asian organic farm lands and 0.05% of the global organic farming. And this is just a 0.65% of the total agricultural lands in the country (Hsieh 2005; Yussefi 2010; Willer and Lernoud 2016). Together with tea, fruits, spices, and few other edible crop categories, vegetables is one of the main commodities under organic agriculture in Sri Lanka. Vegetable crops produced in the developed countries are closely monitored for harmful chemicals. However, in Sri Lanka, heavy doses of insecticides and fungicides are being applied during cultivation making the crop output unsafe to consume (Suriyagoda et al. 2012; Kananke et al. 2014). According to a case study conducted in North Central part of Sri Lanka, farmers apply 155–437 L ha^{-1} of diluted agrochemicals for plant protection in commercial vegetable cultivations. Assuming the standard rate of dilution, this application dosage is three to four times higher than the recommendation (Wijeratne and Weerakkody 2017). Therefore, many people prefer vegetables that have been grown using "low external input" method that falls within organic or ecological farming systems, which do not use chemical pesticides or inorganic fertilizers, and is environmentally friendly, culturally sensitive, socially just and economically viable efficient management system that maintains sustainability.

The organic movement in Sri Lanka started in the 1980s through local nongovernmental organizations (NGOs) with the Philippine Organic Agriculture Movement (Fonseka et al. 2013). An interested group comprising of local NGO representatives, planters, scientists, and environmental officers had drafted a Memorandum of Association to create a movement named Lanka Organic Agriculture Movement (LOAM). During the last two decades, enthusiastic individuals have developed organic farming units, especially in the WZ of Sri Lanka based on acquired knowledge on the benefits of organic farming, owing to increasing demand for export of organically grown products. These units are reported to have a productivity somewhat less than that of the conventional agricultural systems. Vegetables, fruits, and spices grown using these systems bring premium prices, thereby enhancing the economic viability of these production units. The primary use of organic farming methods is seen in homegardening and some smallholder farming operations. However, with sufficient emphasis on research at the initial stages, followed by a well-developed extension program, government awareness can be directed toward organic agriculture (Sangakkara and Katupitiya 2000). Meanwhile research and developments in organic vegetable cultivation have been focused on the use of green manure (i.e., *Gliricidia cipium*), composts, bio-char and bio-fertilizers, and many other organic inputs during the last two decades (Egodawatta et al. 2012; Herath et al. 2015; Rajapakse et al. 2016).

9.2.6 Protected Culture/Hydroponics Systems

Even though protected culture and hydroponics (soil-less culture) has nearly a 100-year long history in the world, it was introduced to vegetable subsector of Sri Lankan agriculture only in the late 1990s. However, its rate of expansion was limited to nearly the first 10 years of its introduction. Most vegetable-growing greenhouses in Sri Lanka (62%) are single span, 100–200 m^2 in size, soft plastic/insect-proof net covered and naturally ventilated. Protected culture in Sri Lanka rarely employs mechanization and automation. Basic greenhouse technologies such as drip-fertigated hydroponics, hybrid planting materials, soilless media, complete fertilizers, intensive canopy management are practiced by all growers with varying intensities. The yield and external quality of protected culture are greater than those of all the other vegetable cropping systems, enabling the produce to capture premium prices in high-price local markets (Weerakkody 2004) despite that Sri Lankan yield levels are at the lower end of the yield range of the leading global producers (150–450 t/ha) such as Mexico, USA, and Canada (Bernal et al. 2010). Protected culture is mainly practiced in Kandy, Matale, NuwaraEliya, and Badulla districts that fall within mid and high elevations in the country. Among the popular crops, bell peppers (*Capsicum annuum*) dominate by comprising 40% of the total extent, followed by salad/green cucumber (*Cucumis sativus*; 19%) and tomato (13%). Meanwhile, lettuce is largely cultivated in circulation-type hydroponics without solid media. There are a number of advantages associated with this method; however, it is an expensive means of producing vegetables. Therefore, hydroponics is used only for crops that can generate a higher income (Kumara et al. 2015).

As discussed above, all cropping systems suffer from specific difficulties and constraints, and few alternative cropping systems are appeared to be expanding due to being advantageous in some aspects over the conventional systems. For example, peri-urban vegetable cultivation and shifting cultivation is gradually wiping out due to scarcity of lands under mounting population pressure. Homegardening has been constrained due to limitations in space and time under urbanization and busy day-schedules of family members in semi-urban areas. Conventional open-field systems are suffering from high cost of inputs, low productivity, labor shortage, and threats from animal pests as a result of prolonged dependency on imported technical inputs and relative growth of other sectors of the economy. In contrast, protected culture and ecological gardening are gathering momentum as a result of growing demand in the high-value markets and the consumer awareness on food safety. However, lack of periodical assessments of the status of these cropping systems has hindered the possible problem identifications and subsequent interventions to resolve them. Assessments done on traditional cropping systems in other South Asian countries (Poudel et al. 1998; Khai et al. 2007) provide necessary inspiration for developing appropriate research strategies for Sri Lankan vegetable-based cropping systems for directing them toward productive and sustainable goals. Therefore, proper analysis of the limiting factors and potential expansions of each cropping system will be needed for improving them and diverting their scales and technologies into more economically and environmentally sustainable status, in order to attain the national objectives of the vegetable subsector in Sri Lankan agriculture and food systems.

9.3 Improvements in Production Technologies

Production technologies include the quality of seeds/planting materials, production of healthy and vigorous transplants (through proper nursery management), irrigation, plant nutrient management, plant protection, and other crop management practices.

9.3.1 Seeds and Planting Materials

About 90% of the seed requirement of up-country vegetables (cool-season vegetable species) are imported (197,348 kg in 2017; SCS 2018). Bean, radish, and tomato seeds are the main up-country vegetables produced locally. Despite the frequent release of new improved varieties and they being multiplied locally by the DOA, an estimated 33% of the national requirement of low country (warm season) vegetable seeds (62,200 kg in 2017) are also imported annually (SCS 2018).

Seed certification program of the DOA follows the ISTA protocols since its inception few decades ago. However, the newly introduced techniques for assuring varietal purity (i.e., molecular assessments) and seed health (i.e., conducive environmental incubation and identification, liquid plating assay, enzyme-linked immunosorbent assay, and polymerase chain reaction (PCR) technology), etc., at the global level (Marcos Filho 2015) are not practically applied so far in Sri Lanka.

Lack of favorable climatic conditions (i.e., long day conditions or low temperature) is the major limiting factor for bolting and subsequent seed set in carrot, beet, lettuce, leeks, cabbage, and few other up-country vegetable species. Even though the response of plant growth regulators (hormones) as specific environmental stimuli is evident in the context of flower initiation in these species, research efforts toward developing alternative stimuli under subnormal conditions have not received sufficient attention. Only few studies on flower induction (i.e., in beet; Kumuduni et al. 2013), extended pollination duration for seed production (Fonseka et al. 2010), fruit thinning and canopy management to improve seed quality (Fonseka et al. 2013; Sooriyapathirana et al. 2013), etc., have been conducted for providing the basic knowledge for further developments in this field. Though molecular, tissue culture, and biochemical technologies have attained a remarkable progress in the rest of the world and the basic analytical facilities have long been established in Sri Lanka, none of them have been utilized to develop seed or vegetative propagation of these exotic vegetable crops to date under situations where the environmental stimuli are absent.

Rapid and uniform seed germination and seedling emergence will ensure better crop establishment, providing a proper platform for subsequent growth and yield formation. Despite the fact the locally produced seeds of low country vegetables (i.e., tomato, capsicum, brinjal, okra, beans (*Phaseolus vulgaris*), gourds, and pumpkin) lose their viability during storage (Fonseka and Fonseka 2009), only limited research findings are available for improving viability through storage and packaging. Mettananda (2005) introduced aluminum foil for packaging okra (*Hibiscus esculentus*) seeds for 11 months under room temperature storage without losing viability, while Sunil and Mabesa (1991) reported of appropriate media and temperature for storing pre-germinated seeds of cucumber. However, based on a quality evaluation done on seed samples collected from range of input markets in Sri Lanka, Wijesinghe et al. (2017) have reported a low degree of adaptation of these storage and packaging technologies in Sri Lanka. The results indicated the loss of moisture as the root cause of most of the issues leading to loss of seed viability during prolonged storage. Apart from this, reports on potential expansion of the local seed industry are confined to few studies (Dissanayake and Weerasena 2000) where high potential of the *Maha* season for producing seeds of the recommended bean varieties in the Dry Zone is reported.

Seed treatments such as priming, pelleting, or film-coating with additives play an important role during germination and seedling emergence. Though seed treatment techniques are quite common in the foreign seed markets (Taylor et al. 1998; Andreoli and de Andrade 2002; Cantliffe 2003), their use in local vegetable seed industry is scarce. Majority of seeds (70–75%) available in the local markets are from "informal" seed sector, and hence the application of such advanced techniques is not feasible. Usually farmers themselves mix seeds with fungicide or insecticide before planting. Sowing of seeds is often done manually resulting in variability in seeding depth, leading to uneven field emergence and subsequent differences in growth and development of the plant stand (Shanmuganathan and Benjamin 1992).

The quality-assured vegetable seed supply, formally shared by the public and private sectors, is around 25–30%, while the informal supply sector (without quality assurance) (i.e. farmers' saved seeds, seeds exchanged between farmers) fulfills the rest of the requirement. Price of the seeds supplied by the formal sector is expensive than the informal sector as the quality assurance process by the former is costly. However, compared to imported "low country vegetable" seeds, locally produced seeds within the formal sector hold a lower share. However, locally produced F1 hybrid vegetable seeds are cheaper than their imported counterparts. Though seeds produced locally are cheap, the quantities available in the market during growing season cannot fulfill the demand in certain crops. The insufficient quantities produced may be attributed to the following: (1) unavailability of basic seed of some vegetable crops in required quantities, making it impossible to multiply them into certified seeds, (2) size of the seed market being not profitable for a given vegetable crop as result of cultivation of large number of vegetable crop varieties, (3) requirement of a large isolation facility due to cross-pollinated nature of vegetable crop species, (4) restrictions in production areas as seed production is normally confined to certain agroclimatic zones, (5) high production costs mainly due to labor shortage and higher wages, (6) fresh market vegetable being more profitable sometimes than the seed production (i.e., hot pepper/chili) resulting in the reluctance of contract growers to engage in seed production, (7) time-consuming seed certification process resulting in delays in getting payments, and (8) release of low-quality seeds (e.g., less viable) due to lapses in the seed certification process.

9.3.2 Nursery Management (Growth Media and Use of Transplants)

The economic boom channeled into the use of "coco peat" (a byproduct of the coconut coir industry) in 1980s and 1990s was followed by a series of studies in a global platform on the versatility of its use as a nursery medium and a soilless (hydroponics) growing system (Nichols and Savidov 2008). Similar results were reported by Sri Lankan researchers by comparing its performances with alternative media and testing the influence of raw material quality on the final products

(Tharanga et al. 2005; Mawalagedera and Weerakkody 2013; Kumarasinghe et al. 2016). However, its further developments in parallel with the western world, aiming at value-added potting media products, have not been done significantly. Even though the local production sector (both vegetables and ornamentals) would not expect such intensifications in the growth media use, as a leading exporter of coco peat to the horticulture industry in the western world, Sri Lanka must spearhead in research and developments of growth media use. Meanwhile few more research reports are available on the possible use of other low-cost, organic media as growing media for nurseries (Somachandra et al. 2012a).

Machine transplanting of vegetable seedlings raised in trays is common in the developed world (Schrader 2000; Parish 2005). In Sri Lanka, majority of the farmers raise their own seedlings on raised beds in an area designated for nurseries. However, vegetable farmers of mid-country Wet Zone used to buy seedlings from commercial nurseries, whereas, farmers who are engaged in protected culture produce their own seedlings in nursery trays. Manipulation of the seedling growth in vegetable nurseries has successfully tried out with the use of spectral filters (Bandara et al. 2004; Ekanayake et al. 2016a, b; Welegama et al. 2016). However, lapses remain in growth control with the use of media and plant nutrition for vegetable crops.

Meanwhile grafting has been practiced for propagation of vegetables in many parts of the world in last three decades as a measure to cope up with the environmental stress conditions. For example, over 40 million grafted vegetables are used annually in the North American greenhouse industry. The technology has now been transformed to robotic control in the new millennium (Kurata 1994; Kubota et al. 2008). However, as at present, this technology has not attracted the interest of the research and extension divisions of Sri Lanka so far.

9.3.3 Irrigation

Use of water-saving or efficient irrigation methods such as boom, drip, and sprinkler irrigation together with reusable or biodegradable plastic mulching techniques are the main source of irrigation in open-field vegetable cultivation in the world today. The benefits of these water-saving irrigation technologies have amply been demonstrated in the literature for many crops including tomatoes (Singandhupe et al. 2003; Hebbar et al. 2004), eggplant (Manjunatha et al. 2004; Aujla et al. 2007), sweet pepper (Antony and Singandhupe 2004), and cabbage (Tiwari et al. 2003) particularly under similar socioeconomic and environmental conditions in the humid tropics.

Except in intensive open-field continuous cropping systems in Kalpitiya and some parts of Nuwara Eliya, micro irrigation is limited in vegetable crop cultivations mainly due to its high initial cost. The reason for the popularity of sprinkler irrigation in the up-country WZ (Nuwara Eliya) and sandy soils in Kalpitiya (in the North Western Province) could have been its cost-effectiveness and low water retention properties of soils, respectively (Weerakkody 2004; Aheeyar et al. 2016). High annual rainfall and the unpredictable changes in rainfall patterns due to climate change have affected vegetable farming systems all over the country, demanding drainage technology rather than irrigation technology at present. However, dry (*Yala*) season and temporary droughts in the wet (*Maha*) season in the vegetable-growing areas in the DZ and IZ, especially in the North Central, North Western, and Eastern Provinces could be benefited from water-saving irrigation methods such as drip, sprinkler, and precision surface irrigation, aided with mulching. Hence, developments of appropriate irrigation technologies for vegetable cropping systems for the DZ and IZ and drainage technologies for vegetable cropping systems in the WZ areas are future challenges for research and developmental programs.

9.3.4 Plant Nutrition

At the global level, the use of chemical or synthetic fertilizers in agriculture has increased from 47 million t to 178 million t during 1965–2010 and reached 182 million t in 2014/2015, of which the global vegetable subsector represents 8.6% (Heffer 2009; Heffer et al. 2017). The resultant yield advantages of increased fertilizer use can be clearly shown by the relative increase in crop productivity during the period in most countries (Hossain and Singh 2000). Meanwhile, the use of chemical fertilizer in agriculture has been much debated in the last two decades for its negative effects on soil and water quality and food safety. As a result, the fertilizer use per unit land during the last two decades is showing a decreasing trend in large agricultural land areas in the world especially in Europe, former USSR, Korea, Japan, and China.

Low fertilizer-use efficiency under improper farming practices and climate change scenario is another challenge faced by the crop sector. As a solution, more efficient and effective forms and application methods of chemical fertilizers such as granules, pellets, slow/controlled release fertilizers, deep placements, nitrification and volatilization inhibitors, super granules, and decision support software-assisted precision application methods have been successfully introduced to agriculture (Chien et al. 2009). Even though large-scale vegetable cropping systems utilize these innovative products and methods, organic or health food market-oriented small- and medium-scale vegetable cropping systems keep faith on improved organic manures and soil amendments. The recent developments in this alternative fertilizer technology are gaining popularity in the world at present (Watson et al. 2002; Willer and Lernoud 2016).

Few decades ago in Sri Lanka, making chemical fertilizer recommendations for crops was the top research priority in plant nutrition. Similar to rest of the world, by visualizing the negative impacts of chemical fertilizer use, the research priorities have altered to their limited or efficient use, through split application, slow-release alternatives, and the use of soil amendments (Wijewardena 1993; Wijewardena and Amarasiri 1993; Marikkar and Lathiff 2000; Nugaliyadde 2000; Wijewardena 2000; Tharmarajah and Kulendran 2001).

Meanwhile the misuse of chemical (synthetic) fertilizers in major vegetable-growing areas, especially in mid-country WZ, is becoming a serious issue as it may lead to pollution of ground water, fixation of nutrients, and increased residues of nitrate and other elements in fresh market vegetables. Several residue analysis studies on leafy vegetables have confirmed the accumulation of heavy metals (i.e., Pb and Cr), exceeding the maximum permissible levels (MPLs; Kananke et al. 2014; Wickramarathne et al. 2016; Dharmasena et al. 2017), which are believed to be health hazards and most probably have entered the environment as impurities of chemical fertilizers.

To alleviate these issues recently, integrated plant nutrient systems (IPNS) and crop- and site-specific fertilizer application approaches have been introduced, based on the results of series of research during late nineties, continuing up to 2014 (Kumuduni et al. 2014; Weerasinghe et al. 2014). It has recently winged into another trend on identifying the specific needs of micronutrients on other crops but very few for vegetable crops (Harris and Puvanitha 2018). Further, the applicability of biochar, a unique soil amendment for maintaining soil fertility by increasing the retention of water (water holding capacity) and plant nutrients, has been excessively studied for many crops, but limited studies have been done for vegetable crops (Ekanayake et al. 2016a, b). This eco-friendly or sustainable approach in plant nutrient management is a further advancement of the IPNS. The next most significant feature in the eco-friendly or sustainable plant nutrient management is the application of biological agents that are capable of releasing plant nutrients (biofertilizers). Plant-house and field research conducted on some of the traditional biofertilizer sources (i.e., *Jeevamutham*) and commercially available bio-fertilizers (Bio-film bio-fertilizer) have proven their positive effects on vegetable crop growth (Dissanayake et al. 2016; Kularathna and Devasinghe 2016; Nawarathne 2017).

Hydroponics is one of the promising strategies for fast-diminishing arable lands in the world where plants can be grown in non-soil media, maintaining a great deal of precision in plant nutrient management. Since the inception of protected culture in Sri Lanka in 1997, hydroponics recommendations have been developed mainly in terms of application dosage, electrical conductivity, and pH management of the fertigation solution for greenhouse tomato, bell peppers, green cucumber, and lettuce (Samarakoon et al. 2006; Weerakkody et al. 2007; Saparamadu et al. 2010; Mawalagedera and Weerakkody 2013). Owing to high cost of fertilizer and their environmental impacts, there are recent trends in developing alternative fertilizer sources for hydroponics in the world (Gore and Sreenivasa 2011; Aboutalebi et al. 2013). Along with this trend, Sri Lankan researchers have developed low-cost fertilizers (Saparamadu et al. 2010), organic amendments in hydroponics (Herath et al. 2008; Mawalagedera et al. 2012) and amendment of micronutrients such as Si into fertigation solutions (Jayawardena et al. 2014; Somapala et al. 2016). However, wide adaptation of these technological advancements is yet to be seen.

9.3.5 Plant Protection

Owing to the high degree of perishability and rapid growth rate, vegetable crops are highly vulnerable to pest attacks and disease infections. Moreover, crop genetic improvements toward the advancements in yield and produce quality have narrowed down the genetic variability of the crop germplasm, making the newly improved varieties more vulnerable to pest and diseases. This situation has favored the utility of chemical pest control since the beginning of green revolution in Sri Lanka, similar to most other countries in the world. Therefore, plant protection researches before 2000 were mainly conducted on developing application protocols of chemical pesticides and improving the efficacy of their use (Wahundeniya 1993).

Despite all scientific study-based recommendations, the overuse and misuse of chemical pesticides have widely been reported in vegetable cultivation in Sri Lanka. Hence, the excessive use of both fertilizer and pesticides (both insecticides and fungicides) in vegetable-cultivated areas could cause environmental and human health hazards. This is evident through farmer surveys as well as residue analysis done recently (Jayakody & Munkittrick 2011; Hadji et al. 2017; Lakshani et al. 2017). The annual average pesticide use in Sri Lanka is about 0.97 kg a.i./ha, while case studies have reported a much higher rate (up to 23.4 kg a.i./ha) in rice-based vegetable cropping system in the DZ of the country (Jayakody & Munkittrick 2011). The majority (85.5%) of the pesticides applied for vegetables are fungicides (Hadji et al. 2017). Although environmentally safer pest and disease management methods such as integrated pest management (IPM) are popular in the world (Kitamura et al. 2004; Sardana et al. 2004), their adoption by the farmers in Sri Lanka is low (Jayasooriya and Aheeyar 2016; Dharmasena et al. 2017).

With the environmental and food safety issues associated with the excessive pesticide use, Sri Lankan researchers were interested in determining their safety margins and devising alternative pest control strategies during the past few decades. Different components of IPM are still under investigation. Among popular plant protection research, screening varieties for resistance to pod borer and foot rot disease of eggplant, soft rot and black rot of cabbage, and bacterial wilt of tomato and tomato yellow leaf curl virus (TYLC) (Kelaniyangoda et al. 1998; Ariyarathne et al. 2004; Rajapakse et al. 2005; Somachandra et al. 2012b; Arakesary et al. 2013; Samarakoon et al. 2014a) are some of the prominent sub areas. In the case of low genetic resistance, external influences such as inoculums of rhizobacteria, grafting, and nonconventional chemical pesticide use have been successfully tested for improving the resistance to some diseases and pests of vegetable crops (Illankoon et al. 2001; Wickramarachchi et al. 2003; Wickramarachchi 2005; Jayawardena et al. 2014).

Among other IPM options, the most developed area is biological control of major pests and diseases. Table 9.3 provides a list of bio-control agents that are successfully tested for controlling major pests and diseases of vegetable crops.

Moreover, agronomy of crops as an option for IPM has a long history. Among them, the application of poultry manure for reducing root-knot nematode attacks (Wahundeniya 1991), effect of companion crops for pest control of tomato (Galanihe et al. 2017), use of polythene sleeves for fruit fly control (Pararajasingham and Wahundeniya 2008), and the use of botanicals such as seed or leaf extracts of neem (*Azaddirachta indica*), garlic, and many other native plant species are significant achievements in alternative pest control strategies within vegetable subsector in Sri Lanka (Wahundeniya 1993). Utilizing a combination of these techniques, overall IPM packages have been tested or introduced for chili and tomato

Table 9.3 Bio-control agents tested against major pest and diseases of vegetable crops

Bio-control agent	Pest	Nature of control	Source
Trichogramma chilonis	Cabbage semilooper (*Trichoplusia ni*)	Egg parasitoid	Singhamuni et al. (2015)
Menochilus sexmaculatus (Coleoptera: Coccinellidae)	*Aphis craccivora* (Hemiptera: Aphididae)	Parasitoid	Priyadarshani et al. (2015)
Paecilomycesfumosoroseus	Cabbage catepillers	Entomophagous fungus	Kudagamage et al. (1996)
Hovefly *(Syrphus*spp.), Lacewing (Micromus spp.) and Lepidopteron (Spagalisepius)	Aphids (of Cole crops and *Solanum* spp.)	Predatory	Mayadunnage et al. (2008)
Burkholderia spinosa and *Bacillus megaterium*	Fusarium wilt and other root borne pathogenic fungi (of tomato)	Entomophagous bacteria	Subasinghe and De Costa (2017)
Bacteriophages	*Ralsotina solanacearum*	Infectious (pathogenic) virus	Kalpage and De Costa (2014)

(Samarakoon et al. 2014b), cabbage (Ariyadasa et al. 2005), and gherkin (*Cucumis anguria var. anguria*) (Bandara et al. 2006). Further, basic studies on the disease identification, general and molecular biology of pests/diseases, degree of virulence, vector control, and other infection pathways have been carried out by many researchers as a prerequisite for developing successful pest control programs. Among them, studies on the biology of cabbage pests (Katipearachchi 2004), studies on hardwood stem borer of *thibbatu* (*Solanum tuberosum*; Hitinayake et al. 2017a), identification of virus-like diseases in spine gourds (*Momordica dioica*; Ranasinghe et al. 2017), phytoplasma disease transmission by bitter gourd (*M. charantia*) seeds (Tennakoon et al. 2017), species richness of fruit fly (*Diptera pepharitidae*; Ranaweera et al. 2017), characterization and virulence of *Xanthomonas campestris* var. *campestris* in Crucifers (Babu et al. 2011), use of PCR in detecting yellow vein mosaic virus (YVMV) infections in okra (Samarasinghe et al. 2009), detection of local isolates of cucumber green mosaic virus (Weerarathne et al. 2009), identification of new mosaic virus from snake gourds and black plant hopper in *Altananthera* (Ariyaratne et al. 2005; Balasingham and Wijayarathnam 2009), molecular and morphological identification of TYLCV in tomato (Samarajeewa et al. 2005), etc., are the significant achievements in plant protection of vegetable crops. However, none of the reports indicate the technological improvements on crop protection with regard to common and highly destructive animal pest attacks that are on the rise for vegetable and many other cropping systems in the recent times.

9.3.6 Crop Manipulation

Plant density management is a very basic research theme in annual crops. However, it is important for intensive space management in protected culture. Research done on greenhouse tomato (Kularathne et al. 2008) has established optimum spacing under semi-intensive technology, while a comparative assessment of vertical space under different staking (crop support methods has been reported for greenhouse cucumber) (Premalatha et al. 2006). Head size of cabbage and size of root/tuber vegetables are market quality parameters in the upmarkets. However, research on agronomic approaches in changing the unit size of harvest is scarce in Sri Lanka. For example, Jayamanne et al. (2015) have improved the head size of cabbage by changing plant spacing.

Staking tomatoes and pole beans (*P. vulgaris*), and trellising Cucurbitaceous crops are widely practiced in Sri Lanka. However, research on the effects of staking and trellising on yield, quality, disease incidence, and ease of harvesting are scarce in contrast to India, where positive effects of staking and trellising were reported by many authors (Singh et al. 2007; Singh and Lal 2003). In Sri Lanka, only few researches have been conducted on new trellising systems for bitter gourds (*Momodica charantia*) and luffa (*Lufa acutangula*) where the recently introduced "fence type" has given a higher yield of quality fruits of both crops (Thavabalachandran 2003; Pararajasingham 2006).

Further, few studies have been reported for maintaining the number of flowers/fruits or seed quality by manipulating sex expression of spine gourd (Hitinayake et al. 2017a, b) and fruit thinning in eggplant and tomato (Sooriyapathirana et al. 2013; Fonseka et al. 2013). In addition, developments in harvesting indices and methods of beans and cabbage have been reported by Champa et al. (2007, 2008). Reports on research and developments in crop manipulations in vegetable crops in Sri Lanka are mostly available for greenhouse crops. The improvements done on crop manipulation of open-field vegetables are comparatively low and limited to few crops and cropping systems.

9.4 Improvements in Produce Quality

9.4.1 Quality and Food Safety

Sri Lanka is ranked 76 out of 113 countries for food quality and safety scoring 49.5 (GFSI 2018). There is a general list of quality parameters usually considered for vegetables. Some of these are external parameters that are priorities in the fresh vegetable market, while the others are internal parameters with much more significance in cooked or processed products. Depending on the edible part of the vegetable, some parameters become more important. Further, the nature of the processed product might decide the importance of internal quality attributes in general or specific for selected crops. In developed countries of which the demand for vegetable and the affordability of customers are high, genetic as well as environmental improvements in the quality of vegetables is a high priority at all times (Weston and Barth 1997).

Lack of concern on quality improvements in breeding or production programs of vegetable crops in Sri Lanka is a consequence of negligence in quality concerns at the market place for vegetables and their processed products. However, as stated earlier under advancements in the input use, few studies have reported the negative trends in internal quality of vegetables due to high levels of heavy metal contamination (33% of the samples) in them (Lakshani et al. 2017). In contrast, Silva et al. (2016) have assured the quality of vegetables against heavy metal contamination.

9.4.2 Bioactive Properties

The antioxidant activity and phenolic and other health-promoting compounds of food have received a great attention in the world in the recent times. Bioactive properties of cultivars and crop wild relatives grown under varying environmental and management conditions have been reported to be highly variable (Velioglu et al. 1998). As a result, the bioactivity has been considered as an internal quality of both fresh and processed vegetable products. Although the research conducted on improving the quality of fruits and vegetables is enormous, recent developments in this context in Sri Lanka is limited to few fruits and vegetables, namely bitter gourd and eggplant (Fonseka et al. 2007; Somawathi et al. 2016). Moreover, several detailed studies have been conducted to investigate the actual phyto-chemicals responsible for the bioactive properties of each vegetable and fruit species (Senarathne et al. 2017).

9.4.3 Dissemination of New Improved Technologies

The importance of technology transfer from a development perspective is nothing new. More than three decades back, Mansfield (1975) stated that, "*One of the fundamental processes that influence the economic performance of nations and firms is technology transfer*". It is well recognized that transfer of technology is at the key to the process of economic growth in both developed and developing countries. Transfer of technology generated by the researchers to farmers is considered to be the responsibility of agriculture extension system. Despite large investments made on technology transfer by both developed and developing countries over the last three decades, anticipated agricultural production is yet to achieve. It is optional that either the technology to be developed locally or be borrowed from other places/countries with similar environmental conditions and socioeconomic situations. The burden on the local extension system is less in introducing packages of borrowed technologies compared to the technologies developed locally. For instance, a new F1 hybrid vegetable variety imported by the private sector after obtaining recommendation of DOA could become popular among farmers than a locally developed F1 hybrid variety of the same crop category released by the DOA. Differences and efficiency of the agricultural technology transfer system and subsequent adoption rate depend firstly on the macro factors such as agroecological, political economic, sociocultural, policy, market intervention, infrastructure, transportation, and communication, while secondly on the institutional factors including research, education and training, input supply, credit, and farmer organizations.

9.5 Overall Impact of Recent Developments and the Way Forward

National Agriculture policy (NAP) of Sri Lanka developed in 2007 has adopted seven major policy statements in horticulture, covering both technological and production aspects. Accordingly, interventions in horticulture at the household, community, and national levels to ensure nutritional food security are the thrust in the research agenda. Although not comparable with regional countries, the recent developments in vegetable breeding in Sri Lanka, as evidenced by release of new varieties, are commendable compared to other subsectors of agriculture. However, the genetic base of vegetable crops is yet to be strengthened through intensive exploration and conservation. The recent research output on alternative pest and disease control strategies as well as alternative plant nutrient management strategies appeared to be progressive compared to other sub areas in agronomic improvements pertaining to vegetable crops. However, these efforts rarely have been converted to practically used IPM or IPNM packages targeting major vegetable crops or cropping systems. As a result, agrochemical use in the vegetable subsector is always on the rise, making the system less cost-effective and environmentally unsustainable. Other than the inherent lapses in field testing for adaptability, the forward march of these new technologies could have been heavily barricaded by their low user-friendliness and long-missing extension links.

The technological improvements in other areas of vegetable crop production such as nursery management, mechanization, micro irrigation, and crop manipulation are generally less attended during last few decades in Sri Lanka. Moreover, the assessment of the local germplasm of vegetable crops for their health-promoting properties and improving these properties further through crop genetics and environmental improvements show lapses, compared to research and developments in vegetable crops in rest of the world today. Overlooking at this new market-value-based genetic and environmental crop improvement could have been the main reason for low productivity of vegetable crops in Sri Lanka. Further to this, fine-tuning the technologies that are specific for upcoming sustainable cropping systems (i.e., organic farming and ecological agriculture) and protected culture needs to be brought to the attention of the policy makers and institutional research committees within the vegetable subsector.

Finally, the burning need of a centrally coordinated collaborative research network, covering major and minor areas of vegetable crops research, can be emphasized for Sri Lanka to upgrade the production technology in par with rest of the Asia within the next 10-year period. In this regard, strengthening of the existing centers for vegetable crops research of the DOA, University system, and other research institutions can be emphasized. The long experience, human resource capital, institutional infrastructure, and stakeholder links of these centers would be a great asset in furthering vegetable crops research and developments in the future.

References

Aboutalebi A, Jahromi MG, Farahi MH (2013) Evaluation of growth and yield of organically-grown basil (*Ocimum basilicum* L.) in soilless culture. J Food Agri Environ 11(1):299–301

Aheeyar M, Manthrithilake H, Pathmaraja S (2016) Drivers of the adoption of farmer-innovated sprinkler irrigation systems; evidence from Kalpitiya, Sri Lanka. In: Proceedings of international perspectives in water resources, Colombo, Sri Lanka, pp 1–14

Andreoli C, de Andrade RV (2002) Integrating matriconditioning with chemical and biological seed treatments to improve vegetable crop stand establishment and yield under tropical conditions. Seed Technol 24:89–99

Antony E, Singandhupe RB (2004) Impact of drip and surface irrigation on growth, yield and WUE of capsicum (*Capsicum annum* L.). Agric Water Manag 65:121–132

Arakesary S, Bavaleeswaram B, Athputhachandran P, Vijayarathnam S (2013) Agronomic evaluation and characterization of local landraces, resistant to shoot and foot borer in brinjal (*S. melongina*). Ann Sri Lanka Dept Agric 15:357–360

Ariyadasa K, Dharmaprema WJ, Samarakoon Y, Nanayakkara A (2005) Adoption of integrated pest management for cabbage (*B. oleraceae* var. *capitata*) cultivation in Matale district. Ann Sri Lanka Dept Agric 7:385–386

Ariyarathne I, Liyanage T, Sagarika M (2004) Virus disease incidence, alternate hosts of pathogens and disease resistance of tomato in Sri Lanka. Ann Sri Lanka Dept Agric 6:13–28

Ariyaratne I, Weerarathne W, Ranathunge R (2005) Identification of new mosaic virus disease of Snake Gourd in Sri Lanka. Ann Sri Lanka Dept Agric 7:13–22

Aujla MS, Thind HS, Buttar GS (2007) Fruit yield and water use efficiency of eggplant (*Solanum melongena* L.) as influenced by different quantities of nitrogen and water applied through drip and furrow irrigation. Sci Hortic 112:142–148

Babu A, Ramanayake R, Hettiarachchi S (2011) Characterization and virulence of *Xanthomonas campestris* pv. campestris isolates from Cruciferaceae crops in Nuwara eliya District of Sri Lanka. Ann Sri Lanka Dept Agric 13:1–12

Balasingham B, Wijayarathnam S (2009) Black Plant Hopper: a newly identified insect pest in Mukunuwenna in Jaffna district. Ann Sri Lanka Dept Agric 11:233–234

Bandara K, Kudagamage C, Senadeera D, Prathapasingha G (2006) Development of an effective integrated pest management system for Melon fly (*Bactroceracucurbitae*) infesting commercial cultivation of Gherkin, Cucurmis sativus. Ann Sri Lanka Dept Agric 8:11–18

Bandara RWDCK Weerakkody WAP, Kumara GKKP (2004) Spectral filters-quality of foliage ornamental plants. Souvenir. 9th Nurseryoysav 2004. Lalbagh Botanical Gardens, Banglore (India). pp 9–10

Bernal LEP, Rumayor-Rodrigez A, Perez-Veyana O, Reyes-Rivas E (2010) Competitiveness of Zactecas (Mexico) protected culture: the fresh tomato industry. Int Food Agribus Manag Rev 13(1):45–64

Cantliffe DJ (2003) Seed enhancements. Acta Hortic 607:53–59

Champa WAH, Palipane KB, Weerakkody WAP, Fernando MD (2007) Maturity indices for harvesting of cabbage (*Brassica olerasiae* L) variety, Green Coronet. Trop Agric Res 19:254–264

Champa WAH, Weerakkody WAP, Palipane KB (2008) Maturity indices for harvesting of Beans (*Phaseolus vulgaris* L) variety, Kentucky Wonder Green in dry zone cultivations in Sri Lanka. Trop Agric Res 20:123–133

Chien SH, Prochnow LI, Cantarella H (2009) Recent developments of fertilizer production and use to improve nutrient efficiency and minimize environmental impacts. Adv Agron 102:267–322

Dharmasena PB, Gunasena N, Jayaweera U (2017) Sustainable management practices for agricultural lands in central highlands of Sri Lanka Part 5—upcountry rain-fed vegetable farming system. FAO Report (GCP/SRL/063/GFF)

Dissanayake D, Seneviratne G, Dissanayaka D, Jinadasa D (2016) Relationship between the quality parameters of biofilm biofertilizers and vegetable crop response. Research Symposium of Rajarata University of Sri Lanka, p 26

Dissanayake DMC, Weerasena SL (2000) Potential of bean seed production in the dry zone. Ann Sri Lanka Dept Agric 2:81–84

Egodawatta WCP, Sangakkara UR, Stamp P (2012) Impact of green manure and mineral fertilizer inputs on soil organic matter and crop productivity in a sloping landscape of Sri Lanka. Field Crop Res 129:21–27

Ekanayake A, Weerasinghe P, Duminda D (2016a) Conjoint use of biochar with compost and inorganic fertilizers on soil quality and plant growth of tomato. Research Symposium of Rajarata University of Sri Lanka, p 24

Ekanayake EMBP, Gunawardena NP, Mawalagedera SMMR, Weerakkody WAP (2016b) Influence of different spectral filters on the colour of foliage of *Codiaeum variegatum* L. Var. Sunny star (croton) during two distinct seasons. In: Proceedings of International Research Symposium, Rajarata University of Sri Lanka, pp 90–93

FAOSTAT (2017) FAO statistics on regional agricultural development. Food and Agriculture Organization of the United Nations, Rome

Fonseka H, Herath K, Devasinghe D, Izuka K (2013) Effect of fruit thinning, canopy management and number of harvests on seed quality of eggplant (*S. melongena*). Ann Sri Lanka Dept Agric 15:217–226

Fonseka H, Warnakulasooriya K, Senanayake G, Fonseka R (2010) Studies on the effect of extended pollination duration on fruit set and seed development and storage temperature on viability and storability of pollen of eggplant (*Solanum melongena* L.) Proceedings of Advances in Genetics and breeding of Capsicum and eggplant, pp 473–482

Fonseka HH, Chandrasekara A, Fonseka M, Wickramasinghe IP, Kumara KPDRSP, Wickramarchchi WNC (2007) Determination of anti-amylase and anti-glucosidase activity of different bittergourd (*Momodica charentia* L.) genotypes. Proceedings of the 1st International Conference on Indigenous Vegetables and Legumes Prospectus for Fighting Poverty, Hunger and Malnutrition. Acta Hortic 752:131–137

Fonseka HHD, Fonseka R (2009) Studies on deterioration and determination of bitter gourd seeds (*Momordica charantia*). Acta Hortic 898:31–38

Galanihe L, Madugalle S, Rajapaksha S, Siriwardena M (2017) Influence of companion crops on sucking pest complex of tomato. Ann Sri Lanka Dept Agric 19(1):27–35

GFSI (2018) Global Food Security Index. https://foodsecurityindex.eiu.com. Accessed 20 Sep 2018

Gore NS, Sreenivasa MN (2011) Influence of liquid organic manures on growth, nutrient content and yield of tomato (*Lycopersicon esculentum*) in the sterilized soil. Karn J Agric Sci 24(2):153–157

Hadji T, Somachandra K, Kamalanie S, Kularathne K, Senenayake N (2017) Small scale farmers' perception on pesticide usage: a case study from intensive vegetable growing areas in Welimada, Sri Lanka. Ann Sri Lanka Dept Agric 19(2):269–273

Harris KD, Puvanitha S (2018) Influence of foliar application of boron and copper on growth and Yield of tomato (*Solanum lycopersicum* cv 'Thilina'). AGRIEAST J Agric Sci 11(2):12–19

Hebbar SS, Ramachandrappa BK, Nanjappa NV, Prabhakar M (2004) Studies on NPK drip fertigation in field grown tomato (*Lycopersicon esculentum* Mill.). Eur J Agron 21:117–127

Heffer P (2009) Assessment of fertilizer use by crop at the global level. International Fertilizer Industry Association, Paris, 75116

Heffer P, Gruere A, Roberts T (2017) Assessment of fertilizer use by crop at the global level. International Fertilizer Industry Association, Paris, 75116

Herath H, Suriyagoda L, Amarasiri K, Premaratne K, Weerakkody WAP (2008) Applicability of straight fertilizers and compost in greenhouse tomato cultivation. Sri Lankan J Agric Sci 45:33–46

Herath I, Kumarathilaka P, Navaratne A, Rajakaruna N, Vithanage M (2015) Immobilization and phytotoxicity reduction of heavy metals in serpentine soil using biochar. J Soils Sediments 15(1):126–138

Hitinayake H, Athapaththu I, Abeysekara W, Madushika KD, Pathirana W, Swarnalatha K (2017a) Management of hardwood stem borer in thibbatu (*Solanum torvum*). Ann Sri Lanka Dept Agric 19(2):350

Hitinayake, H., Sumanarathne, J., Abeysekara, W., Madushika, K., Danushka, W. and Swarnalatha. (2017b). Yield improvement of spin gourd through gynomonoceous hybrid. Ann Sri Lanka Dept Agric, 19(2): 71–78

Hsieh SC (2005) Organic farming for sustainable agriculture in Asia with special reference to Taiwan experience. Research Institute of Tropical Agriculture and International Cooperation, National Pingtung University of Science and Technology, Pingtung, Taiwan, p 65

Hossain M, Singh VP (2000) Fertilizer use in Asian agriculture: implications for sustaining food security and the environment. Nutr Cycl Agroecosyst 57(2):155–169

IASRI (2018) Vegetable research and development in India: role of AICRP on vegetable crops. http://iasri.res.in/aicrpvc. Accessed 18 Jun 2018

Illankoon J, Zoysa I, Wijesekara A (2001) Tomato grafting as a remedy for bacterial wilt in a protected agriculture system. Ann Sri Lanka Dept Agric 3:53–60

Jayakody A, Munkittrick K (2011) Determinants and externalities of pesticides use practices in an intensive mixed vegetable-rice growing catchment in Uma oya in Sri Lanka. Trop Agriculturist 159:61–105

Jayamanne J, Elangeshwaran A, Harris K, Dharmasena H (2015) Market demand for head-size of cabbage and suitable plant spacing for downsizing it to meet the demand. Ann Sri Lanka Dept Agric 17:114–153

Jayasooriya HJ, Aheeyar M (2016) Adoption and factors affecting on adoption of integrated pest management among vegetable farmers in Sri Lanka. Procedia Food Sci 6:208–212

Jayawardena H, Weerahewa H, Saparamadu M, Rathnayaka R (2014) Effect of silicon on anthracnose disease in hydroponically grown *Capsicum annum* l. cv. awlegama. International Research Symposium of Peradeniya University, vol. 18, p 543

Johnson GI (2008) The vegetable industry in tropical Asia (vol. 8, no. 712). AVRDC-World Vegetable Center, Taiwan

Kalpage MD, De Costa DM (2014) Isolation of bacteriophages and determination of their efficiency in controlling *Ralsotina solanacearum*, the bacterial wilt pathogen of tomato. Trop Agric Res 26(1):140–151

Kananke T, Wasnapala J, Gunarathne A (2014) Heavy metal contamination in green leafy vegetables collected from selected market sites of Piliyandala area, Colombo district, Sri Lanka. Am J Food Sci Tech 2(5):139–144

Katipearachchi Y (2004) Biology of Hellula undalis Fab. (Lepidoptera: Pyralidae) on cabbage, *Brassica oleracea* var. capitata L. in the mid country of Sri Lanka. Trop Agriculturist 155:43–48

Kelaniyangoda D, Herath L, Hemachandra B (1998) Sources of Bacterial Wilt (*Ralstonia solanacearum*) disease resistance in tomato. Trop Agriculturist 152:19–29

Khai NM, Ha PQ, Öborn I (2007) Nutrient flows in small-scale peri-urban vegetable farming systems in Southeast Asia—a case study in Hanoi. Agric Ecosyst Environ 122(2):192–202

Kitamura T, Kawai A, Yamada M (2004) Occurrence of insect pests and their natural enemies and establishment of IPM in vegetable fields at West Java, Indonesia, JIRCAS Working Report No. 43, pp 205–210

Kottawa-Arachchi JD, Wijeratne MA (2017) Climate change impacts on biodiversity and ecosystems in Sri Lanka: a review. Nat Conserv Res 2(3):2–22

Kubota C, McClure MA, Kokalis-Burelle N, Bausher MG, Rosskopf EN (2008) Vegetable grafting: history, use, and current technology status in North America. HortScience 43(6):1664–1669

Kudagamage C, Munasingha R, Rajapakse R, Rathnasekara D (1996) Potential of Paecilomyces fumosoroseus in control of cabbage caterpillars. Trop Agriculturist 151:59–62

Kularathna L, Devasinghe D (2016) Effect of effective microorganisms and vermiwash on Lettuce (*Lactuca sativa* L.) in hydroponic cultivation. Int J Chem Environ Biol Sci 4(1):78–81

Kularathne R, Wickramathunge C, Rajasinghe R (2008) Yield improvement of Greenhouse tomatoes (*L. esculentum*) through plant density management. Ann Sri Lanka Dept Agric 10:81–96

Kumara SK, Weerakkody R, Epasinghe S (2015) Viability of controlled environment agriculture for vegetable farmers in Sri Lanka (report No. 179). Hector Kobbakaduwa Agrarian Research and Training Institute, Colombo 7

Kumarasinghe H, Subasinghe S, Ransimala D (2016) Effect of coco peat particle size for the optimum growth of nursery plant of greenhouse vegetables. Trop Agric Res Ext 18(1):51–57

Kumuduni H, Babu A, Nugaliyadde M (2013) Study on flower induction and seed production in beet (*Beta vulgaris*). Ann Sri Lanka Dept Agric 15:363–364

Kumuduni H, Babu A, Nugaliyadde M (2014) Effect of split application of chemical fertilizers combined with organic manure on the yield of carrot (*Daucus carota* L. var sativus). Trop Agriculturist 162:9–24

Kurata K (1994) Cultivation of grafted vegetables II. Development of grafting robots in Japan. HortScience 29(4):240–244

Lakshani P, Rajapaksha M, Sendthuran K (2017) Pestcide residues in selected vegetables in several growing areas by GC/MS using QuEChERS technique. Ann Sri Lanka Dept Agric 19(2):188–208

Manjunatha MV, Rajkumar GR, Hebbara M, Ravishankar G (2004) Effect of drip and surface irrigation on yield and water-production efficiency of brinjal (*Solanum melongena*) in saline Vertisols. Indian J Agric Sci 74:583–587

Mansfield E (1975) International technology transfer: forms, resource requirements, and policies. Am Econom Rev 65(2):372–376

Marcos Filho J (2015) Seed vigor testing: an overview of the past, present and future perspective. Sci Agric 72(4):363–374

Marikkar S, Lathiff M (2000) Effect of biuret content in urea on growth and yield of some selected vegetable crops. Ann Sri Lanka Dept Agric 2:183–188

Mawalagedera SMMR, Weerakkody WAP (2013) Plant growth of coco-peat grown tomato under enhanced fertigation in hot and humid climates. Acta Hortic 1004:151–158

Mawalagedera SMMR, Weerakkody WAP, Premaratne KP (2012) Circulation culture for efficient nutrient uptake and high yield in tomato grown in tropical greenhouses. Trop Agric Res 23(3):104–117

Mayadunnage S, Wijayagunasekara HNP, Hemachandra KS, Nugaliyadda L (2008) Predatory insects of pest insects in a vegetable eco system of the mid country region of Sri Lanka. Trop Agric Res 20:34–41

Mettananda KA (2005) Viability of okra (*Hibiscus esculentus*) seeds under different storage conditions. Ann Sri Lanka Dept Agric 7:191–198

Nawarathne K (2017) Bacterial Bioinoculum: N source for vegetable crops. Ann Sri Lanka Dept Agric 19(2):16–28

Nichols MA, Savidov NA (2008) Recent advances in coir as a growing medium. Acta Hortic 843:333–336

Nugaliyadde M (2000) Effect of fertilizers on growth of bittergourd and root knot nematode, *M. incognita*. Ann Sri Lanka Dept Agric 2:253–262

Pararajasingham N (2006) New trellissing system for bitter gourd (*M. charantia*). Ann Sri Lanka Dept Agric 8:175–182

Pararajasingham N, Wahundeniya I (2008) Influence of sleeving on fruit and seed yield and quality of bitter gourd. Ann Sri Lanka Dept Agric 10:349–350

Parish RL (2005) Current developments in seeders and transplanters for vegetable crops. HortTechnology 15:346–351

Perera KDA (1990) Vegetable production in Sri Lanka. Trop Agric Res Ser 23:9–20

Poudel DD, Midmore DJ, Hargrove WL (1998) An analysis of commercial vegetable farms in relation to sustainability in the uplands of Southeast Asia. Agric Syst 58(1):107–128

Premalatha M, Wahundeniya KB, Weerakkody WAP, Wicramasinghe C (2006) Plant training and spatial arrangement for yield improvements of greenhouse cucumber varieties. Trop Agric Res 18:346–358

Priyadarshani TDC, Hemachandra KS, Sirisena UGAI, Wijayagunasekara HNP (2015) Developmental biology and feeding efficiency of *Menochilus sexmaculatus* (Coleoptera:

Coccinellidae) (Fabricius) reared on *Aphis craccivora* (Hemiptera: Aphididae) (Koch). Trop Agric Res 27(2):115–122
Rajapakse R, Fonseka H, Karunarathne R, Wickramasinghe I (2005) Evaluation of brinjal (*Solanum melongina*) germplasm for foot rot disease. Ann Sri Lanka Dept Agric 7:369–374
Rajapakse RHS, Ratnasekera D, Abeysinghe S (2016) Biopesticides research: current status and future trends in Sri Lanka. In: Agriculturally important microorganisms. Springer, Singapore, pp 219–234
Ranasinghe C, Tennakoon N, Gunasena M, jayasena S, Ekanayake R, Chathurani L, Muthugala D (2017) Virus and virus like diseases in spiny gourd. Ann Sri Lanka Dept Agric 19(2):334–335
Ranaweera NFC, De Silva GAC (1994) Dynamics of vegetable production, distribution and consumption in Asia. Asian Vegetable Research and Development Center (AVRDC), Shanhua, pp 349–357
Ranaweera P, Ranathunga M, Nugaliyadde L, Hemachandra K, Rajapakse ME, Wijesekara L, Edirisinghe E (2017) Abundance of species richness of fruit flies (Diptera: Tephritidae) in major Cucurbit growing areas in Anuradhapura, Kurunegala and Kandy districts in Sri Lanka and Farmenrs' perception on recommended management methods. Ann Sri Lanka Dept Agric 19(2):98–112
SAARC Report (2017) Vegetable production in the South Asia. SAARC Secretariat, Kathmandu
Samarajeewa P, Meegahakumbura M, Rajapaksha R, Gammulla C, Sumanasinghe V (2005) Molecular and morphological identification of tomato yellow leaf curl virus in tomato. Ann Sri Lanka Dept Agric 7:233–244
Samarakoon R, De Costa D, Menike G (2014a) Disease incidences, abundance of insect vectors and beneficial insects of chilli and tomato grown under two crop management practices. In: International Research Symposium of Peradeniya University, vol. 18, no. 2, p 524
Samarakoon S, Balasuriya A, Rajapaksha R, Wickramarachchi W (2014b) Determination of tomato yellow leaf curl virus (TYLCV) resistance in selected Sri Lankan tomato varieties through whitefly inoculation protocol. In: International Research Symposium of Peradeniya University, vol. 18, p 553
Samarakoon UC, Weerasinghe PA, Weerakkody WAP (2006) Effect of EC of the nutrient solution on growth, yield and nutrient uptake of leaf lettuce *(Lactuca sativa)* in stationary culture. Trop Agric Res 18:13–21
Samarasinghe WLG, Samarajeeva PK, Liyanage HM, Gimhani DR (2009) A PCR technique to detect yellow vein mosaic virus (YVMV) infection in okra breeding progenies. Ann Sri Lanka Dept Agric 11:131–140
Sangakkara UR, Katupitiya S (2000) Organic farming in Sri Lanka. Faculty of Agriculture, University of Peradeniya, Sri Lanka
Saparamadu M, Weerakkody WAP, Gunawardhana H, Wijesekera R (2010) A low cost nutrient formulation with a buffer for simplified hydroponics systems. J Hort Forest 2(5):99–103
Sardana HR, Arora S, Singh DK, Kadu LN (2004) Development and validation of adaptable IPM in egg plant, *Solanum melongena* L. in a farmers participatoryapproach. Ind J Plant Protect 32:123–128
Schrader WL (2000) Using transplants in vegetable production. Publication 8013, University of California
SCS (2018) Unpublished report. Seed Certification Service. Department of Agriculture, Peradeniya
Senanayake YDA (2000) Science and technology in agriculture—a concept paper. The First Biennial Conference on Science and Technology. National Science and Technology Commission, Colombo
Senarathne S, Peduruhewa P, Jeewanthi P, Wijesinghe W, Wijerathne D, Karandawala K, Hettiarachchi D (2017) Effect of cooking on ascorbic acid, total polyphenol content and antioxidant activity of selected vegetables. Ann Sri Lanka Dept Agric 19(1):16–26
Shanmuganathan V, Benjamin LR (1992) The influence of sowing depth and seedsize on seedling emergence time and relative growth rate in spring cabbage (*Brassica oleracea* var. capitata). Ann Bot 69:273–276

Silva N, Weerasinghe P, Rathnapriya H (2016) Toxic trace elements in vegetables collected from markets in Kandy district. Trop Agriculturist 164:57–69

Singandhupe RB, Rao GGSN, Patil NG, Brahman PS (2003) Fertigation studies and irrigation scheduling in drip irrigation system in tomato crop (*Lycopersicon esculentum* L.). Eur J Agron 19:327–340

Singh B, Kumar M, Sirohi NPS (2007) Protected cultivation of cucurbits under low-cost protected structures: a sustainable technology for peri-urban areas of Northern India. Acta Hortic (731):267–272

Singh DK, Lal G (2003) Effect of fertilizers and staking on growth and yield of tomato hybrids during summer season. Sci Hort 8:87–94

Singhamuni SAA, Jayasuriya M, Hemachandra KS, Sirisena UGAI (2015) Evaluation of the potential of *Trichogramma chilonis* ishii (Hymenoptera: *Trichogramma tidae*) as a bio-control agent for Trichoplusia ni, cabbage semi-looper. Trop Agric Res 26(2):223–236

Somachandra KP, Gunasekara WMS, Somasiri GAR (2012a) Pinus (*Pinus caribea*) needles as a substrate constituent in soilless media of tomato in protected culture. Ann Sri Lanka Dept Agric 15:375–376

Somachandra KP, Gunasekara WMS, Somasiri GAR (2012b) Response of cabbage varieties to soft rot and black rot diseases. Ann Sri Lanka Dept Agric 14:161–172

Somapala K, Weerahewa D, Thrikawala S (2016) Silicon rich rice hull amended soil enhances anthracnose resistance. Procedia Food Sci 6:190–193

Somawathi KM, Rizliya V, Wickramasinghe HAM, Madhujith WMT (2016) A comparative study on antioxidant and DNA protective activity of different skin coloured brinjal (*Solanum melongena*). Procedia Food Sci 6:116–122

Sooriyapathirana C, Sajeevika C, Fonseka R, Fonseka H (2013) Effects of canopy management and fruit thinning on seed quality of tomato (*Solanum lycopersicum* L) variety *Thilina*. Trop Agric Res 25(2):171–179

Stockdale FA (1921) Scientific research and investigation in regard to the agricultural industries of Ceylon. Trop Agriculturist LVII(3):153–165

Subasinghe SDSK, De Costa DM (2017) Potential of *Burkholderia spinosa* and *Bacillus megaterium* on plant growth promotion and suppression of selected soil-borne diseases of tomato. Proceedings of the 4th symposium of Faculty of Agriculture Undergraduate Research Symposium, Faculty of Agriculture, University of Peradeniya, 12 January, p 114

Sunil G, Mabesa R (1991) Storage techniques for extending the viability of pregerminated cucumber seeds. Trop Agriculturist 147:19–32

Suriyagoda LDB, Ranil RHG, Dissanayake S, Weerakkody WAP (2012) Sustainability of intensive vegetable farming in the upcountry region of Sri Lanka: a situation assessment. Chronica Hort 52(4):14–17

Taylor AG, Allen PS, Bennett MA, Bradford KJ, Burris JS, Misra MK (1998) Seed enhancements. Seed Sci Res 8:245–256

Tennakoon T, Nagasena P, Nandasena K (2017) Possibility of phytoplasma disease transmission through bitter gourd (*M. charantia*) seeds. Ann Sri Lanka Dept Agric 19(2):280–283

Tharanga S, Wathulanda H, Weerakkody W, Gamlath S (2005) Variation of chemical and physical properties of raw coir dust with respect to age, area of origin and the extraction method. Sri Lankan J Agric Sci 42:01–11

Tharmarajah S, Kulendran S (2001) Cucumber response to cattle manure levels and N and K split application frequencies in Sandy Regosols. Ann Sri Lanka Dept Agric 3:275–288

Thavabalachandran M (2003) Effect of different trellising system on growth and yield of Luffa (*L. acutangula*). Ann Sri Lanka Dept Agric 5:241–248

Tiwari KN, Singh A, Mal PK (2003) Effect of drip irrigation on yield of cabbage (*Brassica oleracea* var. *capitata*) under mulch and non-mulch conditions. Agric Water Manag 58:19–28

Velioglu YS, Mazza G, Gao L, Oomah BD (1998) Antioxidant activity and total phenolics in selected fruits, vegetables, and grain products. J Agric Food Chem 46(10):4113–4117

Wahundeniya I (1991) Effect of poultry manure on root knot nematodes (*Meloidogyne spp.*) in tomato (*Lycopersicon esculentum* Mill.). Trop Agriculturist 147:143–154

Wahundeniya I (1993) Control of leaf eating caterpillars in cabbage using an insect growth regulator. Trop Agriculturist 149:81–88
Watson CA, Atkinson D, Gosling P, Jackson LR, Rayns FW (2002) Managing soil fertility in organic farming systems. Soil Use Manag 18(s1):239–247
Weerakkody WAP (2004) Horticulture in Sri Lanka. Chronica Hort 44:23–27
Weerakkody WAP, Mayakaduwa MA, Weerapperuma K (2007) Effect of supply volume and weather based EC adjustments on the growth and yield of greenhouse tomato and bell pepper. Acta Hortic 742:105–111
Weerakkody WAP, Peiris BCN, Gunaratne LHP (2000) Commercial gardening of vegetables in Sri Lanka: present status and environmental constrains. Sri Lankan J Agric Sci 37:135–147
Weerarathne W, Ikiriwatte C, Liyanage T (2009) Detection of local isolates of Cucumber Green Mottle Mosaic Virus and its management. Ann Sri Lanka Dept Agric 11:199–210
Weerasinghe K, Balasooriya A, Ransingha S, Krishantha G, Brahakamanage R, Wijethilke L (2014) Effects of macro and micro nutrients on growth and yield performances of tomato (*Lycopersicon esculentum* Mill.). Int J Environ Ecol Eng 1(10):556
Welegama H, Dissanayake D, Fonseka R, Devasinghe D et al (2016) Seedling emergence and early seedling growth of tomato, brinjal and capsicum on nursery beds with colour polythene. Ann Sri Lanka Dept Agric 18:291
Weston LA, Barth MM (1997) Preharvest factors affecting postharvest quality of vegetables. HortScience 32(5):812–816
Wickramarachchi W (2005) Effect of rhizobacteria on increasing plant growth and Inducing systemic resistance in tomato against early blight disease. Ann Sri Lanka Dept Agric 7:309
Wickramarachchi W, Reddy P, Reddy T, Rao P (2003) Induction of resistance in tomato against early blight disease with non conventional chemicals. Ann Sri Lanka Dept Agric 5:271–280
Wickramarathne M, Maduranga T, Chamara L (2016) Contamination of heavy metals in aquatic vegetables collected from cultivation sites in Sri Lanka. IOSR J Environ Sci Toxicol Food Technol 10(11):76–82
Wijeratne DGKP, Weerakkody WAP (2017) Economic and technical feasibility of upcountry vegetable production in Halmillewa (IL1a). Trop Agric Res 28(4):490–495
Wijesinghe K, Ayoni V, Galaboda J, Perera T, Abeykoon S, Pradeepika K, Liyanage Y (2017) Market level lapses on quality of vegetable seeds in Sri Lanka. Ann Sri Lanka Dept Agric 19(1):242–250
Wijewardena J (1993) Effect of organic and chemical fertilizers on vegetable cultivation in upcountry intermediate zone. Trop Agriculturist 149:1–12
Wijewardena J (2000) Comparison of animal manure sources on potato and vegetable cultivation in UpCountry. Ann Sri Lanka Dept Agric 2:357–370
Wijewardena J, Amarasiri S (1993) Response of vegetable crops in a cropping sequence to K and Mg fertilizers in the upcountry intermediate zone. Trop Agriculturist 149:13–22
Willer H, Lernoud J (eds) (2016) The world of organic agriculture; statistics and emerging trends. Research Institute of Organic Agriculture (FiBL)/I-FOAM, Frick/Bonn
Yussefi M (2010) Organic farming worldwide 2006: overview and main statistics. In: Willer H, Yussefi M (eds) The world of organic agriculture: statistics and emerging trends. IFOAM, Bonn, pp 23–38

Input Intensification in Food Crops Production and Food Security

10

Buddhi Marambe, S. S. B. D. G. Jayawardena, W. M. W. Weerakoon, and Hemantha Wijewardena

Abstract

"Green Revolution" has paid its dividends with an astonishing period of growth of food crop productivity, especially in the developing world, over the past 60 years. Agricultural intensification has been the prime driver of increased per capita food production globally. Sri Lanka is not an exception where Green Revolution took place since late 1950s in the two agrarian structures, namely irrigated and rainfed, with the major objective of increasing the total agricultural production. Food crop production in Sri Lanka has evolved to come up with novel cultivation packages, double cropping, improved seeds, fertilizers and pesticides, mechanization, and institutional arrangements. In this scenario of input intensification, the sustainability of the food production systems has been questioned by many. Hence, transformative changes of agriculture and food systems are the current needs in all countries including Sri Lanka to achieve the sustainable development goals (SDGs), with well-articulated interventions to enhance modernization of the food production sector, increase national level food production, and ensure mechanisms to sustain household and national-level food security.

Keywords

Food production · Agricultural inputs · Modernization of food production sector · Food security

B. Marambe (✉)
Department of Crop Science, Faculty of Agriculture, University of Peradeniya, Peradeniya, Sri Lanka
e-mail: bmarambe@pdn.ac.lk

S. S. B. D. G. Jayawardena · H. Wijewardena
Sri Lanka Council for Agricultural Research Policy, Ministry of Agriculture, Colombo, Sri Lanka

W. M. W. Weerakoon
Department of Agriculture, Peradeniya, Sri Lanka

B. Marambe et al. (eds.), *Agricultural Research for Sustainable Food Systems in Sri Lanka*, https://doi.org/10.1007/978-981-15-2152-2_10

10.1 Introduction

The human population has been able to show an unprecedented growth due to its exceptional ability to interact with and exploit resources in the ecosystem over centuries. This evolution has taken place rapidly and dramatically in the recent past and by the dawn of the twenty-first century, the humans have started consuming about one-third of the total terrestrial net primary production (Imhoff et al. 2004). Plant domestication was a crucial transition in human history that has laid the foundation for origins of agriculture. Since then, the pervasive behavior of human race has substantially modified the earth's landscape throughout the history for agriculture, forestry, and other uses leading to long-lasting impacts on human demography, culture, and the environments. Agriculture has enabled a global shift in many aspects of human life, especially a dramatic increase in population levels. The agricultural land constitutes for about 38% of the total land area and accounts for 70% of all water withdrawals at global scale (https://data.worldbank.org). It is the largest land use on the planet providing livelihood for 38–40% of the world population and contributing to about 30% of the GDP in low-income countries.

Despite the claims that global population will continue to grow and plateau roughly by the middle of this century (Godfray et al. 2010), new reports state that the world population would continue to rise (Table 10.1). However, the population in Asia and south Asia is estimated to contract by 2100 compared to that of 2050. The population projection for Sri Lanka is predicted to show a dramatic reduction after 2035.

Increasing global population would lead to unprecedented demands on agriculture and natural resources (Foley et al. 2011), and make feeding of the growing population a daunting task at least in the next few decades. Higher consumption rates and greater demand for processed food, meat, dairy, and fish (Godfray et al. 2010) would add pressure to the global food system. Moreover, the enhanced competition for land, water, and energy, and many negative effects of food production on the environment (MEA 2005) such as desertification, salinization, soil erosion, and other consequences of unsustainable land management, urbanization, and other human uses (Nellemann et al. 2009), production of biofuels (IPCC 2007), and other biotic factors coupled with the threat of climate change (Marambe et al. 2015a) will continue to negatively affect the global food system.

Table 10.1 Projected population levels in the world (UN 2018)

Region	2017	2020	2025	2030	2035	2040	2050	2100
	Billions							
World	7.6	7.8	8.2	8.6	8.9	9.2	9.8	11.2
Asia	4.5	4.6	4.8	4.9	5.1	5.2	5.3	4.8
South Asia	1.9	1.9	2.0	2.1	2.2	2.3	2.4	2.2
Sri Lanka	*Millions*							
	20.8	21.1	21.3	21.5	21.5	21.4	20.8	15.0

The dominant role that agriculture plays in many developing economies including Sri Lanka and its huge untapped potential has led to major investments focusing on agricultural transformation in recent years. These were aimed at providing the best opportunity for achieving sustainable food systems, sustained economic transformation, enhanced food security, and poverty reduction. Agricultural transformation to catalyze sustainable and inclusive food systems would still be effectively adopted globally as well under Sri Lankan condition due to many persistent constraints such as low investment at farm level, inadequate linkages of farmers to agro-based industries, value chains, and markets. Transforming agriculture in developing economies requires a more holistic approach, innovative thinking and programs, infrastructure and finance.

10.2 The Green Revolution: Success and Impact

"Green Revolution," a term coined by Dr. William S. Gaud who was Director of the US Agency for International Development in 1968 (Swaminathan 1996), is used to symbolize the concerted efforts made to develop high-yielding crop varieties or to denote suitable technologies that enabled crop yields to increase rapidly in Asia and Latin America since late 1950s. This has continued with improvements of cereal varieties and management practices, helping in achieving higher productivity levels even today. The "Green Revolution" was driven by a technology revolution, comprised a package of modern inputs such as irrigation, improved seeds, fertilizers, pesticides, and mechanization, resulting in a dramatic increase in agricultural production.

The concept and the process initially addressed the requirements of rice and wheat and has been referred to the development of high-yielding varieties for a number of other major food crops important to developing countries. The high-yielding varieties were adopted by the global farming community quickly, and by 1980 about 40% of the total cereal area in Asia was planted to modern varieties (World Bank 2007), which reached to about 80% of the cropped area by 2000 (Hazell 2009).

"Green Revolution" has paid its dividends as the developing world witnessed an astonishing period of growth of food crop productivity over the past six decades. This is despite growing land scarcity and rising land values (Pingali 2012). Kush (2001) reported that it took 10,000 years for the world food grain production to reach 1 billion tons in 1960, with only another 40 years to produce 2 billion tons. In the global scenario, despite the populations had more than doubled, the production of cereal crops tripled since 1960s up to early twenty-first century, with only a 30% increase in land area cultivated (Wik et al. 2008). Increasing global population has decreased the per capita availability of arable land from 0.371 ha in 1961 to 0.194 in 2015 alarming of a major constraint faced by the countries to feed its own populations in the years to come. However, "Green Revolution" has continued to provide food for the rapidly increasing human population despite limited agricultural land availability over time. The cereal production per capita has increased from 0.29 t per person in 1962 to 0.39 t per person in 2014 (http://www.fao.org/faostat/en/#home). Further, the population of the low-income countries has doubled from 1966 to 2000, but the food production has increased by 125% during the same period highlighting the tremendous impact of the "Green Revolution."

Sri Lanka has experienced the benefits of "Green Revolution" over the past several decades since gaining independence. The country's attainment of almost complete self-sufficiency in rice production has been a major achievement over the past 60 years. Success was achieved through a combination of factors including the introduction of high-yielding rice varieties, development of irrigation capacity, floor prices, and the use of chemical fertilizers as well as introduction of fertilizer subsidy to promote the use of fertilizer. Improved varieties started replacing traditional varieties since late 1950s. By 1990s, they covered more than 99% of the rice-growing area. The average rice yields increased from about 0.65 t ha^{-1} in the mid-1940s to about 4.5 t ha^{-1} by 2010 (Table 10.2). The total rice production has increased by 14.7-fold in 2010 compared to that in 1940, while the cultivated extent has increased only by 4.8-fold. The productivity of rice (national average production) has also increased by 6.8-fold in 2010 compared that in 1940. This production increase has resulted in a significant reduction in rice imports, which is less than 1% of the current requirement.

Agricultural intensification has dramatically increased over the years, outstripping rates of agricultural expansion (Foley et al. 2011). The dramatic yield increment reported in the global and national crop production mainly owing to "Green Revolution" has been directly tagged to input intensification including seeds, fertilizer, and pesticides, which is a package deal, transforming the agricultural landscape associating it with intensive agriculture (Ramankutty et al. 2018), moving away from the low-input systems practiced traditionally. Agriculture is mainly expanding in the tropics, where an estimated 80% of new croplands are replacing forests (Gibbs et al. 2010), and has doubled the irrigated crop land over the past 50 years.

Table 10.2 Changes in population, cultivated extent, production, and productivity of paddy and import of rice during different decades since 1940s (DCS n.d., various issues)

Year	Population (millions)	Production (million t)	Cultivated extent (million ha)	Yield (t/ha)	Rice imports (% of requirement)
1940	6.0	0.26	0.25	0.65	60
1950	7.5	0.52	0.38	1.56	50
1960	9.9	0.90	0.59	1.86	40
1970	12.5	1.60	0.75	2.63	25
1980	14.7	2.12	0.85	2.94	10
1990	16.3	2.50	0.85	3.18	5
2000	18.5	2.86	0.87	3.86	<1
2010	20.0	4.25	1.10	4.5	<1
Increase over 1940	3.48-fold	14.69-fold	4.84-fold	6.76-fold	

10.3 Agriculture in Sri Lanka: From Being the Granary of the East

Rice (*Oryza sativa* L) being the crop with an enormous symbolism and emotive power (Bandara and Jayasuriya 2007), its self-sufficiency has been a slogan including being a cherished goal of successive governments of the country since gaining independence in 1948. Historically, the royal/state patronage was granted for trading and agriculture as the Feudal system was exclusively charitable to the nation and state economy (Jayasundera 2017). Irrespective of its historical veracity and challenged by the present-day historians (Siriweera 1978), Sri Lanka was supposedly known as "The Granary of the East" during the period of Great King Parakramabahu-I of the Polonnaruwa Kingdom during 1153–1186 AD, for exporting rice to some South East Asian countries and earning foreign exchange. Despite different arguments being put forward, it is believed that the zenith of the ancient hydraulic civilization was reached during King Parakramabahu's period of rule (Seneviratna 1981). However, the rivalries among dynasties, dispute over successions, centralized administration, and incapacitated administrative and political structures failed to keep pace with the expanding economy leading to a political turmoil emanating from invasions, ethnic and religious antagonism, and civil wars. Despite the holding over for a long time against these frequent interruptions, the ancient hydraulic civilization of the dry zone finally collapsed in the thirteenth century (de Silva 1981) due to sudden natural cataclysmic change of the river course of Mahaweli River, resulting in diseases and famine.

Rice was the main economic crop during the pre-colonial era of the country and the irrigated rice lands were used exclusively for wet rice cultivation. Further, the land tenure system was essentially feudal in nature with Chena and dry lands remaining under a communical mode of production (Bandarage 1983). However, with a highly unstable political environment and economy prevailed in the early in the sixteenth century (Knighton 1845), the three European colonial rulers, i.e., the Portuguese (1505–1658), the Dutch (1658–1796), and the British (1815–1948), changed the political and agricultural landscape in the country. The Portuguese, with their main interest on the trade of spices, had a marginal impact on the native production modalities. The Dutch were more interested in maximizing trade surpluses and provided state support to extend rice and coconut cultivation (Wickramasinghe and Cameron 2005). Both Portuguese and Dutch rulers extracted economic profit from the coastal belt of the country while trading cinnamon and other spices; however, they did not have access to open large-scale cash crop plantations in higher altitudes.

The British conquered Sri Lanka under one rule and gained access to the highlands having the highest economic potential to open up for plantation crops. According to Schrikker (2007), the British government decided to promote rice culture only in the peripheral districts while reserving the southwest for the production of cash crops and timber, mainly owing to limited possibilities of growth in the trade of the spices. However, later, the island was ordered by the British government to function as a granary not only to be self-sufficient but also to provide the rest of the Indian empire with rice. However, under the administration of Mr. Robert

Brownrigg (1811–1820), the granary ideal was abandoned and more focus was given to the facilitation of the production of cash crops like coffee. Coffee was the main crop until taken over by tea in 1880s since the former was devastated by the coffee rust (Roland 2007). The labor demand in the plantations with imported South Indian laborers (Abeygunawardane 2014) and the food requirements of plantations remained largely independent of the Sinhalese peasantry. The state therefore had to import rice from India and Burma to fulfill the requirement for the estate labor while maintaining low customs duties on rice (Roland 2007).

The same argument of the then Ceylon being the "Granary of the East" has surfaced during the early nineteenth century where British colonial rulers felt that rice is important not only by being a basic foodstuff for the inhabitants of the island but also an easy good for the government to tax (Schrikker 2007). The Department of Agriculture (DOA) was established in 1912 to undertake research and development in the food crops sector (Mankotte 2012). British rulers initially established the national botanic gardens, basically to identify the plant species adaptable to local conditions, and the DOA was carved out of it. Collection and selection of improved rice varieties and soil classification were the key initial studies carried out by DOA that was significant in paving the way for subsequent research in varietal development and fertilizer recommendations in the food crop sector (Pain 1986). The manpower development for the agricultural sector to conduct cutting edge research was supported by the establishment of the school of agriculture in 1916 and agriculture higher education at the then University of Ceylon in 1947/1948 at Peradeniya (presently the Faculty of Agriculture of the University of Peradeniya), both located in the central province of Sri Lanka.

The overemphasis given to the plantation sector during the colonial era was seen as a major blow to the development of subsistence agriculture in Sri Lanka (Snodgrass 1966). The rice sector of the country remained stagnant at least until 1940s without expansion in cultivated extent and the technologies used. State-sponsored agricultural development programs were started in the 1930s due to the urgency felt for increasing food production in the short run during the Second World War.

Since gaining independence in 1948, the agriculture policy of independent Ceylon was focused on two objectives, i.e., to reach food self-sufficiency, primarily in rice (Bandara and Jayasuriya 2007) and to dismantle the dual agricultural economy of plantation and peasant subsectors (Abegunawardana and Pope III 1986). Though it was an initial attempt to satisfy the political demand of a free-rice ration emanating from the welfare policies of the colonial rule (Pain 1986), these two objectives have continued to evolved further over seven decades to determine the course of the agriculture policy in Sri Lanka to date.

The "Green Revolution" has taken place in Sri Lanka since late 1950s in the two agrarian structures, namely irrigated and rainfed, with the major objective of increasing the total agricultural production. The special focus was to be self-sufficient in rice aiming at improving the living standards of the rural population. The local "Green Revolution" started even before the global community moved along to develop and adopt high-yielding crop varieties. The groundbreaking achievement of releasing the rice variety H-4 helped increasing the then average paddy (de-husked rice) yield to 2.5 t/ha from a mere 0.65 t/ha obtained from traditional rice varieties in 1940s. Development of this variety resulting from a hybridization program has paved the way to the use of present-day rice varieties during the independent Ceylon and then Sri Lanka. The achievement was made by Sri Lankan scientists headed by the well-known rice breeder Dr. Hector Weeraratne, which was accomplished even before the International Rice Research Institute (IRRI; established in 1960) of the Philippines released the so-called miracle seed (IR-8) rice variety for cultivation. The IR-8, though later found to be a failure, was an effort to give high yield potential, photo-insensitivity, and many other desirable characteristics of Japonica rice of Japan and Taiwan, to seeds of Indica varieties from tropical Asia (Farmer 1979).

Overall, Sri Lankan experience of "Green Revolution" can be categorized into two distinct phases. The phase 1 (1958–1969) was evident with the improvement of traditional varieties through hybridization and selection, relatively tall but retained traditional plant architecture (less lodging compared to pure-line selections), resistant to blast disease, moderately responsive to added fertilizer, and with significant improvement of yield compared to pure lines. Double cropping of rice was made possible with the introduction of H series (e.g., H_4, H_7, and H_{10}). The phase 2 was since 1970 where the semi-dwarf gene from IRRI varieties was used in conventional breeding (e.g., IR 262, IR 8) developing semi-dwarf plant architecture to minimize lodging, with high fertilizer responsiveness, development of varieties with different maturity periods (i.e., 2.5, 3, 3½, 4–4.5, and 5–6 months), high tillering capacity and with a higher harvest index (grain:straw ratio). Intensification of agriculture extension in parallel to the technology was one of the significant contributory factors to increase in paddy yields since early 1970s.

10.4 Input Intensification in Agriculture

Innovation is a key driver of productivity growth. In agriculture, innovation often simply takes the form of the utilization of modern inputs and farming practices such as hybrid seeds, mineral fertilizer, crop protection chemicals, and integrated soil and water management practices to address a wide range of production-limiting constraints (Feder et al. 1985; Byerlee 1996). Farmers in Asian and Latin American countries have adopted these technologies during the era of "Green Revolution" and have experienced rapid increases in crop yields over a short period (Pingali 2012).

The global food system has become more dependent on trade, and it has lost resilience and grown more susceptible to shocks and crises (Suweis et al. 2015). Provided that the diets are moderated and crop-based biofuel production is limited (Davis et al. 2014), the conventional technological developments (i.e., agrochemicals and irrigation) can bridge the yield gaps and thus potentially meeting the future global food demand. Further, long-term demographic changes and short-term fluctuations in political and environmental conditions affect the trade dynamics in food distribution around the globe. The global food price crisis in 2007/2008 showed that the countries heavily depended on food imports are at a distinct disadvantage in terms of food security than those who rely on local crop production to meet their dietary demands (Davis et al. 2014). This is important, as Davis et al. (2014) further stated that 76% of countries are not self-sufficient in terms of domestic calorie production.

Unprecedented agricultural land expansions since 1700, and technological innovations that began in the 1930s in the global scale, have enabled more calorie production per capita (Ramankutty et al. 2018). Since 1950s, over a period of a decade, the global land extent devoted to arable agriculture has increased only by ~9% (Pretty 2008), while the grain production has more than doubled. The per capita agricultural production has outpaced population growth as there is 25% more food being produced for each person in the world compared to that of 1960 (Pretty and Barucha 2014). Alexandratos and Bruinsma (2012) reported that, of the total production over the period 1961 to 2005, 77% came from increased yields, 14% from expansion of croplands, and 9% from increase in cropping intensity. Further, Alexandratos and Bruinsma (2012) also predicted that in 2050, 80% of production growth will come from yield growth, and 10% each from cropland expansion and increases in cropping intensity, suggesting that the contribution of cropland expansion to production growth would reduce by approximately 4% in the future.

Over the past six decades, agricultural intensification has been the prime driver of increased per capita food production globally (Pretty 2013). In such efforts improved seeds and agrochemicals (mineral fertilizer and pesticides) have contributed significantly in enhancing crop yields, especially that of cereals. However, the full impact of those agricultural inputs was only realized after the development of high-yielding varieties that were more responsive to plant nutrients, having shorter plant height and stiffer straw to avoid lodging due to the weight of heavier grain heads. Breeding for tropical rice varieties that could mature early and grow at any time of the year providing more crops per year on the same land, and those that are resistant to major insect pests and diseases that would exacerbate under intensive farming conditions have been a priority, while maintaining the desirable cooking and consumption traits. Far reaching changes of the "Green Revolution" has been estimated in the use of modem seeds and chemical fertilizers. The "Green Revolution" agriculture has also relied on the use of pesticides, which are necessary to limit the high levels of pest damages (including weeds, insects, and pathogens) that inevitably would occur with the use of high-yielding varieties and growing a single crop over a large area. The "Green Revolution" in Sri Lanka has evolved with agricultural packages, double cropping, and mechanization and institutional arrangements. The agricultural package consisted of several main elements such as improved and certified seeds, fertilizers, pesticides, and other labor-intensive methods of crop cultivation (Fladby 1983). The inelastic supply of cultivable land being the main constraint on agricultural development in Sri Lanka, the seed-fertilizer technology should increase the productivity of the subsistence sector, thus leading to the high-pay off, input- and management-intensive (Schultz 1968) model of agriculture in Sri Lanka.

The following section intends to present historical trends in the use of agricultural inputs in food crops focusing on seeds, fertilizer, and pesticides, at global and national scenarios. The chapter is not intended to discuss the benefits and negatives of use of such inputs and to compare production and productivity growth with low-input agricultural systems, but to highlight the future of input intensification in modern-day agriculture.

10.5 Seed Sector in Agriculture

10.5.1 Development History, Global Trends, and Trade

Seeds being the first link in the food chain play a significant role in the sustainability of the agri-food system. They are at the very beginning of the food chain, constitute its base and foundation and contributed to agricultural biodiversity (Bonny 2017), and thus is central to rural development and food security. The first evidence of plant domestication reported in the hills over the Tigris river dates back to 9000 BC with hand pollination of date palms by the Assyrians and Babylonians reported in 700 BC. The production of hybrid seeds was initiated in 1920s and became popular with a yield increase in the nineteenth century (Kingsbury 2009; IFPRI 2002). The 1970s saw that genetic engineering making the move of individual genes from one organism to create plant varieties with specific desired qualities (Paarlberg 2010). Crop production has increased by the use of quality seeds of high-yielding varieties and judicious use of fertilizers and water (Singh et al. 2013). With all other factors remaining the same, the use of quality seed of high-yielding varieties could increase crop yield by 15–20% (Agrawal 2011).

Over the years, the market forces have determined the road to agricultural development with the seed industry playing an important role in increasing food production. Commercial seed industry started around the 1740s with the earliest known seed company, "Vilmorin," in France. Since then, the evolution of plant breeding science was an impact of the increasing number of specialized breeding companies established over the past century.

The global seed market increased from US$ 12 billion in 1975 to around US$ 20 billion in 1985, US$ 36.5 billion in 2007 (Bruins 2009) and more than US$ 45 billion in 2011 (ISF 2012). It reached US$ 58 billion in 2016, exhibiting a compound annual growth rate (CAGR) of around 7% during the period 2009–2016, and is expected to grow at a CAGR of more than 6% during 2017–2022 reaching a value of more than US$ 86 billion by 2022 (Anon 2018). The value of genetically modified (GM) seed trade increased more than 25-fold between 1996 and 2007, from US$ 0.28 billion to US$ 7.3 billion (CropLife 2008). Since its commercialization in 1996 with 1.7 million ha of cultivation, GM seed sales grew in its proportion from 9% of commercial seed sales in 2001 to 21% in 2007 and to almost one-third in 2013 (Bonny 2014), but representing only 11% of the cultivated extent. In 2016, the twenty first year of commercialization of GM (biotech) crops, 185.1 million ha of biotech crops were planted by approximately 18 million farmers in 26 countries (ISAAA 2016). In 2016, about 185.1 million ha of biotech crops were planted showing an approximate 110-fold increase in cultivated extent, thus making them the fastest adopted crop technology in the history of modern agriculture and projecting higher market potential for the GM seeds.

Patenting seeds and plant genetic resources have led to their commoditization and commercialization (Sewón 2013). This has made seed production and agricultural production more industrially oriented. The global seed market as at present is highly concentrated with a few international players led by Monsanto, DuPont, and Syngenta (Fig. 10.1) as of 2015.

The past few decades have shown major structural changes in the global seed industry with efforts made for consolidation of seed companies (Lianos et al. 2016; Moss 2013). In 2017, Monsanto merged with Bayer Crop Science, DuPont Pioneer merged with Dow AgroScience, and Syngenta was acquired by ChemChina. The main reasons driving these mergers and acquisitions include the need to secure and expand into new markets and diversify the company portfolios following the decline in prices of crops and cereals. The approval of the above mergers will result in three corporations controlling close to 60% of the global patented seed market.

The seed market has led to a significant shift in farming practices worldwide. Increasing number of farmers are now demanding and purchasing commercially produced high-yielding crop seeds instead of using self-produced seeds. A rapid erosion of seed diversity and increase in the concentration of giant enterprises that control the seed industry were evident over the past few years. UN (2018) reported that 75% of all agricultural biodiversity had disappeared emanating from the introduction of modern varieties that were grown as monocultures. The use of genetic engineering in plant breeding has led to great controversy with respect to the benefits and risks of GMOs (Weasel 2008).

Public sector involvement in seed production has shown a declining trend in many developing countries. Many countries, who have been concentrating on crops

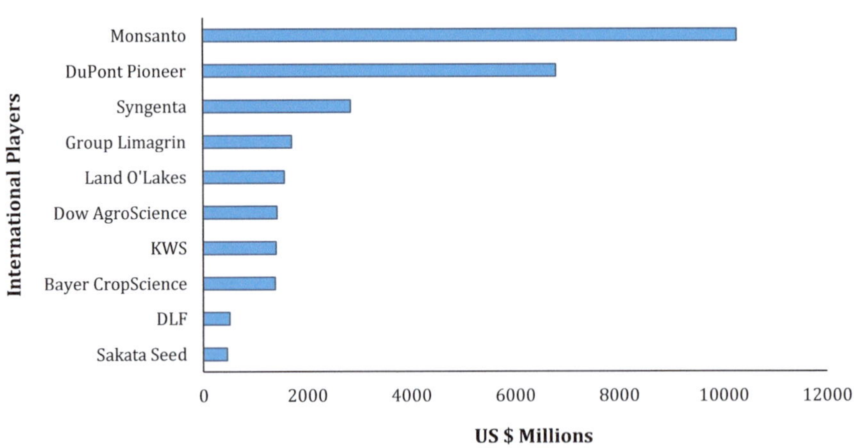

Fig. 10.1 Sales of global top seed companies in 2015 (US$ million) (Zhang 2018; http://news.agropages.com/News/NewsDetail%2D%2D-21186.htm)

for national food security, have either sold the seed production infrastructure to a foreign party (e.g., Malawi) or to private companies (e.g., Uganda), encouraged private seed companies to emerge and compete with the public enterprise (e.g., India and Sri Lanka), and decentralized public enterprises (e.g., Ethiopia). The private seed producers, however, mainly focus on seed of high-value crops such as hybrid cereals and vegetables and industrial crops (Louwaars and de Boef 2011).

10.5.2 Crop Seed Sector in Sri Lanka

10.5.2.1 Transition from Traditional Varieties to Hybrids

Historically, rice economy has played a central role in the overall economic development in Sri Lanka. Archeological excavations have proven that rice cultivation has occurred in the country during 900–600 BC, which was modernized according to the traditional knowledge and farming systems practiced in India, with the establishment of "*Arya*" community Sri Lanka (Withanawasam 2017). Nox (1681) reported of Mauve (*Maa wee*) maturing in 7 months, Hauthiel (*Hathiel*) in 6 months, Honorowal (*Honderawalu*) in 5 months, Heni (*Heenati*) in 4 months, and Aulfankol (*Alpankal*) in 3 months, all growing in puddled muddy paddy fields and *Goda wee* has been cultivated as upland chena. Around 300 traditional rice varieties were being displayed in Kandy Agro Horticultural and industrial exhibition held in 1902 (Senadhira et al. 1980). Although there is no record to state who supplied required seed paddy during the colonial rule, farmers producing their own planting material is the best guess. Further development of traditional rice varieties seems to have not taken place during the colonial era (Withanawasam 2017).

Seed selection of *Hathial* during the early twentieth century could probably be the initiation of rice varietal improvement in the Department of Agriculture (DOA) of the then Ceylon (Withanawasam 2017), followed by pure-line selection in 1920/30s. The first pure-line selection "*Murungakayan*" was later replaced by the old improved variety (OIV) H4 (red pericarp), which was the result of a hybridization program between paddy varieties "*Murungakayan* 302" and "MAS" of Indonesian origin in 1958 during the pre-green revolution era; promotion of the use of chemical fertilizers and machinery for land preparation, and the concept of self-sufficiency in rice resulted in increase in the average paddy yields obtained from traditional varieties (0.65 t ha^{-1}) cultivated in 1940s to 1.73 t ha^{-1}. The development of new high-yielding varieties (HYVs) was commenced by the Department of Agriculture in 1960s to increase the rice productivity further. Five varieties were released in 1971 to replace the traditional types under cultivation in three major maturity classes (Weeraratne 1972). The most significant contribution from rice breeders during early 1970s was the transferring of resistant genes for Blast from TE-TEP, Gall midge from OB678, and Brown Plant Hopper from Ptb 33. Before these genes were incorporated, substantial quantities of pesticides were applied even by using helicopters to control blast disease in Ampara of the Eastern Province of Sri Lanka. The other contributing factor to stabilize rice yields was the development of varieties in the age class of 3½ months by genes obtained from IR262.

Most of these traditional varieties such as *Pachchaperumal*, *Addakari*, *Suwandel*, *Kalu Heeneti*, *Moddaikaruppan*, *Kuruluthuda*, *Murunkan*, and *Periya Velladiyan*, derived from pure-line selection, are still being cultivated by farmers (RRDI 2018). The recent statistics of RRDI (2018) indicate that since 2010, the cultivated extent of traditional varieties in Sri Lanka is only less than 0.83% from total cultivated area of rice of a given year. Area planted to HYVs in the country in 1970/71 was 10.2% (73,600 ha) which increased to 87% (749,700 ha) in 1982–83 of the cultivated extent. Currently about 99% of paddy extent is cultivated to HYVs replacing the traditional varieties and OIVs. The research and development (R&D) program on hybrid rice began in late 1994 and was continued in collaboration with the International Rice Research Institute (IRRI) since 1996 (Abeysekara et al. 2002) resulting in the recommendation of Bg407H with potential yield of 14 t ha^{-1}. Sri Lanka has released rice varieties with various production potentials, resistant to pests and diseases, and tolerant to environmental stresses such as drought (e.g., Bg251). All varieties recommended and released are resistant to blast disease. The achievement of paddy yields at the field level, however, has still been far below from the potential of the HYVs released by the Department of Agriculture (Marambe 2018).

Maize is primarily a rainfed crop in Sri Lanka and is cultivated in the *Maha* season (the main rainy seasons from September to February) in both settle and shifting (Chena) types of highland cultivation. The crop is mainly used as animal feed rather than being a staple food. Sri Lankan researchers have been successful in developing new open-pollinated maize starting from "*Bhadra*" in 1977 to "MI Maize H 02" in 2016 (Table 10.3). The imported hybrids "*Jet*," "*Pacific*," and "*Kafeer*" by the private sector have become popular among farmers. This is mainly due to their higher yields and the buyback of harvest by the private sector. At present, 95% of the maize cultivated extent is covered with imported hybrids. The high rate of adoption of hybrids by farmers increased the yield from 1 t ha^{-1} of OPVs to 3.6 t ha^{-1}. A new hybrid maize variety "MI Maize H1" was released in 2013 yielding 13% more than that of "*Sampath*" (Kumari et al. 2013), the first hybrid maize released in Sri Lanka.

Table 10.3 Maize varieties released by the Department of Agriculture since 1977 (Source: FCRDI 2018)

Variety	Year of release	Yield (mt/ha)
Bhadra	1977	4
Ruwan	1990	4
Muthu	1992	4
Aruna	1992	4
Sampath (hybrid)	2004	4
MI maize H 01 (hybrid)	2013	5.5–6.5
MI maize H 02[a] (hybrid)	2016	5.5–6.5

[a]Moderately tolerant to drought

10.5.3 State and Private Sector Interventions

The seed production and distribution in Sri Lanka was initiated in an organized manner in the late 1950s. Since then, the seed industry matured under the patronage of the state sector and DOA was the sole supplier of seed up to 1984. The national seed certification service (SCS), which was established in 1976, came into operation as a separate unit of the DOA in 1979 and ensured that the quality seeds are distributed among the Sri Lanka farming community. However, with increasing demand for quality seed and planting material, the government of Sri Lanka facilitated the entry of private sector to seed production and supply.

The seed importation was liberalized in 1984 where private sector started importing seeds to fulfill the national requirement. In 1989, the Seedmen's Association was established, and several private sector companies were appointed as leaders to distribute seeds produced by the state organizations. Further, the private sector initiated seed production locally in 1990 where basic seeds were provided to them by the government farms, and the Seed Producers' Association was established in 1996 to strengthen the institutional arrangement in seed production and distribution by the private sector. Seed industry in the country has now become a multi-institutional with both the public and private sector taking part actively. The National Seed Policy (NSP) was approved in 1996 with a view to establish a viable seed industry in Sri Lanka, and the Seed Act was enacted in 2003, which came into effect in 2008. More than 2000 seed handlers have registered under the act and a National Seed Council (NSC) has also been established. The state departments are mandated for varietal improvement and development, while the government of Sri Lanka encourages the private seed companies to develop new varieties.

The DOA produces breeder and foundation seeds, which are multiplied by the contract growers and the private sector. The national seed paddy production program is currently focused on 28 recommended varieties providing approximately 17% of the national requirement. However, basic seeds of 53 varieties of 18 vegetables recommended by DOA are produced in seven government seed farms and meets only 3.8% of total vegetable seed requirement of the country. Seed production and supply of OFCs, except chili and maize, is handled by the public sector.

The certified seed production of paddy, other field crops, and vegetables are illustrated in Figs. 10.2, 10.3, and 10.4. In 2017, the government farms (the farms of the DOA) with its contract grower program produced 38% of certified seed paddy, while the private growers supplied 62% of the requirement (Fig. 10.2). In 2017, the government farms and the contract grower program together produced seed paddy at 18% more than that of 2016 and 373% more than that of 2013. The quantity of seed paddy produced by the private sector reached the maximum in 2013 and declined thereafter (Fig. 10.2). A marked variation in seed paddy production by the private sector was observed over the years due to their investment capacity, some organizations moving out of seed paddy production, and vagaries and uncertainty of the climate. However, the private sector still remained the dominant producer of seed paddy in the country.

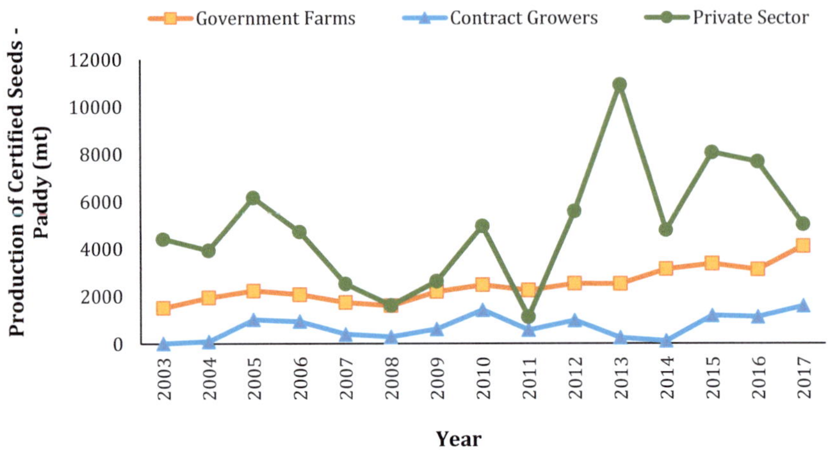

Fig. 10.2 Production of certified seed paddy in Sri Lanka, 2003–2017 (SCPPC (n.d.)—various issues)

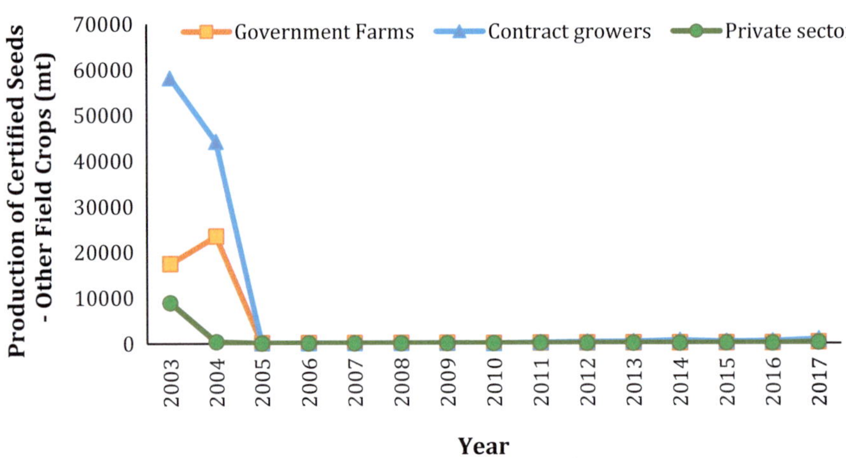

Fig. 10.3 Production of certified seeds of other field crops in Sri Lanka, 2003–2017 (SCPPC (n.d.)—various issues)

Maize, green gram, ground nut, sesame, cowpea, finger millet, horse gram, mustard, black gram, and soybean were included in the OFC seed production program. The registered extent of cultivation for OFC seed production increased by 174% in 2017 compared to that in 2016, as the regions with water scarcity showed a tendency to cultivate OFCs than paddy. The total OFC seed production in 2017 was 1573 mt, which is a 164% increment compared to that in 2016. The seed production program for OFCs was affected after 2003 (Fig. 10.3) mainly due to the importation and demand for hybrid seeds (Fig. 10.5), especially maize, giving higher yields. Yet, more than 85% of the certified seeds of OFC produced in Sri Lanka are provided by the DOA farms and the contract grower system.

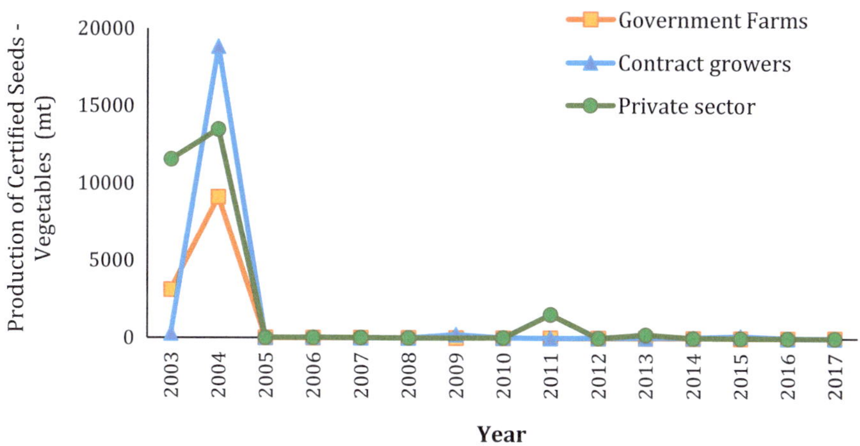

Fig. 10.4 Production of certified seeds of vegetable crops in Sri Lanka, 2003–2017 (SCPPC (n.d.)—various issues)

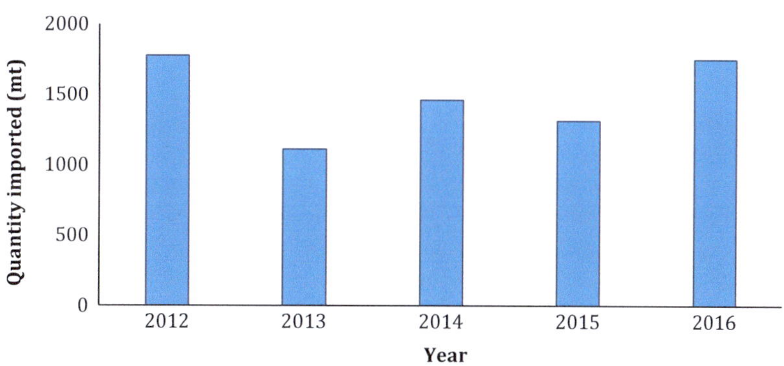

Fig. 10.5 Quantity of hybrid maize seeds imported to Sri Lanka, 2012–2016 (SCPPC (n.d.)—various issues)

The registered extent for vegetable seed production in Sri Lanka in 2017 was 156 ha of which government and contract growers account for 66%. Total extent allocated for vegetable seed production showed a 77% increment in 2017 when compared to that in 2016. Tomato, okra, chili, red onion, big onion, snake gourd, bean, brinjal (eggplant), bitter gourd, radish, luffa, capsicum, yard long bean, cucumber, pumpkin, amaranthus, vegetable cowpea, winged bean, water melon, and papaya (fruit seeds) were included in the seed production program of the country. During 2017, total vegetable seed production was 48 mt, of which 37 mt were certified by DOA (Fig. 10.4), which is a 67% increase compared to 2016.

The seed production programs of the vegetable crops has faced the same fate as that of OFCs, with the increased importation of hybrids by the private sector (Figs. 10.5 and 10.6). The recent statistics indicate that of the total certified vegetable seeds produced locally, where the DOA farms and the contract grower system provide around 88–90%.

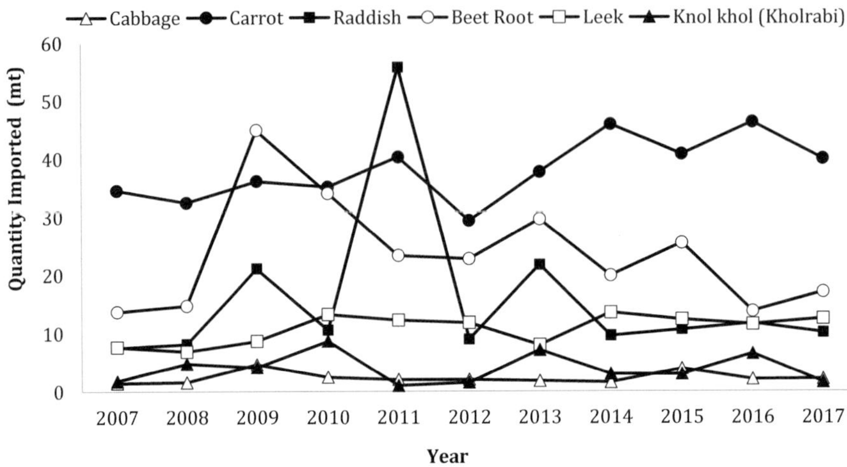

Fig. 10.6 Import of exotic vegetable seeds to Sri Lanka for cultivation, 2007–2017 (SCPPC (n.d.)-various issues)

10.5.4 Seed Imports to Sri Lanka

In Sri Lanka, the procedure for the import of new vegetable seed varieties is governed by the "*Guidelines for The Testing of Vegetable Varieties Imported by The Private Seed Companies (2015)*" published by the Horticultural Crops Research and Development Institute (HORDI) of the Department of Agriculture. Genetically modified organisms (GMOs) are not allowed to be imported to Sri Lanka. The total requirement of exotic varieties of vegetable seeds, including hybrids, is supplied by the private sector through importation. In 2012, the total vegetable seed imports were 3398 mt, and local vegetable seed production was around 43 mt. Over 2300 permits have been issued for seed imports in 2012, of which 30% were for vegetable seeds, and plant and planting materials imports. Sri Lanka annually imports potato, red and big onion, and vegetable (especially upcountry vegetable) seeds worth Sri Lanka Rs. (SLR) 400 million from the USA, India, Netherlands, Thailand, Pakistan, Germany, and Japan.

During the past 15 years, imported maize hybrids developed by internationally renowned companies have been popular among the maize farmers in Sri Lanka, where about 20 maize hybrids were imported for commercial cultivation in 2016 (Fig. 5; Kumari et al. 2017). All the exotic hybrids have been evaluated for adaptability and susceptibility for common pest and diseases before recommendation (Kumari et al. 2015), and the same is practiced for vegetable seeds. Currently, about 95% of the farmers cultivate imported maize hybrids. In 2016, the hybrid seed requirement was around 1000 tons to the value of about Sri Lanka Rupees 1000 million. Seed production of open-pollinated maize varieties (i.e., Ruwan and *Bhadra*-1) are carried out by government seed farms under the DOA and also with contract seed-producing farmers.

A program is now under way to meet 10% of the hybrid seed requirement of "MI Maize H 01" released in 2013 in collaboration with contract farmers, private seed companies, and government institutes (Kumari et al. 2017). The hybrid seed-producing farmers have been encouraged by the provision of parental seeds free of charge and buyback of the F1 seeds at a fixed price. Furthermore, hybrid seed production technology was promoted by field demonstrations and visits, and follow-up is done by extension staff.

Carrot dominated among the exotic vegetable seeds imported to Sri Lanka, mainly to be cultivated in the up-country region, with a stable demand over the years (Fig. 10.6), while pumpkin, okra, and watermelon were the seeds of low country vegetables that were imported during the past decade (Fig. 10.7).

10.5.5 Opportunities and Challenges

Seed production and supply system in Sri Lanka has usually originated from the traditional pattern that has long existed. Today, seed-related issues have become increasingly more complex in the recent past due to more stakeholder involvement (e.g., officials of the state and private sector, scientists, professionals, managers, and farmers). Increasing population growth at a modest rate, anticipated larger holding sizes in agriculture, stronger concerns about the environment, increased use of high-yielding varieties/hybrids, and trained human resources will provide opportunities for the seed industry to prosper. Erosion of plant genetic resources, inadequate seed stock for food security, inconsistent national policies, inadequate numbers of extension experts, imperfect markets and deficient marketing policies, and limited

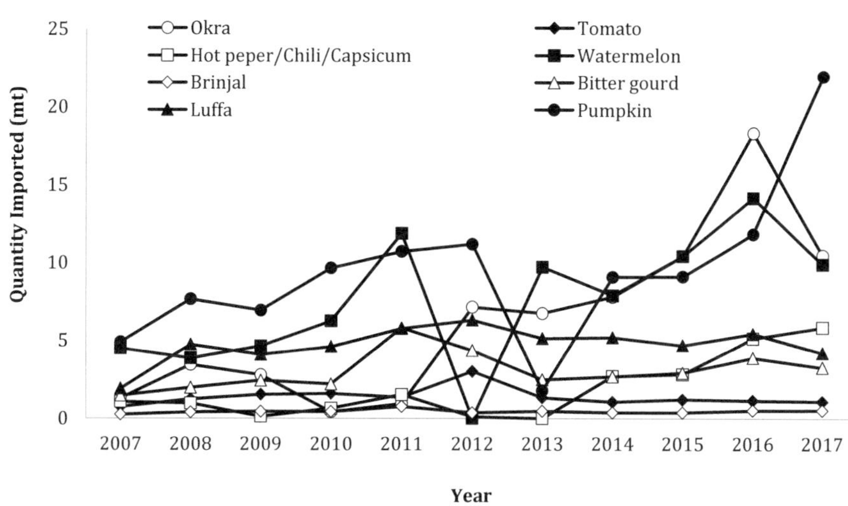

Fig. 10.7 Import of low country vegetable seeds to Sri Lanka for cultivation, 2007–2017 (SCPPC (n.d.)—various issues)

involvement of internationally renowned seed companies are the constraints and challenges that should be tackled soon. The major factors driving the market of *Sri Lankan seed industry* are the government initiatives to develop the seed sector and the growth in demand for vegetables. The factors restraining the growth of the market include the lack of knowledge of quality seeds and ban of genetically modified crops in Sri Lanka. The Ministry of Agriculture appointed a Seed Task Force (STF) to develop strategies to enhance the seed sector in Sri Lanka. The government aimed to develop the seed sector to support development of the agriculture sector, by promoting government and private sector involvement in seed production, implementation of national seed production programs, and initiation of hybrid seed production.

Trends toward consolidation in the seed industry has become increasingly evident in the recent past, as explained elsewhere in this chapter. Mergers may result in efficiency gains; however, higher concentration may also lead to exiting of merged entities from market power and raising prices, along with potential effects in terms of reducing competition in terms of innovation. This may also create a conducive environment providing more opportunities for the seed industry in the future. As stated by Ramankutty et al. (2018), modern innovations to breed climate-smart seeds, improve photosynthetic efficiency and nitrogen fixation, and increase nutritional content and disease resistance could avert yield stagnation, adapt to changes in the growing season and extreme weather events, close the micronutrient gap, and decrease food waste.

10.6 Fertilizer Use in Agriculture

10.6.1 Development History, Global Trends, and Trade

Fertilizer is a critical component of anthropogenic nutrient input in the earth system (Lu and Tian 2013). Being one of the important land management practices and an integral part of "Green Revolution," fertilizer application has contributed to a dramatic increase in production, thus helping to overcome nutrient deficiencies and increase crop yield and soil fertility substantially over the past five decades in a global scale (Erisman et al. 2008).

The first boom in industrial scale fertilizer production for global consumption began with the large-scale mining of guano (sea bird excrement), largely from islands off Peru. Potassium mining in Germany began in the 1860s, coinciding with the manufacture of the phosphate fertilizers from phosphate ores. Industrial fertilizer production showed an exponential growth in the twentieth century with the discovery of the Haber-Bosch process (Galloway et al. 2004) for fixing nitrogen from the atmosphere, which significantly increased the availability of nitrogen fertilizers, which in turn increased the demand for complementary fertilizers with other nutrients. For example, the average use of plant nutrients in 1970 (23.9 kg ha^{-1} of agricultural land) grew rapidly to reach 102.0 kg ha^{-1} by 1995 (Gollin et al. 2005), while the global fertilizer use increased by 500% (over 800% for nitrogen alone) (FAOSTAT (n.d.)—various issues). Excessive use of fertilizer, however, has resulted in many environmental and ecological problems such as air pollution, eutrophication, soil acidification, soil degradation, and crop yield reduction, thus threatening the sustainability of agricultural production in the long run (Lu and Tian 2013).

China has been the fastest growing fertilizer market in past decades, accounting for about 30% of the global fertilizer use. The average application rate of fertilizer for arable land and perennial crops is among the highest in China (approximately 428 kg nutrients ha^{-1}) where the world average stands at 115 kg nutrients ha^{-1}. The fertilizer market has faced an increase in capacity in 2017 with new manufacturing plants in operation, while demand has only moderately expanded in 2017. Uneven global nutrient demand, soft economic prospects, low crop prices, and volatile energy prices were preponderant in 2016, and it is possible that this uncertainty in fertilizer markets could prevail for the next 3–5 years. The world demand for fertilizer-nutrient use as projected by FAO (2017) is shown in Table 10.4. The N, P, and K fertilizer demand in south Asian region for a 5-year period is shown in Table 10.5.

The global value of fertilizer sales around the world in 2014 has been estimated at US$ 172 billion, with nitrogen accounting for more than 50% (IFA 2018). Over the next 5 years (2018–2022), the world fertilizer demand will increase by 1.5% per annum to reach 199 million tons of nutrients, of which 112 million tons is nitrogen, 48 million tons is phosphates, and 38 million tons is potash (Anonymous 2017). The FAO (2017) has also forecasted that the demand for N, P_2O_5, and K_2O to grow annually on an average by 1.5%, 2.25%, and 2.4%, respectively, from 2015 to 2020. Latin America, South Asia and East Asia would account for almost three-fourths of the anticipated increase in fertilizer demand in the next 5 years. In relative terms, Africa, Eastern Europe, and Central Asia are seen as the fastest growing markets. In contrast, little growth, if any, is expected in Western Europe, North America, Northeast Asia, and China.

Nitrogen is likely to remain the most important fertilizer nutrient globally; however, potassium and phosphates have been increasing their market share marginally over the last decade. Demand is expected to expand faster for potash (2.1% annually) than phosphates (1.5% annually) while nitrogen remaining at 1.2% per annum. This reflects the improvement in nitrogen-use efficiency and the need to rebalance fertilization in many regions. Further, a combination of factors such as population growth, changing diets, and changing agricultural practices that are demanding higher agricultural yields per hectare of arable land could have been the main drivers of demand growth over the years.

Table 10.4 Global demand for fertilizer-nutrient use (2015–2020) in '000 mt (FAO 2017)

	Year					
Fertilizer-nutrient	2015	2016	2017[a]	2018[a]	2019[a]	2020[a]
Nitrogen (N)	110,027	111,575	113,607	115,376	117,116	118,763
Phosphorous (P_2O_5)	41,151	41,945	44,120	44,120	45,013	45,858
Potash (K_2O)	32,838	33,149	34,894	34,894	35,978	37,042
Total	184,017	186,668	194,390	194,390	198,107	201,663

[a]Forecasts

Table 10.5 Demand for different fertilizers in south Asia in '000 mt (FAO 2017)

	Year					
Fertilizer	2015	2016	2017[a]	2018[a]	2019[a]	2020[a]
N fertilizer demand	22,273	22,525	23,430	24,002	24,645	25,191
P fertilizer demand	8165	8435	9025	9383	9760	10,107
K fertilizer demand	2958	2991	3226	3407	3612	3812

[a]Forecasts

10.6.2 Fertilizer Use and Imports in Sri Lanka

The National Fertilizer Secretariat (NFS) established in 1979 and functions under the Fertilizer Act No 68 of 1988 is responsible to issue and renew license to import, manufacture, or formulate fertilizer. Currently, Sri Lanka imports chemical fertilizer to meet the requirement, except for Eppawala Rock Phosphate (ERP) and some amounts of Dolomite. A significant increase in the use of chemical fertilizers for crops in Sri Lanka was evident since early 1950s (Wijewardena 2005); however, the widespread use of chemical fertilizer for food crops, especially for paddy, commenced in the year 1960 (Figs. 10.8 and 10.9).

The subsidy scheme focuses on to provide the much needed plant nutrients to the crop and to reap richer harvests from food crops (Ekanayake 2006; Bandara and Jayasuriya 2007; Weerahewa et al. 2010). Prior to 1962, rock phosphate or bone meal and sulfate of ammonia or green manure were the main sources of phosphorus and nitrogen, respectively, used in rice cultivation. Urea became the main source of N from 1967 to 1978, while triple superphosphate (TSP) replaced rock phosphate. Use of ammonium sulfate was recommenced during the period 1990–1994 due to transport restrictions imposed on urea to north and east owing to security reasons. The world price shock and removal of fertilizer subsidy had an impact on fertilizer use in rice in 1996 (Fig. 10.9). Fertilizer subsidy played a major role in promoting mineral fertilizer among farmers, leading to intensification of the use of plant nutrients in agriculture securing the food for the nation.

The paddy fertilizer market grew rapidly (Figs. 10.8 and 10.9) mainly due to the fertilizer subsidy, guaranteed price scheme for paddy, expansion of cultivation of high fertilizer responsive and high-yielding rice varieties (Ekanayake 2006) and enhanced area under irrigation. However, the decline in fertilizer application for paddy observed from 2012 to 2014 could be due to the shift in land use in paddy fields, especially in the wet zone, where majority of the land were either abandoned or made into other nonagricultural uses such as housing construction. Further, there was a reduction in the quantity of subsidized mineral fertilizer, especially urea, with the promotion of organic manures, composts, etc. Further, the new rice fertilizer recommendations for paddy introduced in 2013 did remove the "targeted-yield" approach under which high fertilizer rates were recommended for high-yielding areas. All these would have contributed to the reduction in fertilizer use during 2012–2014 (Dr. Priyantha Weerasinghe—personal communication). The changes in world market price and subsidy policy in Sri Lanka has only made short-term impact on fertilizer use (Figs. 10.8 and 10.9).

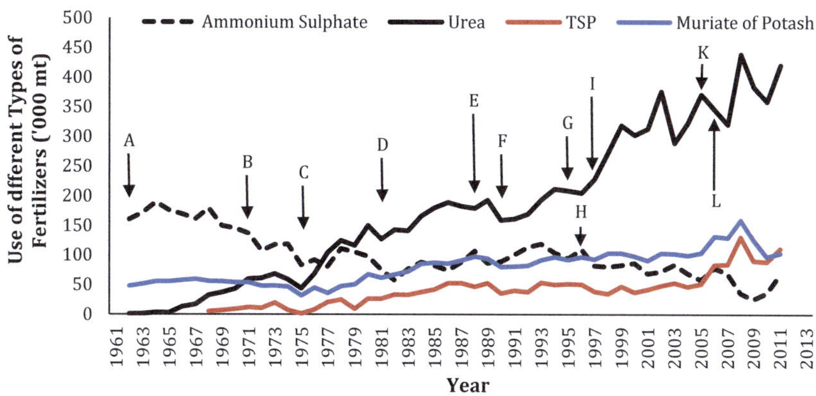

Fig. 10.8 Different types of fertilizer used in Sri Lanka for agricultural purposes 1962–2011 (FAOSTAT (n.d.), various issues). A = Fertilizer subsidy for urea, TSP and MOP introduced for paddy in 1962; B = Banning imposed on private sector to import fertilizer and Ceylon Fertilizer Corporation was given the full responsibility of import and distribute fertilizer in 1971; C = Fertilizer subsidy was introduced for all crops in 1975; D = Subsidy rates were reduced by 3% due to increase in world market prices in 1981; E = Subsidy rates were reduced by 30% due to increase in world market prices in 1988; F = Fertilizer subsidy was completely removed, DOA recommended straight fertilizers for food crops, and ammonium sulfate was recommended due to transport restriction on urea imposed due to security reasons in 1990; G = Fertilizer subsidy was reinstated in 1995; H = The subsidy for ammonium sulfate was removed in 1996; I = Fertilizer subsidy was limited only to urea in 1997; K = Fertilizer subsidy was limited only for N, P, and K straight fertilizers in paddy in 2005; L = Tea, rubber, and coconut smallholder farmers with <2 ha of land became eligible for the fertilizer subsidy in 2006

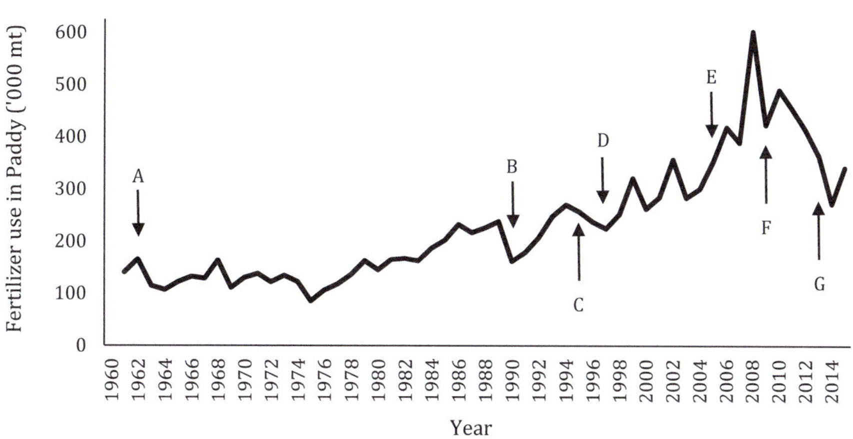

Fig. 10.9 Total fertilizer used in Paddy cultivation in Sri Lanka 1961–2015 (NFS, various issues). A = introduction of fertilizer subsidy for urea, TSP, and MOP in 1962; B = fertilizer subsidy was completely removed in 1990; C = fertilizer subsidy was reinstated for urea, TSP, and MOP in 1995; D = fertilizer subsidy was limited only for urea in 1997; E = fertilizer subsidy was reintroduced for urea, TSP, and MOP in 2005 through the "*Kethata Aruna*" Program until 2015; F = subsidy was coupled with paddy procurement policy in 2009; G = introduction of new fertilizer recommendation for paddy moving away from separate recommendation for high-yielding target areas

In order to promote balanced use of plant nutrients, fertilizer mixtures containing N, P, and K nutrients had been recommended during early stages. The DOA changed its policy in 1990 by recommending straight fertilizers instead of fertilizer mixtures for food crops in order to provide an opportunity for farmers to make necessary adjustments in fertilizer application for optimum yield. Use of straight fertilizer was also promoted to ensure that farmers receive quality fertilizer inputs as mixtures contained many unwanted fillers. However, the use of fertilizer mixtures containing N, P, and K is more popular despite being costly than the straight fertilizers. This may be due to convenience in fertilizer use and lack of available straight fertilizers in the local market.

Most vegetable and potato farmers use high quantities of chemical as well as organic fertilizers. The level of fertilizer applied by the farmers to the vegetables crops is almost two to three times the quantity recommended by the DOA (Wijewardena 2001). This may be due to favorable crop price relationship for vegetable crops. As a result, net returns from vegetable cultivation are much higher than in rice and OFC, making chemical fertilizer a relatively inexpensive input. Various types of liquid fertilizers are also used regularly as foliar fertilizers. In general, fertilizer use for OFC crops is negligible even at present. However, higher rates of fertilizers are used for cash crops such as onion and chili, and even on maize. Excess use of fertilizers has built up high soil P levels in the up-country, which can be environmentally alarming.

The fertilizer subsidy scheme has been the heaviest subsidy in the agriculture sector and a heavy burden on government treasury. This has accounted for 2–2.5% of total government expenditure for the period 2005–2014 (Weerahewa et al. 2010; Rodirigo and Abeysekera 2015), with increase in fertilizer prices in the global scale playing a major contributory role (Fig. 10.10). Low organic matter content in soils has created several problems such as yield decline and yield stagnation in crops such as rice, fruits, and vegetables grown in Sri Lanka. Under such conditions, retention of plant nutrients is low and subsequently decreasing the fertilizer use efficiency. Thus, seasonal application of organic fertilizers is promoted for crop production in Sri Lanka. The Ministry of Agriculture of Sri Lanka has initiated a program to popularize organic fertilizer production and use since 2008 to encourage farmers to produce and use organic fertilizers, and some producers to make large-scale productions on commercial basis. The program is expected to reduce the chemical fertilizer use by 25%. Further, the Ministry of Agriculture of Sri Lanka also promotes adopting Integrated Plant Nutrient Systems (IPNS), a concept that advocates a balanced use of both organic and chemical fertilizers for crop production and is considered the most suitable plant nutrient management system to increase the crop yield while maintaining good soil fertility (Wijewardena and Yapa 1999).

10.6.3 Opportunities and Challenges

While the demand for basic food crops, high-value crops such as fruit and vegetables, animal products, and crops used to produce biofuels is likely to remain strong, it is expected that the increased fertilizer demand to support higher levels of crop

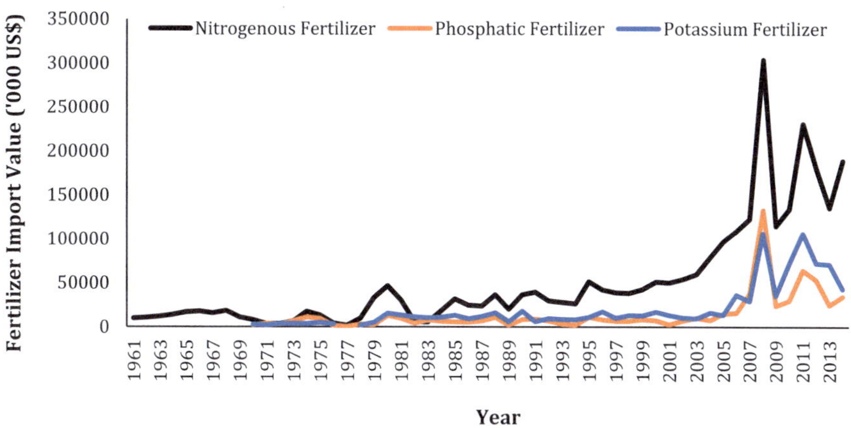

Fig. 10.10 Cost of fertilizer imports to Sri Lanka (FAOSTAT (n.d.), various issues)

production will be adequately supplied at a global scale. The TFI (2017) states that the demand for fertilizer will continue to grow, and industry needs to concentrate on sustainable production of quality-assured products with a timely supply to the end users. The global fertilizer industry was heavily challenged in 2016 with uneven global nutrient demand, soft economic prospects, depressed crop prices, rising market competition, and volatile energy prices. The medium-term outlook for world agriculture remains broadly unchanged compared to 2016, with expectations of relatively flat real prices for most agricultural commodities, reflecting prospects for ample supplies and weakening demand growth. Apart from this, the industry also faces a market condition with persistent oversupply, increasing environmental pressures from nutrient losses, change in consumer demand for more environmentally friendly agricultural systems, and the misuse of fertilizers for unlawful activities are the major challenges to be faced by the fertilizer industry in the future.

In Sri Lanka, the trade policy will continue to protect rice and several other import-competing food crops (e.g., potato and onions) with the use of seasonally varying tariffs and specific duties (Bandara and Jayasuriya 2007). This will affect the domestic input and product prices, and supply conditions. Subsidies for agricultural producer will remain important due to intense social and political pressure. The government envision on low cost and locally produced fertilizer and the drive toward nonchemical agricultural interventions, and social stigma and public cry on increasing incidence of non-communicable diseases such as diabetes and chronic kidney disease of uncertain etiology (CKDu) is likely to change the market dynamics of fertilizer. In Sri Lanka, ad hoc policy changes in the recent past, especially in the fertilizer subsidy by moving from material to cash in 2016 and then returning to material subsidy in 2018, has made it difficult to estimate the fertilizer use and its impact on agricultural production. Avoiding misuse of fertilizers and promotion of Good Agricultural Practices (GAP) including IPNS are challenges to be met and should be high-priority government interventions to safeguard the agricultural production in the country. Sri Lanka is not utilizing its rock phosphate deposit (Eppawala Rock Phosphate—ERP) judiciously but has continued to import P fertilizer. Country should make use this deposit for its own benefit. Globally, P will be in limited supply in the near future, and P-recycling options need to be investigated and made into operation. Owing to possible environmental contaminations and pollution, the country should also move toward using options such as soil and plant test based recommendations and applications of fertilizer for crop production.

10.7 Agro-Pesticide Use

10.7.1 History, Global Trade, and Trends

Crop farming, since inception, has suffered heavily from different types of pests and diseases causing significant losses in yield. Even with the advances in agricultural sciences over the years, losses due to pests and diseases range from 10% to 90%. The first recorded use of insecticides is in about 4500 years ago by Sumerians who used sulfur compounds to control insects and mites. Weeds were mainly controlled by hand weeding, but different "chemical" methods such as use of salt or sea water have been reported (Smith and Secoy 1975). Historically, many inorganic chemicals have been used as pesticides (Smith and Secoy 1976) including Bordeaux mixture, which is still used against various fungal diseases.

Inorganic substances (e.g., sodium chlorate and sulfuric acid) or organic chemicals based on natural sources and few byproducts of coal gas production or other industrial process have been used for pest control until 1940s. Organic chemicals such as nitrophenols, chlorophenols, creosote, naphthalene, and petroleum oils were used for fungal and insect pests, while ammonium sulfate and sodium arsenate as herbicides (Unsworth 2010). High rates of application, lack of selectivity, and phytotoxicity were the major drawbacks in using such products.

The 1940s–1950s experienced an accelerated growth of synthetic pesticides with the discovery of the pesticidal effects of DDT, BHC, aldrin, dieldrin, endrin, chlordane, parathion, captan, and 2,4-D (Unsworth 2010). In 1970s–1980s, the continued interest of research on pesticides, resulted in introduction of herbicides such as glyphosate, and the low use rate of sulfonylurea and imidazolinone (imi), dinitroanilines, aryloxyphenoxypropionate (fop), and cyclohexanediones (dim), insecticides such as the third generation of pyrethroids, and avermectins, benzoylureas, and spray treatment of B_t (*Bacillus thuringiensis*), fungicides belonging to the families of triazole, morpholine, imidazole, pyrimidine, and dicarboxamide. Many of the agrochemicals introduced had a single mode of action, thus making them more selective. Evolution of pesticide resistance also occurred and thus the management strategies were introduced to combat this negative effect. In the 1990s, new herbicides of the families triazolopyrimidine, triketone, and isoxazole, fungicides of families strobilurin and azolone, and insecticides belonging to the families chloronicotinyl, spinosyn, fiprole, and diacylhydrazine were introduced. Many of the new agrochemicals were applied at low rates (i.e., g ha^{-1}), and were more user-friendly and environmentally safe formulations. Today the pest management technologies have evolved and further expanded to include genetically engineered crops such as herbicide-tolerant soybeans, corn, canola, and cotton (CropLife 2002). Use of Integrated Pest Management (IPM) systems, which discourage the development of pest-resistant populations and reduce the use of agrochemicals, has also become more widespread. Such changes have altered the nature of pest control and have the potential to reduce and/or change the nature of agrochemicals used.

The recent studies have indicated that both the global cost/benefit and the volume of pesticides use have increased during 1990–2007, and decreased thereafter (Zhang 2018). The global pesticide usage by category is illustrated in Fig. 10.11. The global pesticide market was 28 billion US$ in 1999, 58.46 Billion US$ in 2015, and is expected to reach an estimated $81.1 billion by 2021 (http://www.lucintel.com/global-pesticide-industry-2016-2021.aspx). The global pesticide market is forecasted to grow at a CAGR of 4.4% by value from 2016 to 2021. The global volume market for pesticides is projected to reach 3.2 million tons by 2019, with a CAGR for volume growth being 6.1% between 2014 and 2020 (https://www.agprofessional.com/article/study-shows-global-pesticide-market-reach-81-billion-five-years). The Asia Pacific Crop Protection market generated a revenue of 7.9 billion in 2016 and is anticipated to contribute 19.1 billion US$ by 2025, growing at a CAGR of 12.2% (https://www.inkwoodresearch.com/reports/asia-pacific-crop-protection-market/). The crop protection market is rapidly increasing due to the need of food for continuously rising population, a decrease in arable land, need for higher crop yield, and increasing purchasing power.

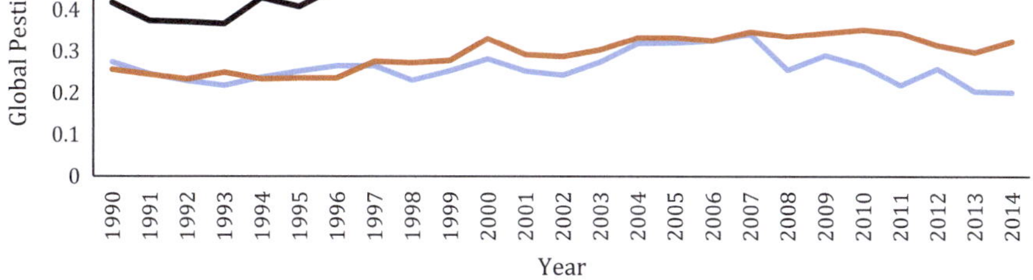

Fig. 10.11 Global use of pesticide in agriculture (Source: FAOSTAT (n.d.), various issues)

10.7.2 Pesticide Use in Sri Lanka

The first historical records of pesticide use in agriculture is the large-scale experimentation insecticides in tea plantations in 1950s (Thirugnanasuntharan 1987). The first reference to any herbicide trial in Tea was in 1949 with the use of 2-4D (Somaratne 1988), while fungicides were used to control the Blister Blight Leaf disease in 1946 (Arulpragasam et al. 1987).

Liberal import of pesticides to Sri Lanka was done up to 1962 where import restrictions were imposed up to the value of Sri Lanka Rs 5.23 million in 1963 by the Revenue Controller of then Ceylon. The DOA also recognized the need for regulation of supply, distribution, and use of pesticides and submitted a draft bill entitled "Act on Poisons used in Agriculture" to the legal draftsman in 1964, which unfortunately seemed to have been shelved. However, in 1972 with the intension of reviving the legislation on pesticide regulation, the Department of Agriculture with the assistance from the Food and Agriculture Organization (FAO) of United Nations (UN) on establishment of national infrastructure for the control of pesticides. The FAO submitted the report to the Government of Sri Lanka in 1972 with another report titled "Pesticide Control" submitted in 1976 recommending enacting of a pesticide bill. The introduction of more liberal economic policies after 1977 demonstrated the urgent need for pesticide control legislation where the import of pesticides to the country increased dramatically by 237%, from 2166 tons in 1976 to 5144 tons in 1979. Resulting from unabated continuous lobbying by the DOA and other stakeholders, finally Sri Lanka government enacted the Control of Pesticides Act No. 33 of 1980 with effect from 5th September 1980 with the assistance from the FAO, which was later amended as the Control of Pesticides (Amendment) Act No 6 of 1994.

Pesticides are either imported to Sri Lanka in bulk for repacking or brought in as concentrates (technical grade material) for formulation locally prior to marketing. The Sri Lanka Customs Ordinance No 17 of 1869 and Sri Lanka Customs (Amendment) Act No 9 of 2013 have classified all pesticide as licensed items, requiring an import license for every consignment. The Controller of Imports issues the import license only on a written approval by the Registrar of Pesticides under sect. 17 of the Act. The approvals are issued on consignment basis after assuring the product quality, etc. The Registrar cannot issue the import approvals unless the product is duly registered under the Act.

The Pesticides Control Act No. 33 of 1980 has contributed much to making the use of agricultural pesticides safe to the user particularly by creating the necessary background for the introduction of pesticides with lower mammalian toxicity. By the Act, WHO class 1A and IB products are banned to be imported and use. Most pesticide products that are allowed to be imported today fall under WHO class II or III. Several pesticides have been banned or in restricted use in Sri Lanka due to various reasons as depicted in Table 10.6.

Though the significance of agro-pesticides in food crop production in Sri Lanka cannot be ignored, the import of pesticides has continued to be a drain of foreign exchange to the country. The value of the total volume of pesticides imported to Sri

Table 10.6 Pesticides banned and severely restricted in Sri Lanka

Year of ban (regulatory)	Year of ban (legal)	Name of the pesticide
1970	2001[a]	Endrin[e]
1976	2001[a]	DDT[e]
1980	2001[a]	Chlordimeform
1980	2001[a]	Dieldrin[e]
1980	2001[a]	Phosphamidon
1980	2001[a]	Thalium sulfate
1984	2001[a]	2,4,5-T
1984	2001[a]	Ethyl-parathion
1984	2001[a]	Methyl-parathion
1986	2001[a]	Aldrin[e]
1986	2001[a]	Lindane[e]
1987	2001[a]	HCH (mixed isomers)[e]
1987	2001[a]	Mercury compounds
1988	2001[a]	Arsenic (arsenites and arsenates)
1988	2001[a]	Heptachlor[e]
1988	2001[a]	Leptophos
1989	2001[a]	Captafol
1990	2001[a]	1,3-Dichlorpropane
1990	2001[a]	Aldicarb
1990	2001[a]	Quintozene (PCNB)
1994	2001[a]	Pentachlorophenol[e]
1994	2001[a]	Chlordane[e]
1995	2001[a]	Methamidophos
1995	2001[a]	Monocrotophos (60% SL—restricted use to control coconut weevil)
1998	2001[a]	Endosulfan (35% EC)[e]
2008	2014[b]	Paraquat (20% SL)
2011	2014[b]	Paraquat (6.5% SL)
2011	2014[b]	Dimethoate(40% EC)
2011	2014[b]	Fenthion (50% EC)
2011	2014[b]	Cyromazine (75% WP)
2012	2014[b]	Alachlor(36% EC)
2013	2014[c]	Prapnil (36% EC)
2013	2014[c]	Carbofuran (3% GR)
2013	2014[c]	Carbaryl (85% WP)
2013	2014[c]	Chlorpyrifos (20% EC and 40% EC)
2014	2014[c]	Glyphosate (36% SL)
2015	2015[d]	Glyphosate (36% SL)

[a]Ban of registration by the government extraordinary gazette No. 1190/24 of 29.06.2001 under the Control of Pesticides Act No. 33 of 1980

[b]Ban of registration by the government extraordinary gazette No. 1854/47 dated 21.03.2014 under the Control of Pesticides Act No. 33 of 1980

[c]Regional restriction for sale, offer for sale and use as per the government extraordinary gazette No. 1894/4 of 22.12.2014 under the Control of Pesticides Act No. 33 of 1980

(continued)

Table 10.6 (continued)

[d]Ban of importation by the gazette extraordinary No. 1813/14 of 05.06.2015 under the Import and Export (Control) Act No. 01 of 1969 and gazette extraordinary No. 1937/35 of 23.10.2015. The latter was rescinded by the gazette extraordinary No. 2079/37 of 11.07.2018

[e]Classic organochlorine pesticides (POPs pesticide) listed under the Stockholm Convention

Lanka was SLR 0.2 million in 1970, 1.3 million in 1972, 37.9 million in 1979 (Weeraratne 1983), and 1350 million rupees in 2006 (12.3 m US$). There are approximately 114 active substances and nearly 440 agricultural pesticides in commercial use (Lakshani et al. 2017).

Pesticide imports to Sri Lanka dropped by about 30% in 2014 compared to that of 2013 (Fig. 10.12) due to stringent control over high volume pesticides such as chlorpyrifos, carbaryl, carbofuran, and propanil. A further reduction of about 25% was observed in 2015 due to the ban imposed on paraquat in 2014 and glyphosate in 2015. The latter accounted for 25% of herbicide formulation and 18% of all pesticide formulations imported to the country (Department of Agriculture 2015). However, the total active ingredients imported to Sri Lanka (i.e., including active ingredients in all formulations) showed a declining trend in herbicide and insecticides (Fig. 10.12) with fungicides maintaining a steady status over the past 25 years.

10.7.3 Opportunities and Challenges

The future in the global pesticide industry looks promising with opportunities in insecticide, fungicide, and herbicide markets, with herbicide segment remaining as the largest market. Increasing demand for weed management and food security is expected to drive the pesticides demand in the future. Limited availability of arable land, demand for healthy diet, and priority for food safety are the major challenges faced by the pesticide industry or agrochemical industry as a whole. Innovation and new product development for long-lasting crop protection, advanced plant health benefits, simplifying the application of using pesticides, development of nano-pesticides and bio-pesticides to reduce environmental pollution and development of formulated technology for weed control in different food systems are the priority intervention needed to uplift the industry from its present scale. Concerns over chemical-induced non-communicable diseases, potential carcinogenicity of agrochemicals, and trends toward organic farming are the challenges faced by the industry and use of pesticides in agriculture in the medium and long term.

Government interventions to ban and/or restrict the use of popular insecticides and herbicides in 2014 and 2015 have had a significant negative impact on the pest control, pesticide use, and trade in Sri Lanka. Claims of misuse of pesticides alleged contribution of pesticides in causing non-communicable diseases such as CKDu and other human health hazards (Marambe et al. 2015b; Marambe 2018), fear of being potentially carcinogenic, use of some pesticides to commit suicide, environmental concerns of the general public, and a move to search for nonchemical pest control methods are some major concerns, which is likely to continue in the

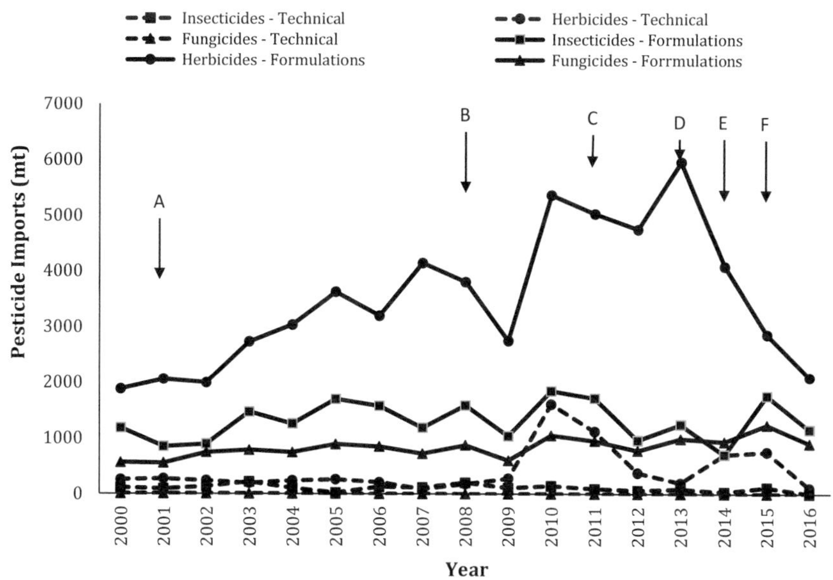

Fig. 10.12 Pesticide imported to Sri Lanka (ROP, 2018); A = legal ban of large number of 23 pesticides; B = regulatory ban of Paraquat (20% SL); C = regulatory ban of Paraquat (6.5% SL), Dimethoate (40% EC), Fenthion (50% EC), and Cyromazine (75% WP); D = regional restriction of sale on Propanil (36% EC), Carbofuran (3% GR), Carbaryl (85% WP), Chlorpyrifos (20% EC and 40% EC); E = regional restriction of sale on Glyphosate (36% SL); F = regulatory and legal ban on importation of Glyphosate (36% EC)

foreseeable future. Decision making based on scientifically valid information is crucial to overcome such negative impacts on agricultural systems of Sri Lanka. Use of bio-pesticides/bio-herbicides should be encouraged in Sri Lanka based on well-designed scientific research.

10.8 Synthesis

Producing more food to feed a growing population amidst a rapidly diminishing arable land and (rural) labor force, demanding for cultivation of energy crops for a potentially huge bioenergy market, contributing to develop agriculture-dependent economies in a sustainable manner, adopting efficient and sustainable production methods, and adapting to climate change are the multifaceted challenges faced by global agriculture in the twenty-first century. Approximately 90% of the global growth in crop production, and 80% growth in developing countries, is expected to be a result of higher yields and increased cropping intensity, while the remainder coming from land expansion. Pressure on renewable water resources and energy sources would continue to be severe affecting the anticipated growth rate of agriculture.

"Green Revolution" has fulfilled the task of increasing productivity and total production to feed a large majority of the world population toward the end of the twentieth century and in the twenty-first century. However, the success achieved was at huge cost on inputs and the environment. The agricultural inputs such as seeds and agrochemicals (fertilizer and pesticides) have dramatically increased their use at global and national scales since the adoption of Green Revolution practices in 1960s. Rapid adoption of high-yielding crop varieties, hybrids and GM crops, and other technologies have made agriculture the supporting livelihood of farming communities all over the world, and Sri Lanka is not an exception. However, negative impacts on traditional germplasm, indiscriminate and misuse of agrochemicals affecting human and environmental health have been rising concerns over the years emanating from agro-input intensification in agricultural systems. Breeding for climate resilient varieties using such important characteristics inherited by the traditional crop germplasm and crop wild relatives, IPNS with a judicious balance of chemical fertilizer, organic matter and bio-fertilizer, and integrated pest management that include host-plant resistance, functional biodiversity, biological control including use of bio-pesticides and judicious use of agro-pesticides are the current and future needs aiming at sustainable development and sustainable food security of the country.

Transformative changes of agriculture and food systems are the current needs in all countries, but with different priorities to align with different economies. There is no doubt that agriculture must change to meet the rising demand, to contribute more effectively to the reduction of poverty and malnutrition, but should become more ecologically sustainable. Such a transformation will be a crucial to achieve many of the Sustainable Development Goals (SDGs) by 2030. The government of Sri Lanka should aim at well-articulated interventions to increase national-level food production, and ensure mechanisms to sustain household and national-level food security. Better understanding on the risks, improving resource use efficiencies, target-oriented public–private partnerships, better use of state and private investments, coupled with sound and coherent agricultural policies would help the country to reach the targets in the sustainable development agenda.

Acknowledgments The authors acknowledge support extended by Dr. J.A. Sumith (Registrar of Pesticides, Department of Agriculture), Dr. Lakmini Ashraff (Additional Director, Seed Certification and Plant Protection Service—SCPPC, Department of Agriculture), and Mr. B.H. Rohan Induka (Assistant Director, National Fertilizer Secretariat, Ministry of Agriculture) in providing statistical information.

References

Abegunawardana P, Pope CA III (1986) Impacts of price and yield variation on profitability of management plans of subsistence farmers in Sri Lanka. Agric Adm 22:197–204

Abeygunawardane DV (2014) Why is the subsidy on fertilizers for rice in Sri Lanka continued? A thesis submitted for the degree of Doctor of Philosophy at The University of Queensland, Australia

Abeysekara SW, Jayawardena SN, Kiriwaththuduwage S, Abeysiriwardena DSZ (2002) Hybrid rice research and development in Sri Lanka. In: Virmani SS, Mao CX, Hardy B (eds) Hybrid rice for food security, poverty alleviation, and environmental protection, pp 341–354. Proceedings of the 4th International Symposium on Hybrid Rice, Hanoi, Vietnam. Los Baños (Philippines): International Rice Research Institute

Agrawal PK (2011) Climate change and its impact on agriculture. In: *Souvenir-Indian Seed Congress*, 22–23 Feb, Hyderabad, p 67–76

Alexandratos N, Bruinsma J (2012) World agriculture towards 2030/2050: the 2012 revision. Food and Agriculture Organization of the United Nations, Rome

Anon (2018) $58 Billion Seeds Market: Global Industry Trends, Share, Size, Growth, Opportunity and Forecast 2017–2022 – Research and Markets. https://www.businesswire.com/news/home/20170421005206/en/58-Billion-Seeds-Market-Global-Industry-Trends. Accessed 24 June 2018

Anonymous (2017) Fertilizers face growing market imbalance. In: Insight. Gulf Petrochemicals and Chemicals Association, 41, 10–11

Arulpragasam PV, Addaickan S, Kulatunga SM (1987) Recent developments in the chemical control of blister blight leaf disease of tea. Effectiveness of EBI fungicides. Sri Lanka J Tea Sci 56:22–34

Bandara J, Jayasuriya S (2007) Distortions to agricultural incentives in Sri Lanka: main report (English). Agricultural distortions working paper no. 31. World Bank, Washington, DC

Bandarage A (1983) Colonialism in Sri Lanka: the political economy of the Kandyan highlands, 1833–1886. Mouton, Berlin

Bonny S (2014) Taking stock of the genetically modified seed sector worldwide: market, stakeholders, and prices. Food Sec 4:525–540

Bonny S (2017) Corporate concentration and technological change in the global seed industry. Sustainability 9:1–25

Bruins M (2009) The evolution and contribution of plant breeding to global agriculture. In: *Proceedings of the Second World Seed Conference*, pp. 18–29. September 8–10, Food and agriculture Organization of the United Nations, Rome

Byerlee D (1996) Modern varieties, productivity, and sustainability: recent experience and emerging challenges. World Dev 24(4):697–718

CropLife (2002) A history of crop protection and Pest control in our society. CropLife Canada. http://www.croplife.ca/english/pdf/Analyzing2003/T1History.pdf. Accessed 4 June 2018

CropLife (2008) Facts and figures – the status of global agriculture. CropLife International, Brussels, Belgium. http://www.croplife.org/library/attachments/eab5d7c1-5b80-492b-b52bd3de8819eb77/2/Facts%20And%20Figures%202009.pdf. Accessed 4 June 2018

Davis KF, D'Odorico P, Rulli MC (2014) Moderating diets to feed the future. Earth's Future 2:559–565

DCS (n.d.) (various issues): Annual reports. Sri Lanka: Department of Census and Statistics

de Silva KM (1981) A history of Sri Lanka. University of California Press, Berkeley

Department of Agriculture (2015) Socio-economic and statistics division. Department of Agriculture, Peradeniya

Ekanayake HKJ (2006) Impact of fertilizer subsidy in paddy cultivation in Sri Lanka. Staff Studies. (Central Bank of Sri Lanka) 36(1 & 2):73–92

Erisman JW, Sutton MA, Galloway JN, Klimont Z, Winiwarter W (2008) How a century of ammonia synthesis changed the world. Nat Geosci 1:636–639

FAO (2017) World fertilizer trend and outlook 2020. Food and Agriculture Organization of the United Nations, Rome, p 27

FAOSTAT (n.d.) – various issues. http://www.fao.org/faostat/en/

Farmer BH (1979) Green revolution. The Macmillan Press Ltd., London

FCRDI (2018) Annual Report. Field Crops Research and Development Institute, Mahailuppallama

Feder G, Just RE, Zilberman D (1985) Adoption of agricultural innovations in developing countries: a survey. Econ Dev Cult Chang 33:255–298

Fladby B (1983) Household viability and economic differentiation in Gama, Sri Lanka. Bergon occasional paper in social anthropology, no. 28. University of Bergon, Bergon
Foley JA, Ramankutti N, Brauman KA, Cassidy CS, Gerber JS et al (2011) Solutions for a cultivated planet. Nature 478:337–342
Galloway JN, Dentener FJ, Capone DG, Boyer EW, Howarth RW, Seitzinger SP et al (2004) Nitrogen cycles: past, present and future. Biogeochemistry 70:153–226
Gibbs HK, Ruesch AS, Achard F, Clayton MK, Holmgren P, Ramankutty N et al (2010) Tropical forests were the primary sources of new agricultural land in the 1980s and 1990s. Proc Natl Acad Sci USA 107:16732–16737
Godfray HCJ, Beddington JR, Crute IR, Haddad L, Lawrence D, Muir JF, Pretty J et al (2010) Food security: the challenge of feeding 9 billion people. Nature 327:812–818
Gollin D, Morris M, Byerlee D (2005) Technology adoption in intensive post-green revolution systems. Am J Agric Econ 87:1310–1316
Hazell P (2009) The Asian Green Revolution. IFPRI Discussion Paper 00911. IFPRI, Washington D.C.
IFA (2018) International fertilizer association. https://www.fertilizer.org/En/Home/En/Default.aspx?hkey=6c3f9638-dbf5-4ec7-84b3-d6a993e7db8a
IFPRI (2002): Green revolution – curse or blessing?. International Food Policy Research Institute, http://www.ifpri.org/sites/default/files/pubs/pubs/ib/ib11.pdf
Imhoff ML, Bounoua L, Ricketts T, Loucks C, Harriss R, Lawrence WT (2004) Global patterns in human consumption of net primary production. Nature 429:870–873
IPCC (2007) Climate Change 2007 – Impacts, Adaptation and Vulnerability Contribution of Working Group II to the Fourth Assessment Report of the IPCC. Intergovernmental Panel for Climate Change. Cambridge University Press
ISAAA (2016) Annual Report. USA: international service for the acquisition of agri-biotech applications
ISF (2012) Annual Report. Switzerland: International Seed Federation
Jayasundera M (2017) Ancient Sri Lankan economic systems. Financial Mark Inst Risks 1:24–27
Knighton W (1845) The history of Ceylon from the earliest period to the present time. Longman, Brown, Green & Longmans
Kingsbury N (2009) Hybrid. The University of Chicago Press, Chicago & London
Kumari WMR, Gunawardhana GA, Perera AM, Bandara HMS, Paththinige SS (2013) Evaluation of quality protein maize (Opeque 2) as substitutes for conventional maize cultivars in Sri Lanka. Ann Dept Agric 15:37–49
Kumari WMR, Gunawardhana KNC, Priyantha MGDL, Bandara HMS, Dassanayake DMJK (2015) Performance of imported maize hybrids under local condition in Sri Lanka. Ann Dept Agric 17:350–354
Kumari WMR, Karunathilake WAK, Weerakoon WMW, Chandrasiri WACK, Abeysinghe AMBN (2017) Scaling-up of proven Technology for Maize Improvement through participatory approach in Sri Lanka. In: Pandey PR, Koirala KB (eds) Best practices of maize production Technologies in South Asia. SAARC Agriculture Centre and CIMMYT, Bangladesh, pp 111–128
Kush G (2001) Green revolution: the way forward. Nature Reviews – Genetics 2:815–822
Lakshani PWY, Rajapaksha MKLK, Sendthuran K (2017) Pesticide residues in selected vegetables in several growing areas by GC/MS using QuEChERS technique. Ann Sri Lanka Dept Agric 19:188–208
Lianos I, Katalevsky D, & Ivanov A (2016) The global seed market, competition law and intellectual property rights: untying the Gordian knot. Research Paper Series: 2/2016, Centre for Law, Economics and Society
Louwaars PN, de Boef WS (2011) Integrated seed sector development in Africa: a conceptual framework for creating coherence between practices, progrms, and policies. J Crop Improv, 26: 39–59
Lu C, Tian H (2013) Net greenhouse gas balance in response to nitrogen enrichment: perspectives from a coupled biogeochemical model. Glob Chang Biol 19:571–588

Mankotte K (2012) Department of Agriculture in retrospect (1912–2011). Department of Agriculture, Peradeniya

Marambe B (2018) Present status of the food crop sector in Sri Lanka. In: Modernization of Agriculture and Industry: Challenges for Sri Lanka Sri Lanka Economics Association (In Press)

Marambe B, Punyawardena R, Silva P, Premalal S, Rathnabharathie V, Kekulandala B et al (2015a) Climate, climate risk, and food security in Sri Lanka: need for strengthening adaptation strategies. In: Leal W (ed) Handbook of climate change adaptation. Springer-Verlag, Berlin Heidelberg, pp 1959–1789

Marambe B, Abeysekara ASK, Herath HMS (2015b) Weeds and weed management in agricultural and natural ecosystems: an overview of the Sri Lankan context. In: Rao VS, Yaduraju NT, Chandrasena NR, Hassan G, Sharma AR (eds) Weed Science in the Asian-Pacific Region, chapter 9. Asian-Pacific Weed Science Society and Indian Society of Weed Science, Hyderabad, pp 213–240

MEA (2005) Millennium ecosystem assessment. Summary for decision makers. Island Press, Washington, DC

Moss DL (2013) Competition, intellectual property rights, and transgenic seed. SDL Rev 58:543

Nellemann C, MacDevette M, Manders T, Eickhout B, Svihus B, Prins A, Kaltenborn B (eds) (2009) The Environmental Food Crisis. The environment's role in averting future food crises. A UNEP rapid response assessment, Arendal, UNDP

NFS (various issues) Unpublished Reports. National Feertilizer Secretariat, Ministry of Agriculture, Battaramulla, Sri Lanka

Nox R (1681) An historic relation of the island of Ceylon in the east indies. Richard Chiswell (Publ), London

Paarlberg R (2010) Food politics: what everyone needs to know. Oxford University Press, New York, NY, p 218

Pain A (1986) Agricultural research in Sri Lanka: an historical account. Mod Asian Stud 20:755–778

Pingali PL (2012) Green revolution: impacts, limits, and the path ahead. PNAS 109(31):12302–12308

Pretty J (2008) Agricultural sustainability: concepts, principles and evidence. Philos Trans R Soc B: Biol Sci 363:447–465

Pretty J (2013) The consumption of a finite planet: Well-being, convergence, divergence and the nascent green economy. Environ Resour Econ 55:475–499

Pretty J, Barucha ZP (2014) Sustainable intensification in agricultural systems. Ann Bot 114:1571–1596

Ramankutty N, Mehrabi Z, Waha K, Jarvis L, Kremen C, Herrero M et al (2018) Trends in global agricultural land use: implications for environmental health and food security. Annu Rev Plant Biol 69:789–815

Rodirigo C, & Abeysekera L (2015) Determinants of fertilizer demand: evidence from Sri Lanka. In: Proceedings of the International Postgraduate Research Conference 2015. University of Kelaniya, Kelaniya, Sri Lanka, p 22

Roland W (2007) Indian labour immigration and British labour policy in nineteenth-century Ceylon. Mod Asian Stud 41:575–602

ROP (2018). Unpublished Report. Office of the Registrar of Pesticides, Getambe, Sri Lank

RRDI (2018) Unpublished data. Rice Research and Development Institute, Batalagoda, Sri Lanka

Schrikker AF (2007) Dutch and British colonial intervention in Sri Lanka 1780–1815: expansion and reform. Brill Publ. 272 pp.

Schultz TW (1968) Economic growth and agriculture. McGraw-Hill, New York

SCPPC (n.d.) (various issues): Annual Report. Seed certification and plant protection service, Department of Agriculture, Peradeniya

Senadhira D, Dhanapala MP, Sandanayaka CA (1980) Progress of Rice varietal improvement in the dry and intermediate zones of Sri Lanka. Rice symposium 80. Department of Agriculture, Peradeniya

Seneviratna A (1981) The hydraulic civilization of the ancient Rajarata in Sri Lanka. International Irrigation Management Institute (IIMI), Colombo

Sewón M (2013) The web of global seed trade - commercialization and commoditization of plant genetic resources. University of Applied Science, Novia
Singh RP, Vara Prasad PV, Raja Reddy K (2013) Impact of changing climate and climate variability on seed production and seed production. Adv Agron 118:49–110
Siriweera WI (1978) Agriculture in Mediaeval Sri Lanka. Ceylon Hist J 15:34–50
Smith AE, Secoy DM (1975) Forerunners of pesticides in classical Greece and Rome. J Agric Food Chem 23:1050–1055
Smith AE, Secoy DM (1976) A compendium of inorganic substances used in European pest control before 1850. J Agric Food Chem 24:1180–1186
Snodgrass DR (1966) Ceylon: an export economy in transition. Yale University Press, Illinois
Somaratne A (1988) The use of herbicides in tea in Sri Lanka. In: Sri Lanka journal of tea science conference. Tea Research Institute of Sri Lanka, Talawakelle, pp 143–154
Suweis S, Carr JA, Maritan A, Rinaldo A, D'Odorico P (2015) Resilience and reactivity of global food security. PNAS 112:6902–6907
Swaminathan MS (1996) Sustainable agriculture towards an Evergreen revolution. Konark Publishers Private Limited, Delhi
TFI (2017) State of the fertilizer industry. The Fertilizer Institute, Washington, DC
Thirugnanasuntharan K (1987) Feeding rates and survival of live wood termites in woody plants. Sri Lanka J Tea Sci 56:67–82
UN (2018) World population prospects: the 2017 revision. Population Division of the Department of Economic and Social
Unsworth J (2010) History of pesticide use. International Union of Pure and Applied Chemists. https://agrochemicals.iupac.org/index.php?option=com_sobi2&sobi2Task=sobi2Details&catid=3&sobi2Id=31
Weasel LH (2008) Food fray: inside the controversy over genetically modified food: AMACOM. p 256
Weerahewa J, Kodithuwakku S, Ariyawardana A (2010) The fertilizer subsidy programme in Sri Lanka. In: Pinstrup-Anderson P (ed) Case study no. 7–11. Food policy for developing countries: the role of government in the global food system. Cornell University, New York
Weeraratne CS (1983) Pesticides – an overview with particular reference to Sri Lanka. Econ Rev 8:4–12
Weeraratne H (1972) Progress of rice breeding in Ceylon since 1960. In: Rice breeding. International Rice Research Institute, Philippines, pp 137–140
Wickramasinghe A, & Cameron DC (2005). British Capital, Ceylonese Land, Indian Labour: The Imperialism and Colonialism of Evolution of Tea Plantations in Sri Lanka, Paper presented to Critical Management Studies Conference (Management and Organizational History), Cambridge, 4–6 July
Wijewardena JDH (2001) Fertilizer and soil amendments use on potato in relation to soil fertility in rice based cropping systems of up-country of Sri Lanka. Ann Sri Lanka Dept Agric 3:353–363
Wijewardena JDH (2005) Improvement of plant nutrient management for better farmer livelihood, food security and environment in Sri Lanka. Proceedings of a Regional Workshop. Rome: Food and Agriculture Organization of United Nations
Wijewardena JDH, Yapa UWSP (1999) Effect of combined use of animal manure and chemical fertilizer on potato and vegetable cultivation in the upcountry of Sri Lanka. Sri Lankan J Agric Sci 3:70–82
Wik M, Pingali P, Broca S (2008) Global agricultural performance: past tends and future prospects. In: Background paper for the world development report 2008. World Bank, Washington DC
Withanawasam DM (2017) Heritage of heirloom rice varieties of Sri Lanka. Department of Agriculture, Peradeniya. (https://www.doa.gov.lk/rrdi/images/Mydoc/publication1.pdf)
World Bank (2007) World development report 2008: agriculture for development. World Bank, Washington, DC
Zhang W-J (2018) Global pesticide use: profile, trend, cost/benefit and more. Proc Int Acad Ecol Environ Sci 8:1–27

A Cross Section of Century-Long Experiences in Entomological Research in Crop Sectors: Directions for Future Research

11

K. S. Hemachandra

Abstract

Commercial crop cultivation in Sri Lanka became evident in the colonial times along with the shift of sustainable mixed cultivation systems to monoculture system. Insect pest problems also became evident with the change of cultivation system drawing the attention of growers. Studies focusing on pest management were first conducted with tea and coffee, followed by other plantation crops. Well planned, formal research studies commenced and results were published upon the establishment of Department of Agriculture in 1912 followed by other research stations for tea, rubber, and coconut. Biology and ecology of the insect pest species, pest behavior, and practices for population management were the interest of studies, and the recommendations were implemented and monitored. Most of the plant protection strategies that we practiced today were initiated at the early part of the twentieth century. Nonchemical pest management practices such as use of resistant varieties, shifting of planting dates, use of physical barriers against insect activities, management of alternate hosts, and classical biological control have been attempted to manage the target pest populations. Insecticide application was the interest by the mid of the last century, which has continued to date. The possible non-target effects and development of resistance in insect population were studied along with the chemical control of insect pests. Integrated pest management was the approach in many crops, along with modern extension strategies such as farmer field school. Of the insect pest management practices, certain strategies have not been fully utilized, and future research should be aimed for effective utilization of such pest management tools. Biological control, especially the conservation biocontrol and augmentative biocontrol, including microbial control, has a great potential to be used in the Sri Lankan context. Use of resistant varieties is a classic pest management approach

K. S. Hemachandra (✉)
Department of Agricultural Biology, Faculty of Agriculture, University of Peradeniya, Peradeniya, Sri Lanka

B. Marambe et al. (eds.), *Agricultural Research for Sustainable Food Systems in Sri Lanka*, https://doi.org/10.1007/978-981-15-2152-2_11

but yet has a great potential to be used even in modern agriculture. Collaborative research to produce resistant varieties with good agronomic characters and high yield is a timely need and an essential tool to manage the insect pests. Novel insecticides with low mammalian toxicity are encouraged with improved safety procedures and modern application technologies. Overall, the new directions of entomological research should be under the considerations of sustainability, and environment friendliness.

Keywords
Tea · Diversity · Biocontrol · Botanicals · Insecticides

11.1 Introduction

Crop cultivation in Sri Lanka has a long history, and it has continued for centuries as a sustainable system. Upon the establishment of commercial monoculture systems, sustainability of the systems was disturbed and entomological problems appeared along with other issues in crop management practices. The Department of Agriculture (DOA) was established in 1912 (Anon 2000) followed by the Tea Research Institute (TRI), Coconut Research Institute (CRI), Rubber Research Institute (RRI), and Sugarcane Research Institute (SRI) to address the issues in agriculture including the pest management. The researchers studied different aspects of entomology including the biology, ecology, natural enemies of insect pest species, and management strategies including chemical, biological, and regulatory control, and the use of botanicals. The research findings were published in many different periodicals such as annual reports, administrative reports, and in journals such as Tropical Agriculturist, Ceylon Journal of Science (Biological Sciences), Sri Lanka Journal of Food and Agriculture, Journal of Science Council, Journal of National Science Foundation, Tropical Agricultural Research and Extension, and Tropical Agricultural Research and in proceedings of symposia. These sources were examined as per the availability in major libraries to understand the trends of entomological research in the past. The information gathered in this exercise is far below a comprehensive coverage of the literature but sufficient enough to understand the research trends in the past and therefore can be considered as a cross section of the research findings in agricultural entomology. The objective of this review was to portray the past entomological research highlighting the research area and make suggestions or identify the directions for future entomological research.

In this article, I reviewed the important land marks in commodity research, using tea as an example, because tea crop has been the subject of research with many clear examples. The research studies carried out in the areas of insect diversity, population management, and insect physiology were also reviewed. The entomological research directed toward the industrial entomology such as bee keeping, medical and veterinary entomology, and forest entomology were excluded to make the review more focus on agricultural entomology. Further, the research related to mite pests, vector control, and use of biocontrol agents in weed management were purposefully excluded in this review to make a better focus on insect pest management.

11.2 Entomological Research on Commodity Crops: Tea

Along with foreign invasion, traditionally practiced mixed farming system was disturbed and commercial monoculture systems were appeared in different parts of the country. Tea plantation was first established in 1867 in Loolkandura estate, and soon spread into other areas with the failure of coffee crop due to coffee rust disease (*Hemileia vastatrix*) in 1869.

Dr. Edward Green was the first government entomologist and mainly responsible for managing the tea pests. His work on "Insect pests of tea plant" (Green 1890) was a valuable resource for the tea planters in Sri Lanka as well as in nearby countries. The Tea Research Institute (TRI) was established in 1925, and the entomological research conducted during 1925–2000 was reviewed by Vitarana (2000). The first entomologist in the TRI was Mr. S.S. Light, and he worked in collaboration with the DOA at Peradeniya. The two major tea insects: tea tortrix and tea shot-hole borer had been described by using the specimens collected in Sri Lankan plantations. Tea tortrix (Lepidoptera; Tortricidae) was first described by Nietner as *Capua menicana* in 1861 as a coffee pest, and the name was later revised as *Capua coffearia,* followed by *Homona coffearia.* The tea shot-hole borer (*Xyleborus fornicatus*; Coleoptera: Scolytidae) was described in 1868, using a beetle collected in Sri Lanka.

As the tea tortrix and shot-hole borer have become the limiting factors of tea cultivation, the TRI started to gather information on life histories, distributions, host ranges, parasitoids, and population dynamics of these pest insects (King 1936), which are key information needed for the formulation of effective management practices. As per the management practices of tea tortrix, hand collection of egg masses (King 1935) has been suggested, but labor cost involved was a concern. Top pruning and burning of pruned parts has been another effective physical technique (Hutson 1925, 1932). Establishment of wind belt and flight breaks were ideas to control tea tortrix (Jardine 1919b). Green (1906) reported the parasitism of tea tortrix, revealing the possibility of managing the insect pest population, which can be now explained as a biological control program. Local egg parasitoid, *Trichogramma erosicornis* Westw., has been recognized as a biocontrol agent (Light 1929), and an attempt for mass production using an alternate host *Sitotroga cerealella* has been made (Light 1929). An egg parasitoid, *Trichogramma minutum*, was imported and released after augmentation (King 1930), which can be explained as classical biological control approach. *Macrocentrus homonae* was introduced to Sri Lanka, along with several other parasitoid species in 1935 from Java, Indonesia by Dr. G.E.J. Wixou. The required level of control of tea tortrix was achieved and the attempt was successful (King 1937, 1939), and the success of biocontrol program has been continued to date. Biology and behavior of the parasitoid were studied by

Gadd (1946), including population dynamics and synchronization of host and parasitoid populations, which could lead to the better use of biocontrol agents. The monitoring of the population has revealed evidences for the spread of the parasitoid into the different geographical locations where the elimination of native parasitoids by the exotic species has been realized and reported by Austin (1957). The impact of hyperparasitoids on *M. homonae* population has been well studied (Cranham 1962). Among hyperparasitoids, *Ceraphron fijiensis*, *Hemiteles* sp., and *Goryphus variibalteatus* were common in tea plantations.

The association between tea tortrix outbreak and application of insecticides has been realized where the chemicals that have been recommended and applied for the control of shot-hole borer have subsequently been restricted on tea cultivation. Among the chemical control, classic chemicals had been tried on tea, such as lead chromate (Jardine 1919a), tobacco and soap spray, DDT, trichlorophon, and methomyl. Use of sex pheromones has also been an attempt to control tea tortrix (Sivapalan and Vitarana 1975). The concept of integrated pest management was also adopted to manage the tea tortrix population in tea (Vitarana 2000). Over the years, the development of population management strategies of tea tortrix has branched into many directions over the past and achieved a good sustainable population management system.

Borer complex in tea was studied by Rutherford (1914), who has found many species. The two species found in tea were *Xyleborus fornicates* and *X. compactus*. The shot-hole borer (*X. fornicates*; SHB) has been a serious pest in tea and therefore was declared as a pest under the provision of Plant Protection Ordinance in 1907, to restrict the spread of the pest among the plantations. This is a fine example for use of regulatory pest control even at the beginning of the last century. The host range of shot-hole borer was studied (Cranham 1963a), and 49 plant species have been found to support the completion of its life cycle (Danthanarayana 1968). Extensive studies on biology, ecology, life cycle, gallery formation, and effect on tea yield have been conducted by the staff of the TRI over the years. The relationship between the borer and the fungus *Monacrosporium ambrossium* was first revealed by Gadd and Loos (1947).

The flight behavior of the beetle has been studied using "Johnson suction trap" (Calanido 1964a, b). Distribution studies have revealed that SHB prefers the tea cultivation at the elevation around 600 m. The borer attack on tea plants leads to a complex situation including the yield loss, die back of branches, soil erosion, poor frame of the bush, and wood rot (Cranham 1963b). Biological control of shot-hole bore was first attempted by introducing a predatory beetle species imported from United Kingdom (Austin 1956), but the predator has been a bit larger than the borer galleries having no access to the shot-hole borer broods. A chalcid wasp, *Perniphora robusta*, was imported from Switzerland and introduced to Sri Lankan tea fields in 1969, and another chalcid wasp *Tricholas xylocleptis* and a braconid wasp *Heterospilus alter* were imported from Switzerland and released in Sri Lankan tea fields in 1970 (Sivapalan 1971). An entomopathogenic fungus, *Beauveria bassiana*, has been tested to manage the borer population (Senaratne 1984) without promising results in managing the borer population. Biological control of SHB was not successful due to various biological reasons such as cryptic habitat of SHB, poor efficacy of available natural enemies, and poor viability of entomopathogenic spores. This has been the case even for today.

Chemical control of SHB initiated with the application of resin, soap, and fish oil as paint. In 1950, synthetic insecticides became available and those were used as a spray and an effective borer control was achieved. Sixty-four insecticides including organocholorines, carbamates, and pyrethroids have been evaluated (Vitarana 2000), and promising chemicals were sprayed to manage the borer population.

Clonal resistance was also an approach for borer control, and a reasonable effort has been taken to screen the clones where TRI 3022, 3023, 3044, 3047, and 3073 have shown resistance against the borer attack. Some work has been done to understand the biochemical mechanism of the resistance. Karunaratne et al. (2008) reported the involvement of volatiles of the tea plant, namely eugenol, hexanol, α- and β-pinene, geraniol, and methyl salicylate, to attract SHB adults.

Similar studies have been conducted with the focus on population management with other tea pest species such as live wood termites (Hutson 1923; Jepson 1926), scavenging termites, nettle grub, geometrid caterpillars, mealybugs, ants, white grubs, red borer, lobster caterpillar, red slug, and bag worms. Depending on the economic significance of the attack of these pest species, research studies have been conducted at different depths. Except the live wood termites, other species are occasional pests and the attack on tea plant has minor significance. Nevertheless, mite attack on tea is significant under dry weather as it facilitates the rapid buildup of the population level, and the natural control factors are not adequate to keep the mite population under control. Several species of mites have been reported and red spider mite, scarlet mite, purple mite, and yellow mite are important species among them (Vitarana 2000).

As a well-established research station, the staff of TRI has gathered most of the biological and ecological information of tea pests. The use of pheromones for mass trapping and mating disruption has been the recent research interest, which has been logically focused. This aspect can be further studied using modern technologies. Expansion of distribution of pest species is a concern with the warming environment and might affect the growth rates of the pest. Screening for resistant tea clones is still remaining as a key area of research at the TRI, and it is one of the ways out to manage the insect population. However, the mechanism of resistance is not well understood, hence requires further attention in planning future research programs. A rapid method of screening tea clones using molecular tools is a current need. Live wood termite is still a problem, and there is no efficient method to detect the infestation early.

11.3 Research on Insect Diversity

Over the twentieth century, entomological research has been focused on different study areas, including insect diversity. The diversity of scale insects in Sri Lanka was first studied by Dr. Green with his own interest, and a book titled "Coccidae of

Ceylon" was published in four volumes (Green 1896–1922). This reference material contains taxonomic and morphological descriptions of scale insects that Green collected in different geographical areas of Sri Lanka. Some of the voucher specimens and the original art drawings have been still conserved in the insect collection at the Horticulture Crop Research and Development Institute (HORDI), at Gannoruwa in Peradeniya, Sri Lanka. Coccidae of Ceylon is a standard text book and a valuable resource for the coccid taxonomist even today. The diversity of beetles in Sri Lanka was first studied by Weather Horn (1871–1883) (Wijesekara and Wijesinghe 2003), followed by George, M.R. Henry. Fauna of British India, including Ceylon and Burma, is a series of books, which addressed the major arthropod orders examining the diversity of different insect groups. Dragonflies (Odonata) have been an interest to Lieftinck (1955) followed by Fonseka (2000) and Sumanapala (2017).

In addition to the residential scientists, some visiting naturalists have done some studies by collecting insect specimens in Sri Lanka (Wijesekara and Wijesinghe 2003). Fred Keiser, a Swiss naturalist, studied the dipterans in Sri Lanka in 1953–1954. Paul A. Remy is another visiting naturalist who studied the soil macro arthropods in 1959. Professor W. Fernando studied the aquatic insects in temporary and artificial habitats (Fernando 1959a), aquatic hemiptera (Fernando 1959a), parasitic hymenoptera (Fernando 1959b), major insect groups such as cockroach (Fernando 1959d), and many other insect groups with minor significance as agricultural insects such as collembolans. The hemipteran diversity has been studied by the curator of Colombo museum, Mr. G.P.D. Karunarathne. The diversity of blister beetle in Sri Lanka has been studied by Mohameidsaid (1973). A long-term collaborative work with Smithsonian Institute, USA, has been a significant land mark on studies of insect diversity, and the work "Biosystematic studies in Ceylonese insects" was led by Mr. Karl Krombein during 1969–1980 (Wijesekara and Wijesinghe 2003). At present, the National Museum has a deposit of more than 100,000 specimens of dry-mounted and wet-preserved insects. Mr. Claude Besvchet and Mr. Ivan Lobl were Swiss visitors who studied the soil insects in 1970, while the aphid diversity in Peradeniya University Park has been studied by Wijerathna and Edirisinghe (1995) where eight species were reported with two new country reports. These species have been associated with 21 grass species. Edirisinghe (1994) reported 17 species of aphids in the same locality associated with 52 host plant species. The diversity of agromyzid flies has been studied by Wijesekara (2002) in which four new country records along with 34 other locally available species were reported. The bee diversity in Knuckle mountain range has been studied by Karunaratne and Edirisinghe (2008a), and the highest diversity has been found in low elevation associated with home gardens where plant diversity is much higher than at the higher altitudes of Knuckles. The bee diversity in Peradeniya University Park has been studied by Karunaratne and Edirisinghe (2002), in which 48 species in 21 genera in four families have been reported. A taxonomic key to identify the bee species has been given (Karunaratne and Edirisinghe 2008b). Diversity of tiger beetles has been studied by Dangalle et al. (2014) in which the occurrence of *Cicindella waterhousei* has been reported.

The diversity of thrips in Sri Lanka has been reviewed by Tillekaratne et al. (2011) in which details of 78 species have been given. Thrips in common vegetable crops have been studied by Tillekaratne et al. (2008). The diversity of mealybugs (Sirisena et al. 2013a, b), leafhoppers (Gnaneswaran et al. 2008a, b, 2010), fruit fly (Ranaweera et al. 2017), termites (Ekanayake and Karunarathne 2011; Hemachandra et al. 2011), and soft scales (Sirisena et al. 2016) have been studied, in which new country records along with already reported species have been documented. The diversity of ants in Sri Lanka was well studied by Dias (2014) and her coworkers. Butterfly fauna has been well documented by Poorten and van Poorten (2016). The diversity of natural enemies in Sri Lanka has been studied by many different scientists in different habitats with the focus on promotion of biological control, and the details of such studies are reported under biological control in this review.

In addition to the study of diversity of common/interesting groups of insects, the diversity of insect fauna associated with different crops has been studied. After Green (1906), the tea studies on insect fauna has been followed up by the subsequent entomologists in TRI such as Rutherford (1914), Stockdale (1920), Hutson (1921, 1922), and Light (1928a, b). Similarly the insect fauna of other commercial crops has been studied including rubber (Petch 1921), coconut (Hutson 1922, 1927; De Mel 1927, 1928), paddy (Fernando et al. 1954), sugarcane (Rajendra 1979), and cotton (Green 1909). Further, new pest species have emerged in association with different crops and those species have been recorded; black planthopper (*Harmalia heitensis*; Balasingham and Vijayaratnam 2009), *Helopeltis antonii* in guava (Jayawardene et al. 2007), banana leaf roller (*Erionota thrax*; Premawardhane et al. 2016), jewel beetle (*Sphenoptera indica*) in ground nut (Rajeshkanna et al. 2016), amaranthus stem borer (*Hypolixus truncatelus*; Rajeshkanna et al. 2017), durian stem borer (*Batocera rufomaculata*; Atapaththu et al. 2015), and *Leptocorisa acuta* in rice (Mandanayake et al. 2014) are some examples that have been recorded in recent literature. Wijeratne (2000) compiled the records on insects feeding on plants providing a catalog for Sri Lankan crop pests. Studies on insect biodiversity have expanded our knowledge more toward the insect taxonomy and phylogenetic relationships. Many universities around the world maintain insect collections, insect museums. The Natural History Museum in London plays an invaluable role in the area of insect biodiversity. The research in biodiversity assessment could be reoriented as the diversity in ecosystems, thereby the findings on insect diversity could be better use in sustainable management of the ecosystems. This aspect could be strengthened focusing on the roles played by different insect groups. Further, an official pest list is not yet available for Sri Lanka and is an important tool to handle the plant quarantine issues that arise in the international trade. The information available through insect diversity studies could be the base for preparation of an official pest list. There is a greater need to establish a national collection of insects in the form of an insect museum as the current collections are staggered and poorly maintained.

11.4 Biology and Ecology of Pest Species

Along with the fauna, the biology and ecology of major pest species have been studied, providing the essential details needed for the formulation of management strategies. The occurrence of insect pest outbreaks has been continuously reported in the Administrative Reports of the DOA. The arthropod community in rice ecosystem has been studied by Bambaradeniya and Edirisinghe (2008) in which 282 insect species belonging to 90 families and 17 orders have been reported. Similar results have been reported by Bambaradeniya et al. (1998) and Edirisinghe and Bambaradeniya (2006). The diversity of pest species in rice ecosystem has been studied by Rajendram and Devarajah (1990) who reported eight major and minor pest species and five predatory species. The studies on the insect pest control have been well reported even during the mid of the twentieth century, such as for paddy bug (Fernando et al. 1957), paddy stem fly (Fernando and Manickavasagar 1957), paddy pentatomid bug (Alwis 1941), and paddy pest control (Fernando et al. 1954). Temporal variation of insect pest populations in rice ecosystem has also been studied (Uvarov 1930; Mandanayake et al. 2015), and entomological research in paddy crop has been well reviewed in the Rice Congresses held in 1980, 1990, 2000, and 2010 in Sri Lanka. The insect pests associated with sugarcane which include 26 species in different taxonomic groups have been well reported by Rajendra (1979) together with the associated parasitoids and predators. The biology and control of pink mealybug in sugarcane has been studied (Rajendra 1974).

Vegetable crops are heavily infested by many insect species. Hence, the demand for the control of vegetable pests is much more intense than the insect pest control in perennial crops. Along with this demand, the biology and ecology of vegetable pests have been well studied and reported. The biology of *Diaphania indica*, a leaf-eating lepidopteran insect in cucurbit vegetables, has been studied by Ganehiarachchi (1997). Another cucurbit insect gall midge (*Lasioptera chichindae*) has been studied along with the control practices (Faheemah and Sulaiman 1990). The biology of flea beetle (*Scelodonta strigicollis*) in grapevine has been studied (Jeyaseelan and Mikunthan 2004) to facilitate the formulation of control measures. The biology of *Hellula undalis*, one of the cabbage-feeding insects, has been studied by Ketipearachchi (2004). Cutworms have been a serious problem in potato cultivation and their biology has been reported. Brinjal is a widely grown crop and its pest situations together with the biology have been studied (Thevasagayam and Canagaslingham 1961). Insect pests of dhal (*Cajanus cajan*) have been studied by Thevasagayam and Canagaslingham (1960). Some other species such as *Bagrada picta* (Hutson 1935), cotton steiner (Manikavasagar 1955), paddle leg bug (Fernando 1957), and potato pests (Fernando and Manickavasagar 1958) have been well studied. The recent entomological research studies in vegetable crops have been well reported in the reports produced by different research institutes of the DOA for the centenary celebrations (Anon 2000). Further, almost all insect pest species attacking vegetable and other crops cultivated in Sri Lanka have been well studied by the scientists and reported in scattered places. This information is valuable in the formulation of management practices.

Cacao has been an important commercial crop and its pest species have been studied in the early part of the century. Cacao capsid bug (*Helopeltis ceylonensis*; De Silva 1961b; Fernando and Manickavasagar 1956) and *Empoasca devastans* have been studied in relation to its control (Fernando 1959c). Interest on coffee berry borer control is evident with the research conducted by Hutson (1936). The recent work on management of insect pests in agriculture export crops such as coffee, cacao, pepper, cardamom, cinnamon, clove, and ginger have been well reported in the Proceedings of the Annual Symposium of Minor Export Crops (ASMEC) conducted by the Department of Export Agriculture in Sri Lanka. Most of these researches have focused on management of the pest populations using nonchemical strategies.

Apart from the insect pest species attacking crops, the insect species associated with grain and grain product have been studied (Ganesalingam 1976, 1977) where six insect species have been identified as the common pests in grain stores while another five species as less common minor pests. Stored grain pests (Easter 1954), potato tuber moth (Manickavasagar 1953), and *Ephestia cautella* (Fernando 1939) have been well studied in relation to their biology and control.

Mosquitoes have been an interesting group of insects as the vectors of human diseases and the diversity of mosquitoes has been studied by Amerasinghe (1990), providing a key to identify the *Anophiline* female mosquitoes and the larvae (Amerasinghe 1992). Mosquito diversity has been extended to natural forest area, and 34 species belonging to 10 genera have been reported (Kulasekera and Amerasinghe 1990). Further, the physio-chemical characterization of mosquito breeding habitats in irrigated lands has been studied (Amerasinghe et al. 1995) to facilitate the management of mosquito populations.

Biology and ecology of pest species are key information in the formulation of control strategies. This has been well understood even at the beginning of the century, which is evident with early entomological research. Even in the recent past, researchers have paid attention to study the biology and ecology of insect species that have become pests lately. It is important to study the insect ecology and biology with a special focus on population management strategies. The nonchemical pest control strategies are generally based on biology and ecology of the pest species. The behaviors of insect pest species have been poorly studied but important in the formulation of management strategies or to improve the existing management strategies.

11.5 Management of Insect Pest Populations

With respect to the management of insect pest populations in crops grown in Sri Lanka, many different approaches have been attempted and different levels of success have been achieved. Among those approaches, biological control, use of resistant varieties, botanicals, and insecticides have been highlighted in the literature.

11.5.1 Biological Control of Insect Pests

Biological control has been an interest for the researchers even in 1880s with the success on cottony cushion scale management using a predatory beetle *Rodolia cardinalis* in California in USA on citrus. The interest on biological control was evident with the work of Myers (1929) and Neave (1927). Management of plantation crop pests through biological control has been an interest because of many favorable biological and ecological factors. In Sri Lanka, management of tea tortrix has been well achieved using *Macrocentrus homonae*, which was started in 1935 while the coconut leaf miner was well managed using *Demokia javanica*, which was initiated in 1970s. The coconut caterpillar, too, has been managed using a set of parasitoids *Brachemaria nephantidis*, *Bracon hebator*, *Eriborus trochanteratus*, and *Trichospullus pupivora*. More recently, the papaya mealybug population has been managed using a parasitoid species *Acerophagous papayae*. The management of leaf miner (*Liriomyza huidobrensis*) was also attempted by introducing a braconid parasitoid *Dygliphus isaea*. All these attempts were classical biological control, and the insect pest species concerned were exotic species. Many of those introductions were made by observing the foreign experiences and were successful, and required level of control has been achieved. In addition, many researchers became interested in biological control with the realization of health and environmental hazards of application of synthetic insecticides.

The first step toward the biological control is to identify the naturally, locally available parasitoid and predatory species in Sri Lankan cropping systems. These data are staggered and a few of the key research work are highlighted in this review. De Silva (1961a) presented a very comprehensive list of local parasitoids and predators of insect pests in Sri Lanka. In this compilation, 200 insect pest species have been listed, together with their parasitoids and predators. The author, however, considered the list as a preliminary list which indicates that there could be many more natural enemies that are not included in the list. Ketipearachchi (2002) listed 80 species of parasitoids of crop pests, including hypoparasitoids. The parasitoid or hypoparasitoids have been collected through laboratory rearing of the host species. Rajendra (1979) listed the parasitoids and predators associated with sugarcane insects. Senadhira (1967) published a bibliographical guide to the animal parasites, which include the insect parasitoids and predators of insect pests in Sri Lanka. Rajendram and Devarajah (1990) reported 10 species of predators in the rice ecosystem. Fellowes and Amarasena (1977) reported the parasitoid species associated with grain legumes in Mahaillupullama area. Mayadunnage et al. (2009) reported the syrphid species associated with aphids. The diversity of coccinellid species in vegetable ecosystem in the mid country area (Mayadunnage et al. 2007) and predatory insects of vegetable insects (Mayadunnage et al. 2008) have been studied. The parasitoids of brinjal pod and shoot borer (Nagalingam et al. 2006), and leaf miner *Liriomyza sativa* (Nagalingam et al. 2007) and leaf miners (Nagalingam et al. 2008) in vegetables have been studied. Egg parasitoids of rice leaf folder (Hemachandra and Perera 2016; Perera et al. 2015; Hemachandra et al. 2014c) and cabbage caterpillars (Hemachandra et al. 2015) have been studied. Parasitoids and predators of mealybugs have been studied by Perera et al. (2011) and Rathnamalala et al. (2014), respectively. Samarakoon et al. (2017) studied the parasitoid fauna in rice ecosystem. The parasitoids and predators of fruit fly have been studied (Bandara and Billah 2015).

Although the parasitoid fauna of insect species or ecosystems were well documented, the attempts to use them in biocontrol programs are comparatively limited. Perera and Hemachandra (2014) studied the longevity, fecundity, and oviposition of *Trichogramma bactrae* with the interest of mass rearing of the egg parasitoids. Singhamuni et al. (2015a) attempted to mass rear egg parasitoids *Trichogramma chilonis* and *Tricogramma achaeae* (Hymenoptera: Trichogrammatidae) on *Corcyra cephalonica* eggs. Further, Singhamuni et al. (2015b) evaluated the potential of egg parasitoid *Trichogramma chilonis* (Hymenoptera: Trichogrammatidae) as a biocontrol agent for *Trichoplusia ni*, the Cabbage Semi-looper. The interactions among egg parasitoids and cabbage caterpillars were studied (Singhamuni and Hemachandra 2013). Ganesalingam (1966) studied the effect of environmental factors on parasitism of *Nezara viridula* by *Telenomus basalis*, which is a scelionid egg parasitoid.

Some work has been carried out to understand the biology and ecology of parasitoids, and the findings are quite helpful to enhance the efficacy of biocontrol. Mandanayake et al. (2016) studied weed floral diversity in rice, which support the better survival of parasitoid adults. The biology of *Acerophagous papayae* has been studied to enhance the mass rearing of parasitoids (Premarathne et al. 2017). The levels of parasitism of papaya mealybug in different ecological zones have been examined (Galanihe et al. 2015).

Implementation of a successful biological program is quite challenging, and the related constrains were reviewed by Ahangama and Gilstrap (2007). However, considering the benefits and the sustainability, the biological control of crop pests should be attempted through natural enemy conservation and augmentative release, especially using egg parasitoids. Use of egg parasitoids is more important as the pest is destroyed at egg stage, before the larva damages to the crop. Moreover, less use of pesticide in the crop system encourages the development of natural enemy fauna providing a suppression of insect pest populations. Mass release of predators and parasitoids to manage insect pest populations has been the practice in protected agriculture as well as in open fields, and it is an area that Sri Lankan researchers should pay more attention. This approach has been developed as commercial ventures in many countries. There is a timely need to establish a central place to do the insect rearing with modern facilities, which is useful to enhance the biocontrol research. Ecological engineering is a new area of research that needs the attention of Sri Lankan researchers. In this approach, the ecosystem is managed in favor of natural enemies, expecting enhanced suppression of insect pests. Microbial control of insect pests, too, is an underutilized area in Sri Lanka which needs the attention.

11.5.2 Use of Botanicals in Insect Pest Management

Use of plant-based products has been an interest for decades. Traditionally, farmers have used different plant parts and plant exudations to repel insects. Burning of cashew (*Aanacarium occidentale*) shell is a well-known practice to repel adult mosquitoes in rural homesteads. Use of *kappetiya* (*Croton lacciferus*) leaves as a mulch in betel cultivation, along with the planting of turmeric (*Curcuma longa*), has been a practice over the years to manage the pests of betel. The interest on plant-based pest control products became less with the use of synthetic insecticides, which are more effective, reliable, and convenient. However, upon the realization of effect of pesticides on human and environmental health, the interest of botanicals was raised again among researchers (Dubey 2010). The pesticidal properties of 55 medicinal plants (Hewage et al. 1997), bark of wood apple (Bandara et al. 1989), and several Sri Lankan plants (Perera et al. 1995) have been examined using different target insects. Repellent properties of essential oils against cockroach were examined by Paranagama and Ekanayake (2004). The latex of *Euphorbia antiquorum* has insecticidal properties (De Silva et al. 2008). *Ocimum* spp., *Lantana camara*, and neem (*Azadirachta india*) seed powder have been found effective to manage the damage of maize stem borer (*Chilo partellus*; Gunewardena et al. 2017). Insecticidal properties of rhizomes of wild ginger (*Zingiber purpureum*) have been proven against *Calasobruchus maculatus*. Use of botanicals together with insecticides (chitin biosynthesis inhibitors) has been examined against cabbage-leaf-eating caterpillars and found promising (Bandara and Kudagamage 1989). A botanical mixture of garlic, ginger, and green chili is effective against the thrips in chili (Rajeshkanna et al. 2017). Balasingham et al. (2003) proved that neem and garlic (*Allium sativum*) extracts are effective in controlling thrips. It appears that many of the tested plants had the pesticidal properties. However, development of such plant extracts as a usable commercial formulation is the need, and research toward developing such formulation is essential. The registration of plant-based products is still challenging as the unavailability of eco-toxicological data of plant-based materials. Though the plant-based products are natural, it does not mean that the extracts are safe and no effects on human and animal health. Therefore, testing on toxicity and potential environmental and health effect is necessary.

11.5.3 Plant Resistance in Insect Pest Management

Use of host-plant resistance or resistant varieties has been an effective strategy to manage the insect pest population. Therefore, the research on host-plant resistance has been evident over the years for the short-term crops and perennial crops such as tea. The tea clones TRI 2023, 2027, 3013, 3014, 3018, and 3025 have been introduced as tolerant clones against the shot-hole borer (Walgama 2012). A significant progress has been made in breeding rice varieties that has resistance against brown planthopper (BPH—*Nilaparvata lugens*; Nugaliyadde et al. 2001), gall midge, and thrips. Rice variety Bg379-2 is the first variety developed in Sri Lanka possessing resistance to BPH, and the variety has derived BPH resistance from an introduced traditional rice variety Ptb 33 from Pathambi (India) (Nugaliyadde and Wilkins 2000). The basis for resistance against BPH has been studied using *Oryza nivara* and *O. eichingeri* (Madurangi et al. 2011); antixenosis and antibiosis effects of *O. nivara* accessions against BPH have been reported (Madurangi et al. 2013). Screening of rice varieties in the greenhouse and in the fields has been discussed in relation to oviposition, honeydew production, and nymph survival (Kudagamage and Nugaliyadde 1982). Use of molecular tools in selection of rice lines against BPH resistance has been reported (Jayathilaka et al. 2014). The occurrence of biotypes and reaction of rice varieties with new biotypes have been demonstrated using gallmidge (*Orseolia oryzae*) as the test insects. Varietal resistance against rice leaf folders (*Cnaphalocrocis medinalis* and *Marasmia patnalis*) has been done by Dhanapala and Claridge (1990). Screening of cowpea (*Vigna unguiculata*) varieties against bean fly (*Ophiomyia phaseoli*) has been done by Dharmasena and Fernando (1988). Different levels of resistance of mung bean (*Vigna radiata*) varieties against *Calasobruchus* spp. has been reported (Dharmasena and Subasinghe 1986). Attempts to control vegetable pests through the use of host-plant resistance are limited over the last 50 years. Paddy varieties resistant to thrips, BPH, and gall midge have been bred, and it is a policy of the Rice Research and Development Institute (RRDI) of Sri Lanka to release only those varieties that show resistance, at least moderate resistance, to those three insect pests. It is an appreciable move toward sustainable pest management and should be adopted in other crop groups as well. The insect-resistant vegetable varieties, at least for key pest species, are rarely available. Currently, most of the vegetable seeds are imported and distributed by the private companies, and no adequate attention has been paid for the levels of insect resistance. Use of host-plant resistance is well compatible with IPM and has many advantages to be used in pest management. Therefore, it is important that plant breeders and entomologist work together to breed varieties that are resistant for insect damages. The breeding programs should be continued according to a national plan irrespective to the change of institutional infrastructure. New technology and molecular tools could be used for rapid screening of varieties for insect resistance.

11.5.4 Insect Pest Management Using Insecticides

History of the use of insecticide in pest control goes back to the colonial time (Jardine 1919a). Sodium fluoride has been used in cutworm control (Anon 1928) and calcium cyanide to control *Helopeltis* in tea (Brittain and Shaw 1926). Since 1977, the use of insecticides has been extensive due to their availability in the country, with the adoption of open economic policies. The plant protectionists have also encouraged the farmers to use insecticides in insect pest control as it is an easy way out of managing the most of the insect pest problems in agriculture.

Use of insecticides to control insect pests has become popular and an effective strategy. However, the insecticide application is associated with many health and environment consequences. Screening of insecticides to control the most troublesome insect pests in major crops has been done in the recent past according to a routine protocol, which is a mandatory requirement imposed by the Registrar of pesticides. Evaluation of different insecticides against specific pest species such as etofenprox against brinjal fruit and shoot borer (Marasinghe et al. 2017b), carbosulfan 200 g/l SL, diazinone 5% GR, fipronil 0.3% GR, virtako 40% WG against yellow stem borer (Rajeshkanna and Sumankali 2015), emamectin benzoate and metaflumizone against onion caterpillar, *Spodoptera* sp. (Gunawardena et al. 2008), carbosulfan, and diazinone for rice bug control (Sarathchandra et al. 2016) has been done recently. Insecticides for the control of leaf miner (Wahundeniya 2001) and banana weevil (Galanihe et al. 2017b) have been tested and recommended. Dharmasena (1993) screened several insecticides to control the cowpea pod borer (*Maruca testulalis*). Bean fly is a seedling pest of bean and requires special attention, where seed treatment has been successful in its control (Wijesekara and Abeytunga 1983). Control of maize stem borer (*Chilo partellus*) has been achieved using the insecticides carbofuran and diazinon.

Insect growth regulator-type insecticides show low mammalian toxicity. Hence, use of such insecticides on leafy vegetables like cabbages is logical (Wahundeniya 1993) to control leaf-eating insects. Screening of classic insecticides to control cabbage leaf caterpillars has been reported (Jesudasann and Yogaratnam 1985). Pyrethroid insecticides came to the market in 1980s, after testing of such products in Sri Lanka (David and Peries 1985).

Methyl bromide has been used for fumigation of stores and export commodities; however, the current regulations have restricted the use of this chemical for quarantine purposes. Therefore, there is a need of alternate chemicals such as phosphine, which is used for fumigation of grain and grain products. Wijayaratne et al. (2009) examined the use of smoke, generated with partial combustion of paddy husk to manage stored product insects. Warshamana et al. (2016) tested the effectiveness of ethyl formate (16.3 w/w) to control mealybugs in pineapple (*Ananas comosus*) through fumigation and found effective. Further, efforts have been made to introduce liquid phosphine (ECO$_2$FUME: a cylinderized gas formulation with 2% phosphine and 98% carbon dioxide w/w), which is recognized as a safer fumigant for quarantine and pre-shipment (QPS) fumigation and non-quarantine purposes.

Development of resistance in insect populations to insecticides is a concern, and management of resistance development in insect populations is essential. Mechanisms of resistance development have been reviewed (Karunaratne 1998). Resistance development in insect populations may lead to the failure of control using insecticides. This has been demonstrated with whitefly population (Marasinghe et al. 2017a). Damayanthi and Karunaratne (2005) studied the biochemical characteristics of insect pests of vegetables and predatory coccinellids in relation to the development of insecticide resistance. Further, Karunaratne and Weerakoon (2007) reported the involvement of metabolic and insensitive acetylcholinesterase mechanism in insect pests and predators of rice ecosystem.

Exposure of farmers and the consumers to insecticides has been a concern, and there are studies such as chlorpyrifos ingestion with diet (Marasinghe et al. 2015), vegetable contamination with chlorpyriphos (Menike et al. 2012), pesticide residues in vegetables (Lakshani et al. 2017), organophosphate residues in food commodities (Marasinghe et al. 2011), and pesticides in water (Aponso et al. 2003). Heavy use of insecticides in vegetables is a great concern as the high risk of ingestion of residues. Therefore, Hariharan and Yamini (2007) studied the pesticide usage in vegetables in the Vavunia district of Sri Lanka and reported the types of products used by the farmers. Improper disposal of pesticide containers poses a threat on contamination of water and soil. Therefore, Prasanna (2015) and Nishantha et al. (2016) assessed the status of the problem in relation to the vegetable farmers and reported that pesticides are poorly handled and requires the pest monitoring system and improvement of pesticide application technology.

Effect of pesticides on non-target organisms is also a major concern. Iroshani and Mohotti (2011) studied the effect of agrochemicals on soil micro-arthropods and found the relatively less abundance of soil micro-arthropods in agricultural land. Susceptibility of natural enemies in rice ecosystem for insecticides has been studied by Karunaratne et al. (2007). The effect of pesticide on biodiversity has been studied indirectly by comparing the arthropod fauna in conventional and pesticide-free fields (Shalika and Hemachandra 2008; Mandanayake et al. 2013; Hemachandra et al. 2013, 2014a, b) and found the poor diversity and abundance of arthropods in conventionally managed agricultural fields.

It is important that synthetic insecticides should be available to manage the insect populations when it reached epidemic or intolerable levels. Hence, more effective and safe to handle chemicals should be screened on the need basis. Resistance development in insect populations is a major issue to be dealt with and thus insecticides with different modes of action should be made available to the farmers. More attention on effect of synthetic chemicals on human and environmental health should be paid, and research should be extended toward the examination of the relation between human health and pesticides. It appears that not much attention has been paid on improving insecticide application technology, which is a requirement for efficient use of pesticides.

11.5.5 Nonchemical Methods in Insect Pest Management

There has been an interest on non-chemical pest control even in the early part of the twentieth century. The work in the latter part of the twentieth century is more intense probably due to the realization of the negative impacts of synthetic insecticides. The concept of IPM has been well placed in 1980s with the support of DOA and other projects funded by the NGO and INGOs. Integrated approach for rice pest control has been implemented (Nugaliyadde et al. 2000) with special reference to management of BPH (Nugaliyadde et al. 2001). Integrated control of fruit fly has been formulated (Bandara et al. 2017) by using many strategies such as bagging of fruits (Pushpakumari and Mahagollage 2016), attraction of males using plant base attractants (Gunawardhana 2017), and attraction of flies with protein baits (Bandara et al. 2017) and field sanitations.

There are many nonchemical strategies that have been tested and recommended at different time periods. Timely pruning (Pyper 1928) and application of manure (Jepson 1926; Gadd 1923; Gadd and Jardine 1923) have been proposed to manage shot-hole borer populations in tea. Asynchronization of peak pest population with susceptible stage of the crop through changing the planting date (Subasinghe and Amarasena 1988) is another strategy to manage *Chilo partellus* in maize. Jesudasann and Yogaratnam (1984) studied the population fluctuations of diamond backmoth (*Plutella xylostella*) and its parasitoids and suggested to minimize the application of insecticides during the off peak time. Use of companion crops such as okra, leeks, and mint has been an interest to manage the sucking pests such as whitefly in tomato (Galanihe et al. 2017a). Manipulation of host plants in the ecosystem is another approach of nonchemical pest control. Association of weedy hosts and rice plant bug has been studied by Nugaliyadde et al. (2001), and host plant preference of papaya mealy bug has been studied by Galanihe et al. (2015). The level of nitrogen fertilizer has an effect on pest population of maize stem borer (Gunawardena and Madugalla 2014). Use of poultry manure in chili crop has an effect on thrips population (Vijayaratnam et al. 2008). Use of organic and inorganic mulch to control aphids in tomato has been recorded (Senanayake et al. 2014). Pheromone-based trapping and mate disruption is an approach to use in population management of potato tuber moth (Nugaliyadde et al. 2014). Leafy vegetables with resistance to insects and control the insects through integrated approach have been reported (Wahundeniya et al. 2005). Flight activity of *Liriomyza huidobrensis* is useful in designing traps for monitoring of the insect populations or for mass trapping. The aspect of nonchemical pest control is very important in developing sustainable pest management program. These aspects should be studied in detail in the future for better use of locally available, renewable resources to develop pest control programs.

11.6 Research on Regulatory Pest Control

History of regulatory pest control goes back to 1901, where Insect Pest and Quarantine Ordinance was placed, followed by Plant Pest Ordinance in 1907, Water Hyacinth Ordinance No. 4 in 1909, Plant Protection Ordinance No. 10 in 1924, and the Plant Protection Act No. 35 of 1999. Movement of plants infested with SHB has been restricted as early as in 1920s. Similar applications have been practiced throughout Sri Lanka when pest epidemics were experienced, e.g., coconut leaf miner, coconut mite, and Weligama coconut leaf wilt disease. Restriction on importing planting materials through imposing appropriate permit conditions has assisted in preventing the accidental entry of quarantine species. However, regulations laid out in 1981 are being implemented, and no regulations were gazetted in relation to the Plant Protection Act No. 35 of 1999. Along with the regulatory control, Warshamana et al. (2012) reported the newly introduced quarantine pests or quarantine pest species detected at the port. Kulatunga et al. (2012) reported the non-compliances to the import/export conditions. Hewage et al. (2007) reported the pest infestations of export foliage consignments. Pest species detected at the ports have been reported (De Silva and Weerasinghe 2007). In order to provide a speedy service, it is important to have pest identification guides combined with new technologies. Development of electronic keys, sharing of knowledge and means of speedy communication with National Plant Protection Organizations (NPPOs) of other countries are also essential to provide a better quarantine service.

11.7 Research on Insect Physiology

The local research in the area of insect physiology and insect behavior are limited and scattered. Functional characteristics (Modder 1967a), activity level of carbohydrate (Modder 1967a, b, c), and glucose diminishing process of alimentary canal of cockroach (Modder 1967c) have been studied. Transformation of alimentary canal through larva, pupa to adult has been studied in *Ptinus tectulus* (Ganeslingam 1982a). Oosorption of *P. tectulus* has also been examined (Ganeslingam 1982b). The oviposition cycle and the possible stimulus for oviposition of *Dermastus maculatus* have been examined (Ladduwahetty 1967). Sensory physiology of coconut black beetle larvae has been examined by Costa and Ganesalingam (1967). The structure and development of testes of *Sitophilus zeamais* have been studied by Ganesalingam (1975). Not many recent studies were found focusing on insect physiology, probably due to lack of funding into the subject area. Much of the research has been focused on applied research, expecting the application of results at short run for the benefits of the farming community.

11.8 Entomological Research on Knowledge Generation

Many research studies have been conducted to generate new knowledge, which is important for the progression of the subject. However, the results of those studies may or may not be directly applied in pest management or for the benefit of the farmers. Dharmaratnam and Sabaratnam (2002) studied the elytral pattern of *Ephilachna* species and concluded that separation of species is more reliable when the characteristics of genitalia and elytral shape are used. The number of black spots on elytra is not a reliable character. Rearing of insects on artificial diets is another aspect of entomology research. This has been attempted with fruit fly (*Bactocera dorsata*; Ranaweera et al. 2017) and diamond backmoth (Perera and Senanayake 2014).

11.9 Conclusions/Future Directions

Management of insect populations to prevent or minimize the yield loss and yield quality loss associated with the insect attack is the prime focus on research, conducted during the last century. The depth of the research and number of research on a pest species depend on the economic significance of the pest or the gravity of the problem. Biological and ecological aspects of the pest species have been studied through research in early and the later part of the twentieth century to some extent. It is quite justifiable because of the need of biological and ecological information for planning of effective control measures. The work toward IPM is appreciable, and more attention has been paid toward the nonchemical pest control methods. In order to get the maximum use of nonchemical methods, further understanding on ecology and behavioral aspects of the insect pest species is needed. As an example, it is widely accepted that management of alternate host plants in the ecosystem is important to interfere with the growth of pest populations. However, in many cases, a clear understanding on the host range of insect pests is not evident. Subsequently, alternative host control has not been fully utilized to manage the pest population. Mating disruption and mass trapping using pheromones have not been fully utilized in the Sri Lankan cropping systems, which have a great potential to use to manage some insect species. Among nonchemical methods, use of plant resistance and biological control including the microbial pest control are primarily important for sustainable pest management. Biocontrol practices that can be implemented at farmer levels should be promoted based on research findings. Commercial biocontrol such as augmentative release is not practiced due to poor facilitation and lack of service providers.

There is no adequate market to attract private sector service providers. Research and development work toward this direction is needed. It is also important to look for new insecticides, which are safer on environment and human health, paying attention on different modes of action to manage the development of resistance. There is an increasing demand for agriculture produce, which are free of pesticide residues. There is no sustainable system in place to ensure that the marketed products are safe for the consumption. This aspect requires the attention of entomologists as well as the toxicologists. There is an increasing interest among farmers and extension officers on plant-based crop production products. However, these products have not been properly tested for non-targets, fate in the environment, etc. Consequently, there is a technical barrier for their registration as a crop protection product. Area-wide pest management is an interest, and no enough attention has been paid on this area of research. Better understanding of this aspect is important in managing active insects, which fly over long distances, especially to manage the re-invasions. This aspect would become more important upon the consideration of possible changes of climate and weather patterns. Further, new communication technologies are poorly used in delivering the crop protection information to the farmers. Hence, there is a dire need in developing appropriate technologies in collaboration with the expertise on information and communication technology (ICT), extension officers, and entomology researchers. Directing of future research into these areas is very useful to formulate sustainable, economically viable pest management systems in all crop systems.

References

Ahangama D, Gilstrap FE (2007) Constraints to the implementation of biological control in Sri Lanka. Biocontrol Sci Tech 17(8):773–795

Alwis ED (1941) The paddy pentatomid bug, *Sctinphara* (Podps) *lurida* Burm. Trop Agric 46:217–220

Amerasinghe EP (1990) A guide to the identification of the Anopheline mosquitoes (Diptera:Culicidae) of Sri Lanka I. Adult females. Ceylon J Sci Biol Sci 21:1–16

Amerasinghe EP (1992) A guide to the identification of the Anopheline mosquitoes (Diptera:Culicidae) of Sri Lanka II. Larvae. Ceylon J Sci Biol Sci 22:1–13

Amerasinghe EP, Indrajith NG, Ariyasena TG (1995) Physico-chemical characteristics of mosquito breeding habitats in an irrigation development area in Sri Lanka. Ceylon J Sci Biol Sci 24:13–29

Anon (1928) Sodium fluoride and cutworms: new methods of fighting the pest. Trop Agric 70:115–116

Anon (2000) Siyawasaka abimanaya. Department of Agriculture, Peradeniya

Aponso GLM, Magamage C, Ekanayake WM, Manuweera GK (2003) Analysis of water for pesticides in two major agricultural areas of the dry zone. Ann Sri Lanka Dept Agric 5:7–22

Atapaththu IK, Mahananda MLLC, Dewage DSKP, Rajapakse RPKC (2015) Stem beetles in durian cultivation. Ann Sri Lanka Dept Agric 17:339

Austin GD (1956) Historical review of shot-hole borer investigations. Proc. symposium on shot-hole borer. Tea Quart 27:97–102

Austin GD (1957) Report of the entomologist. TRI Bull 39:52–53

Balasingham B, Vijayaratnam S (2009) Black planthopper (*Harmalia heitensis* Tho): a newly identified insect pest in mukunuwenna [*Alternanthera sessilis* (L) DC] in the Jaffna district. Ann Sri Lanka Dept Agric 11:233

Balasingham B, Vijeratnam S, Jeganathan K (2003) Efficacy of botanical pesticides and chemical prothiofos against thrips, *Scirtothrips dorsalis* in chilli. Ann Sri Lanka Dept Agric 5:321–323

Bambaradeniya CNB, Edirisinghe JP (2008) Composition, structure and dynamics of arthropod communities in rice agro-ecosystem. Ceylon J Sci Biol Sci 37:23–48

Bambaradeniya CNB, Fonseka KT, Ambagahawatte CL (1998) A preliminary study of fauna and flora of a rice field in Kandy, Sri Lanka. Ceylon J Sci Biol Sci 25:1–22

Bandara BMR, Gunathilake AAL, Sotheeswarna S, Wijeratne EMK (1989) Insecticidal properties and an active constituent of *Limonia acidissim*. Natl Sci Council Sri Lanka 17(2):237–239

Bandara KANP, Billah MK (2015) Parasitoids/predators and alternative hosts of economically important fruit fly pests of low, mid and high elevations of Sri Lanka. Ann Sri Lanka Dept Agric 17:303–309

Bandara KANP, Kudagamage C (1989) Potential of chitin inhibitors and botanical insecticides for the control of cabbage caterpillars. Trop Agric 145:125–129

Bandara KANP, Kumari LGS, Kumari ASP, Magollage MNE (2017) Efficient, user and environment friendly protein bait trap to replace spot application in area-wide management of fruit flies. Ann Sri Lanka Dept Agric 17:319–320

Brittain WH, Shaw WS (1926) Preliminary report on the application of calcium cyanide dust to the control of *Helopeltis* in tea. Trop Agric 67:209–214

Calanido D (1964a) Report of research officer, Entomology Unit, Hantana Substation. In: TRI annual report. Tea Research Institute of Sri Lanka, Thalawakele, pp 91–94

Calanido D (1964b) Studies of the population ecology of the shot-hole borer, *Xyloborus fornicates* Eichh. In tea in Ceylon. Tea Quart 35:1–51

Costa HH, Ganesalingam VK (1967) The sensory physiology of the larvae of the rhinoceros beetle, *Oryctes rhinoceros* L. Ceylon J Sci Biol Sci 7:42–50

Cranham JE (1962) Report of the entomologist. In: TRI annual report. Tea Research Institute of Sri Lanka, Thalawakele, pp 71–80

Cranham JE (1963a) Report of the entomologist. In: TRI annual report. Tea Research Institute of Sri Lanka, Thalawakele, pp 50–69

Cranham JE (1963b) Shot-hole borer biology and control. Notes for planters. Tea Quart 34:127–143

Damayanthi BT, Karunaratne SHPP (2005) Biochemical characterization of insecticide resistance in insect pests of vegetables and predatory ladybird beetles. J Natl Sci Found Sri Lanka 33:115–122

Dangalle CD, Pallewatta N, Vogler AP (2014) Inferring population history of tiger beetle species of Sri Lanka using mitochondrial DNA sequences. J Natl Sci Found Sri Lanka 43:47–63

Danthanarayana W (1968) The distribution and the host range of shot-hole borer. Tea Quart 39:61–69

David I, Peries R (1985) Synthetic pyrethroid insecticides and their potential use in Sri Lanka. Trop Agric 141:65–73

De Mel CNEJ (1927) Pest and diseases of coconut in the North-Western province. Trop Agric 68:252–256

De Mel CNEJ (1928) Notes on coconut pests. Trop Agric 70:85–86

De Silva MD (1961a) A preliminary list of the native parasites and predators of insect pests in Ceylon. Trop Agric 117:115–141

De Silva MD (1961b) Biology of *Helopeltis ceylonsis* De Silva (Heteroptera-Miridae), a major pest of cacao in Ceylon. Trop Agric 117:149–156

De Silva RSY, Weerasinghe ASP (2007) Pests detected in agricultural commodities imported to Sri Lanka. Ann Sri Lanka Dept Agric 2:75–79

De Silva WAPP, Manuweera GK, Karunaratne SHPP (2008) Insecticidal activity of *Euphorbia antiquorum* L. latex and its preliminary chemical analysis. J Natl Sci Found Sri Lanka 36:15–23

Dhanapala MG, Claridge MF (1990) Resistance of rice (*Oryza sativa*) to two species of rice leaf folders, *Canphalocrocis medinalis* Guenee and *Marasmia patnalis* Bradley (Lepidoptera:Pyralidae). Trop Agric 146:13–20

Dharmaratnam M, & Sabaratnam M (2002) Identification of *Epilachna* spp. found in Sri Lanka. Proceedings of Annual Symposium of Sri Lanka Association of Advancement of Science, 58, D10

Dharmasena CMD (1993) Efficacy of insecticides on cowpea pod borer *Maruca testulalis* GREYER (Lep: Pyralidae). Trop Agric 149:101–108

Dharmasena CMD, Fernando MHJP (1988) Resistance of cowpea varieties to bean fly *Ophiomyia phaseoli* (Tryon) DIP: Agromyzidae. Trop Agric 144:155–161

Dharmasena CMD, Subasinghe SMC (1986) Resistance of mung (*Vigna radiate*) to *Callosobruchus* spp. (col: Bruchidea). Trop Agric 142:1–6

Dias RKS (2014) Ants of Sri Lanka. Biodiversity Secretariat of Ministry of Environment and Renewable Resources, Colombo

Dubey NK (2010) Natural products in plant pest management. CABI Publishing, London

Easter SS (1954) Infestation control in stored rice. Trop Agric 60:217–219

Edirisinghe JP (1994) Annotated list of aphids of Peradeniya campus, Sri Lanka. Ceylon J Sci Biol Sci 23:25–29

Edirisinghe JP, Bambaradeniya CNB (2006) Rice fields: an ecosystem rich in biodiversity. J Natl Sci Found Sri Lanka 34:57–59

Ekanayake EWMTD, Karunarathne WAIP (2011) Termite assemblages in lower Hanthana forest and variation in worker mandible structure with food type. In: Proceeding of the Peradeniya University International Research Sessions, Sri Lanka. University of Peradeniya, Peradeniya, p 16, 146

Faheemah S, Sulaiman M (1990) The cucurbit gall midge, *Lasioptera chichindae* Grover (Diptera:Cecidomylldae) and its possible control. Trop Agric 146:67–78

Fellowes RW, Amarasena J (1977) Natural parasites of some grain legume pests in the dry zone. Trop Agric 133:83–90

Fernando CH (1959a) Some observations on aquatic insects found in temporary and artificial habitats in Ceylon. Ceylon J Sci Biol Sci 2:1–4

Fernando EFW (1959b) Two new species of parasitic hymenoptera from pests of tea in Ceylon. Ceylon J Sci Biol Sci 2:249–252

Fernando HE (1957) The biology and control of *Leptoglossus membranaceus* Fabricus (Fam. Coreidae Ord. Hamiptera). Trop Agric 113:107–118

Fernando HE (1959c) Studies of *Empoasca devastans* Dist. (Aam. Jassidae, Ord. Hemiptera), a new pest of cacao causing defoliation, and its control. Trop Agric 115:121–144

Fernando HE, Elikewela Y, Manickavasagar P (1957) Resistance to benzene hexachloride of the rice bug *Leptocorisa varicornis* F. (Fam. Coreidae Ord. Hemiptera) and its insecticidal control. Trop Agric 113:229–244

Fernando HE, Manickavasagar P (1956) Economic damage and control of cacao capsid *Helopeltis* sp. (Fam. Capsidae Ord. Hemiptera) in Ceylon. Trop Agric 112:25–36

Fernando HE, Manickavasagar P (1957) Investigation on the control of Paddy stem fly *Atherigona* sp. (Fam. Anthomyiidae Ord: Diptera). Trop Agric 113:341–345

Fernando HE, Manickavasagar P (1958) Investigation on potato insects and their control with special reference to *Dorylus orientalis*. Trop Agric 114:127–139

Fernando HE, Weerawardane GV, Manickavasagar P (1954) Paddy pest control in Ceylon: paddy pests and their control with chemicals. Trop Agric 110:159–174

Fernando M (1939) The sources of *Ephestia* infestation of stored cacao in Ceylon. Trop Agric 92:141–155

Fernando W (1959d) Two new cockroaches of the genus Sphecophila from Ceylon. Ceylon J Sci Biol Sci 2:244–248

Fonseka TD (2000) The dragonflies of Sri Lanka. WHT Publications, Sri Lanka

Gadd CH (1923) The effect of certain manurial substances on shot-hole borer of tea. Trop Agric 60:75–77

Gadd CH (1946) *Macrocentrus homonae,* a polyembryonic parasite of tea tortrix *Homona coffearia* Neitn. Ceylon J Sci Biol Sci 23:67–79

Gadd CH, Jardine NK (1923) The control of shot-hole borer in tea by the use of general measures. Trop Agric 60:299–304

Gadd CH, Loos CA (1947) The ambrosia fungus of *Xyleborus fornicates* Eichh. Br Mycol Soc 31:13–18

Galanihe LD, Madugalle S, Rajapaksha ST, Siriwardena MI (2017a) Influence of companion crops on sucking pest complex of tomato. Ann Sri Lanka Dept Agric 19:27–35

Galanihe LD, Ranatunga RMMS, Rajapaksha SP (2015) Current status of papaya mealybug, *Paracoccus marginatus* (Hemiptera: Pseudococcidae) parasitism by *Acerophagous papaya* in selected areas of Sri Lanka. Ann Sri Lanka Dept Agric 17:187–190

Galanihe LD, Vithana KVAIK, Rajapaksha ST (2017b) Efficacy of granular formulations of insecticides for the control of Banana Pseudostem weevil, *Odoiporus longicollis* (Coleoptera: Curculionidae). Ann Sri Lanka Dept Agric 19:365

Ganehiarachchi SASM (1997) Aspects of the biology of *Diaphania indica* (Lepidoptera: Pyralidae). Ceylon J Sci Biol Sci 25:203–209

Ganesalingam VK (1966) Some environment factors influencing parasitization of the eggs of *Nezara viridula* L (Pentatomidae) by *Telenomus basalis* Wollaston (Hymenoptera: Scelionidae). Ceylon J Sci Biol Sci 6:1–14

Ganesalingam VK (1975) Structure and development of the testes in the adult *Sitophilus zeamais* Mostschulsky (Coleoptera: Curculionidae). Ceylon J Sci Biol Sci 11:70–76

Ganesalingam VK (1976) A study of insects in four rice stores in the Kandy districts in Sri Lanka. Ceylon J Sci Biol Sci 12:30–46

Ganesalingam VK (1977) Insects in paddy and rice in storage in Kandy districts. Ceylon J Sci Biol Sci 12:169–176

Ganeslingam VK (1982a) Transformation of the alimentary canal of the larva, pupa and adult of *Pinups tectus* Boieldieu (Colecoptera:Ptinidae). Ceylon J Sci Biol Sci 15:136–141

Ganeslingam VK (1982b) Some observation of oosorption in *Ptinus tectus* Boieldieu (Colecoptera:Ptinidae). Ceylon J Sci Biol Sci 15:142–154

Gnaneswaran R, Hemachandra KS, Ahangama D, Wijayagunasekara HNP, Wahundeniya I (2008a) Species of *Nephotettix* Matsumara (Hemiptera: Auchenorrhyncha: Cicadellidae) in Sri Lanka. Trop Agric Res 20:414–418

Gnaneswaran R, Hemachandra KS, Viraktamath CA (2008b) Leafhopper fauna (Hemiptera: Auchenorrhynhca: Cicadellidae) in Peradeniya, Sri Lanka. Proceedings of the International Conference on Biodiversity Conservation Management (BIOCAM) 2008, 3–6 February, in Cochin, India, 110–113

Gnaneswaran R, Hemachandra KS, Wijayagunasekara HNP, Ahangama D (2010) A key for identification of *Deltocephaline* genera (Hemiptera: Cicadellidae: Deltocephalinae) associated with vegetable ecosystem in Sri Lanka. Trop Agric Res 21(4):421–426

Green EE (1890) Insect pests of the tea plant. Colombo Independent Press, p 104

Green EE (1896–1922) Coccidae of Ceylon. Dulau & Co. Ltd, London

Green EE (1906) Entomological notes, report of a meeting of the Maskeliya planters' association. Trop Agric 26:301–308

Green EE (1909) Insects associated with cotton plants in Ceylon. Trop Agric 33:318–321

Gunawardena KNC, Madugalla SRK (2014) Effect of different nitrogen fertilizer levels on stem borer (*Chilo partellus* Swinshoe) (Lepidopetra: Pyralidae) damage in maize. Ann Sri Lanka Dept Agric 16:105–114

Gunawardena KNC, Madugalla SRK, Bandara HMS, Gunawardena NAM (2008) Efficacy of emamectin benzoate and metaflumizone for the control of onion caterpillar (*Spodoptera* spp.) (Lepidoptera:Noctuidae). Ann Sri Lanka Dept Agric 10:315–321

Gunawardhana GCA (2017) An ecofriendly method to control fruit fly. Ann Sri Lanka Dept Agric 19:358

Gunewardena KNC, Wickramasooriya IPSD, Jayanthai KP (2017) Management of maize stem borer *Chilo partelluss* Winhoe (Lepidoptera: Pyralidae) using selected plant powders with insecticidal properties. Ann Sri Lanka Dept Agric 18:118–120

Hariharan R, Yamini P (2007) Pesticide usage in vegetable cultivation in Vavunia district. Ann Sri Lanka Dept Agric 9:223–226

Hemachandra II, Edirisinghe JP, Karunarathne WAIP, Gunatilake CVS (2011) Diversity and abundance of termites in a mahogany plantation in the Gannoruwa hills. Proceeding of the Peradeniya University International Research Sessions, 16, 143

Hemachandra KS, Bandara BPUR, Weerakkody WAP, Wasantha JKS, Sirisena UGAI, Priyadarshanie TDC, Soorasena JM (2013) Diversity and abundance of ground beetles (Carabidae: Coleoptera) in eco-friendly home gardens in comparison with conventional home gardens in southern dry zone of Sri Lanka. In: Proceedings of the International Symposium on Agriculture and Environment. Faculty of Agriculture, Ruhuna University of Sri Lanka, Kamburupitiya, pp 257–260

Hemachandra KS, Bandara BPUR, Weerakkody WAP, Wasantha JKS, Sirisena UGAI, Priyadarshanie TDC, Soorasena JM (2014a) Comparison of soil dwelling insect fauna in eco-friendly versus conventional home gardens at selected localities in Hambantota district. In: Proceedings of the 1st Ruhuna International Science and Technology Conference. Ruhuna University of Sri Lanka, Kamburupitiya, pp 22–23

Hemachandra KS, Bandara BPUR, Weerakkody WAP, Wasantha JKS, Sirisena UGAI, Priyadarshanie TDC, Soorasena JM (2014b) Comparison of diversity and abundance of hymenopteran parasitoids in ecofriendly home gardens and conventional home gardens in Hambanthota district of Sri Lanka. In: Proceedings of the Peradeniya University International Research Sessions, 18. University of Peradeniya, Peradeniya, p 434

Hemachandra KS, Perera MCD (2016) Assessment of egg parasitoid guild of rice leaf folder, *Cnaphlocrocis medinalis* (Guenee): a search for biocontrol agents. J Natl Sci Found 44(2):185–192

Hemachandra KS, Polaszek A, Singhamuni SAA (2014c) Egg parasitoids of *Trichoplusia ni: Trichogramma achaeae* Nagaraja and Nagarkatti 1969, and *Trichogramma chilonis* Ishii (Hymenoptera: Trichogrammatidae) in Sri Lanka. In: Proceedings of the 1st Ruhuna

International Science and Technology Conference. Ruhuna University of Sri Lanka, Kamburupitiya, p 86

Hemachandra KS, Polaszek A, Sirisena UGAI, Singhamunni SAA, Perera MCD (2015) Trichogrammatid egg parasitoids (Hymenoptera:Trichogrammatidae) of rice leaf folder and cabbage caterpillars in Sri Lanka. Sri Lanka J Food Agric 1(1):1–7

Hewage CM, Bandara KANP, Karunaratne V, Bandara BMR, Wijesundara DSA (1997) Insecticidal activity of some medicinal plants of Sri Lanka. J Natl Sci Counc Sri Lanka 25(3):141–150

Hewage LC, Ekanayake HMRK, Fernando KNS, Nimalananda NPH, Fernando WMR, Weerasinghe RU (2007) Insect, mite, diseases infestation in export foliage nurseries in Sri Lanka. Ann Sri Lanka Dept Agric 9:227–232

Hutson JC (1921) Scale insects and mite pests on tea. Trop Agric 56:378–380

Hutson JC (1922) Investigation into shot-hole borers of tea. Trop Agric 59:24–27

Hutson JC (1923) Notes on termites attacking tea in Ceylon. Trop Agric 60:83–87

Hutson JC (1925) Notes on a caterpillar pest of dadap, *Tarasama dorsalis* Wlk. Trop Agric 65:27–32

Hutson JC (1927) Recent work on some pests of economic crops. Trop Agric 68:220–228

Hutson JC (1932) Insect pests and green manuring. Trop Agric 79:129–146

Hutson JC (1935) The painted Bagrada bug (*Bagrata picta*). Trop Agric 85:191–193

Hutson JC (1936) The coffee berry borer in Ceylon (*Stephanoderes hampei* Ferr.). Trop Agric 86:379–383

Iroshani RGS, Mohotti K (2011) Impact of agro chemicals on soil micro arthropods density in different farming systems: a case study at Mandaramnuwara. Ann Sri Lanka Dept Agric 13:217–218

Jardine NK (1919a) Field experiments with anti-tortrix fluids. Bull Dept Agric Ceylon 46:4

Jardine NK (1919b) Tortrix flight breaks. Bull Dept Agric Ceylon 45:4

Jayathilaka KPOS, Kottearachchi NS, Kakulandara DS, Madanayake MARA (2014) Progress towards the marker assisted selection of rice lines for aroma and BPH resistance. Trop Agric 162:45–62

Jayawardene MMS, Piyasena HDAS, Perera KDA (2007) *Helopeltis antonii* (Hemiptera: Miridae): a new pest of guava (*Psidium guajava*) in Sri Lanka. Ann Sri Lanka Dept Agric 9:239–244

Jepson FP (1926) Some preliminary notes on tea termites. Trop Agric 67:67–79

Jesudasann D, Yogaratnam V (1984) Seasonal population fluctuation of diamondback moth (*Plutella xylostella* L.) and its larval parasitoids in the uplands of Sri Lanka. Trop Agric 140:27–36

Jesudasann D, Yogaratnam V (1985) Evaluation of pesticides for controlling the pest complex on cabbage in the uplands of Sri Lanka. Trop Agric 141:1–12

Jeyaseelan B, Mikunthan G (2004) Biology of the flea beetle, *Scelodonata strgicollis* Mots. (Coleoptera:Emuolphidae) on grapevine in the Jaffna district. Ceylon J Sci Biol Sci 32:35–37

Karunaratne SHPP (1998) Insecticide resistance in insects: a review. Ceylon J Sci Biol Sci 25:72–99

Karunaratne SHPP, Weerakoon KC (2007) Involvement of metabolic and insensitive acetylcholinesterase mechanisms in insecticide resistance of rice insect pests and predatory populations from Batalagoda, Sri Lanka. J Natl Sci Found 35(2):103–108

Karunaratne SHPP, Weerakoon KC, Nugaliyadda L, Manuweera GK (2007) Susceptibility of rice insect pests and their natural enemies to commonly used insecticides. J Natl Sci Found Sri Lanka 35:97–102

Karunaratne SW, Kumar V, Pettersson J, Kumar NS (2008) Response of the shot-hole borer of tea, *Xyleborus fornicatus* (Coleoptera:Scolytidae) to conspecifics and plant semiochemicals. Acta Agric Scand Sect B – Soil Plant Sci 58(4):345–351

Karunaratne WAIP, Edirisinghe JP (2002) Bee diversity and floral hosts in selected habitats of the Peradeniya University Park. Ceylon J Sci Biol Sci 30:21–36

Karunaratne WAIP, Edirisinghe JP (2008a) Diversity of bees at different altitudes in the knuckles forest reserve. Ceylon J Sci Biol Sci 37:61–72

Karunaratne WAIP, Edirisinghe JP (2008b) Keys for the identification of common bees of Sri Lanka. J Natl Sci Found Sri Lanka 36:69–89
Ketipearachchi Y (2002) Hymenpteran parasitoids and hyperparasitoids of crop pests at Aralaganwila in the north central province of Sri Lanka. Ann Sri Lanka Dept Agric 4:293–306
Ketipearachchi Y (2004) Biology of *Hellula undalis* fab. (Lepidoptera:Pyralidae) on cabbage, *Brassica oleracea* var. Capitata L. in the mid country of Sri Lanka. Trop Agric 155:43–48
King CBR (1930) *Trichogramma,* an egg parasite. Tea Quart 3:61–62
King CBR (1935) Report of the entomologist. TRI Bull 12:14–15
King CBR (1936) Report of the entomologist. TRI Bull 13:35–40
King CBR (1937) The tea tortrix problem. Tea Quart 10:46–53
King CBR (1939) Tea tortrix control. Tea Quart 12:86–91
Kudagamage C, Nugaliyadde L (1982) Laboratory and field investigation of Brown planthopper (*Nilaparvata lugens* Stal.) resistance in rice varieties. Trop Agric 138:149–158
Kulasekera VL, Amerasinghe FP (1990) A survey of the mosquito fauna (Diptera: Culicidae) of Sinharaja forest, Sri Lanka. Ceylon J Sci Biol Sci 21:55–74
Kulatunga WMDH, Silva SGR, Warshamana IK, Rajapaksha RWPM (2012) A study to investigate non-compliance on a consignment of wheat bran pellets. Ann Sri Lanka Dept Agric 14:265–271
Ladduwahetty AM (1967) Oviposition cycles and the stimulus for oviposition in *Dermestes maculatus* DeGeer (Coleoptera:Dermestidae). Ceylon J Sci Biol Sci 7:128–138
Lakshani PWY, Rajapaksha MKLK, Sendthuran K (2017) Pesticide residues in selected vegetables in several growing areas by GC/MS using QuEChERS technique. Ann Sri Lanka Dept Agric 19:188–208
Lieftinck MA (1955) A synopsis of the dragonflies (Odonata) of Ceylon. Zoologische Mededeelingen Leiden 34:67–87
Light SS (1928a) Report. TRI Bull 2:25–34
Light SS (1928b) Pests of tea nurseries – a warning. Tea Quart 1:19–22
Light SS (1929) Tea tortrix egg parasite culture. Trop Agric 72:176–177
Madurangi SAP, Ratnasekara D, Senanayake SGJN, Samarasinghe WLG, Hemachandra PV (2013) Antixenosis and antibiosis effects of *Oryza nivara* accessions harbouring bph2 gene on brown plant hopper (*Nilaparavata lugens* (Stal)). J Natl Sci Found Sri Lanka 41:147–154
Madurangi SAP, Samarasinghe WLG, Senanayake SGJN, Hemachandra PV, Ratnasekara D (2011) Resistance of *Oryza nivara* and *Oryza eichingeri* derived lines to brown planthopper, *Nilaparvata lugens* (Stal). J Natl Sci Found Sri Lanka 39:175–181
Mandanayake MARA, Amarakon AMA, Sirisena UGAI, Hemachandra KS, Wilson MR, Kahawaththa UC (2014) Occurrence of *Leptocorisa acuta* (Thunberg) (Hemiptera: Alydidae) in Sri Lanka. Ann Sri Lanka Dept Agric 16:323–326
Mandanayake MARA, Kekulandara KMHGNSB, Wijayagunasekara HNP, Balasuriya A, Kahawaththa UC (2013) A comparative study on arthropod faunal diversity in organic and conventional rice ecosystems at Batalagoda, Sri Lanka. Ann Sri Lanka Dept Agric 15:169–181
Mandanayake MARA, Priyantha MGDL, Bentota AP, Sirisena UGAI, Bandara AMKR, Kahawaththa UC (2016) The effect of flowering plants to enhance natural enemy population in fields. Ann Sri Lanka Dept Agric 18:177–180
Mandanayake MARA, Sirisena UGAI, Bandara AMKR, Bentota AP, Hemachandra KS, Liyanage HLK, Rathnayake WMUK, Kahawaththa UC (2015) Temporal variations of insect pest populations of rice and their relationship with meteorological variables. Ann Sri Lanka Dept Agric 17:56–58
Manickavasagar P (1953) The potato tuber moth, *Gnorimoscma operculella* (Zeller). (order Lepidoptera). Trop Agric 109:118–121
Manikavasagar P (1955) A preliminary study on the control of red bug of bandakka. Trop Agric 111:28–31
Marasinghe JP, Connel D, Yu Q (2015) Health risk on dietary intake of chlorpyrifos with an average diet in Sri Lanka. Ann Sri Lanka Dept Agric 17:28–32

Marasinghe JP, Hemachandra KS, Nugaliyadde L, Karunaratne SHPP (2017a) Control failure of Sri Lankan whitefly (*Bemisia tabaci* Genn.) is due to high resistance development against recommended insecticides. J Natl Sci Found 45(1):23–31

Marasinghe JP, Madugalle S, Rangana KMP, Ranaweera PH (2017b) Bioefficacy of etofenprox against brinjal shoot and pod borer, *Leucinodes orbanalis* Guenee (Lepidoptera:Crambidae). Ann Sri Lanka Dept Agric 19:284–288

Marasinghe JP, Magamage C, Shiromi MGD, Aravinna AGP (2011) Organophosphate residues in food commodities in Sri Lanka: review. Ann Sri Lanka Dept Agric 13:81–93

Mayadunnage S, Wijayagunasekara HNP, Hemachandra KS, Nugaliyadde L (2007) Predatory coccinellids (Coleoptera: Coccinellidae) of vegetable insect pests: a survey in mid country of Sri Lanka. Trop Agric Res 19:69–77

Mayadunnage S, Wijayagunasekara HNP, Hemachandra KS, Nugaliyadde L (2008) Predatory insects of pests in a vegetable eco system of the mid country region of Sri Lanka. Trop Agric Res 20:34–41

Mayadunnage S, Wijayagunasekara HNP, Hemachandra KS, Nugaliyadde L (2009) Occurrence of Aphidophagous Syrphids in aphid colonies on cabbage (*Brassica oleracea*) and their parasitoids. Trop Agric Res 21:99–109

Menike AMW, Shanthini R, Kalpage CS, Karunaratne DGGP, Kankanamge A (2012) Chlorpyrifos contamination of fresh water in a commercial vegetable cultivation area in Sri Lanka and factors affecting contamination. J Natl Sci Found Sri Lanka 40:333–344

Modder WWD (1967a) Functional characteristics of the alimentary canal of the cockroach, *Blaberus craniifer* Burmeister. Ceylon J Sci Biol Sci 7:1–8

Modder WWD (1967b) The distribution and activity levels of carbohydrates in the alimentary canal of the cockroach, *Blaberus craniifer* Burmeister. Ceylon J Sci Biol Sci 7:9–24

Modder WWD (1967c) Report of a glucose diminishing process in the gut of the cockroach, *Blaberus craniifer* Burmeister. Ceylon J Sci Biol Sci 7:25–31

Mohameidsaid MS (1973) Blister beetles (Meloidae) of Sri Lanka. Ceylon J Sci Biol Sci 13:203–252

Myers JG (1929) The principles of biological control. Trop Agric 73:43–47

Nagalingam T, Wijayagunasekara HNP, Hemachandra KS. Parasitoids of vegetable leaf miner in the mid country areas of Sri Lanka. Entomological Society of Manitoba, 13–14 November 2008

Nagalingam T, Wijayagunasekara HNP, Hemachandra KS, Nugaliyadde L (2006) Association of Parasitoid, *Trathala flavoorbitalis* with brinjal pod and shoot borer, *Leucinodes orbonalis* in the mid country of Sri Lanka. Sri Lanka J Agric Sci 43:124–132

Nagalingam T, Wijayagunasekara HNP, Hemachandra KS, Nugaliyadde L (2007) Parasitoids of *Liriomyza sativae* Blanchard (Diptera: Agromyzidae) in the mid country of Sri Lanka. Trop Agric Res 19:59–68

Neave SA (1927) The control of insect pests by means of parasites. Trop Agric 69:230–232

Nishantha KMDWP, Sandika AL, Babu AGC, Hettiarachchi HASN, Pushpanjali K, Abey-Tilakeratna PD, Nugaliyadde MM (2016) Farmers' knowledge and attitudes on pesticide usage in vegetable cultivation in Sri Lanka. Trop Agric Res Ext 19(2):244–252

Nugaliyadde L, Abeyasiriwardena DSDEZ, Samanmalee LGA, Pathirana R, Wilkins RM (2001) Inheritance of resistance in rice to brown planthopper: its implications on rice varietal improvement in Sri Lanka. Ann Sri Lanka Dept Agric 3:167–175

Nugaliyadde L, Dissanayake N, Mitrasena J (2000) Advances of pest and diseases management of rice in Sri Lanka: a review. Ann Sri Lanka Dept Agric 2:409–422

Nugaliyadde L, Wilkins RM (2000) Influence of some rice varieties on the feeding behaviour of *Nilaparvata lugens*. Ann Sri Lanka Dept Agric 2:239–251

Nugaliyadde MM, Bender C, Wijewardena EP, Bandara HM, Maduranga PKS, Gunasinghe SW, Nugaliyadde L (2014) A pheromone based management method for potato tuber moth (*Phthorimaea operculella*). Ann Sri Lanka Dept Agric 16:347–251

Paranagama PA, Ekanayake EMDS (2004) Repellant properties from essential oil of *Alpinia calcarata* Rosc. Against the American cockroach, *Periplanta Americana*. Ceylon J Sci Biol Sci 32:1–12

Perera MCD, Hemachandra KS (2014) Study of longevity, fecundity and oviposition of *Trichogrammatoidea bactrae* Nagaraja (Hymenoptera: Trichogrammatidae) to facilitate mass rearing. Trop Agric Res 25(4):502–509

Perera MCD, Hemachandra KS, Sirisena UGAI (2015) *Trichogramma bactrae* (Hymenoptera: Trichogrammatidae): a potential biocontrol agent of rice leaf folder. Trop Agric Res 26(3):537–546

Perera MCD, Wijayagunasekara HNP, Hemachandra KS (2011) Diversity of parasitoids of mealybug (Hemiptera: Pseudococcidae) in some selected areas of Sri Lanka. Proceedings of Undergraduate Research, vol 2. Department of Agricultural Biology, Peradeniya, pp 9–12

Perera MTMDR, Senanayake N (2014) Evaluation of low cost artificial diet to rear diamondback moth, *Plutella xylostella* L. Ann Sri Lanka Dept Agric 16:195–204

Perera PACR, Kumar V, Perera L, Hemachandra KS, Ranaweera J (1995) Pesticides from Sri Lankan plants for the control of coconut pests. Proceedings of the Annual Symposium of Sri Lanka Association for Advancement of Science, 51, 107–108

Petch T (1921) Pests of *Hevea*. In: Petch T (ed) Diseases and pests in rubber tree. Macmillan, London, pp 223–248

Poorten GM, van Poorten NE (2016) The butterfly fauna of Sri Lanka. Lepodon Books, Toronto

Prasanna J (2015) Pesticide storage, disposal and use by the vegetable farmers in the Muthur divisional secretariat, division of Trincomalee district. Ann Sri Lanka Dept Agric 17:332–335

Premarathne MAGCN, Dissanayake KGPC, Nishantha KMDWP, Hemachandra KS, Jayasinghe WH (2017) Fecundity and oviposition behavior of *Acerophagous papaya* (Hymenoptera: Encyrtidae): a parasitoid of papaya mealybug, *Paracoccus marginatus* (Hemiptera: Pseudococcidae). Ann Sri Lanka Dept Agric 19:168–175

Premawardhane KAJC, Kumuduni HK, Wijesinghe WMCN (2016) Banana leaf roller, a new pest to Sri Lanka. Ann Sri Lanka Dept Agric 18:275

Pushpakumari AS, Mahagollage MGNE (2016) Fruit bagging technique for improving the fruit yield and quality by reducing the fruit fly damage in rose apple. Ann Sri Lanka Dept Agric 18:280

Pyper G (1928) Branch canker (woodrot) and its connection with season of pruning, style of pruning and shothole borer infection. Trop Agric 70:281–287

Rajendra A (1974) The biology and control of *Saccharicoccus sacchari* Ckll. (Hom: Pseudococcidae) the pink mealy bug of cane in Sri Lanka. Ceylon J Sci Biol Sci 11:23–28

Rajendra A (1979) Insect pests attacking sugarcane in Sri Lanka and the parasites and predators recorded on them. Ceylon J Sci Biol Sci 13:29–36

Rajendram GF, Devarajah FR (1990) Survey of some rice insect pests and their predators in three districts of Sri Lanka. Ceylon J Sci Biol Sci 18:79–92

Rajeshkanna S, Keerthiga G, Jayasinghe WH, Hemachandra KS (2017) Efficacy of green chilli, ginger and garlic (3G) solution for the control of Chilli Thrips (*Scirtothrips dorsalis* Hood). Ann Sri Lanka Dept Agric 19:255–259

Rajeshkanna S, Mukunthini S, Jogeswaran T, Nugaliyadde L (2016) Jewel beetle (*Sphenoptera indica*): a new pest of groundnut in the northern provinces of Sri Lanka. Ann Sri Lanka Dept Agric 18:295

Rajeshkanna S, Sumankali T (2015) Effects of selected insecticides to control of yellow stem borer in rice in Sri Lanka. Ann Sri Lanka Dept Agric 17:336–339

Ranaweera PH, Ranathunga M, Nugaliyadde L, Hemachandra KS, Rajapakshe M, Erabadupitiya HRUT, Wijesekara LK, Edirisinghe EMRD (2017) Abundance and species richness of fruit flies (Diptera:Tephritidae) in major cucurbit growing areas in Anuradhapura, Kurunegala and Kandy districts in Sri Lanka and farmers perception on recommended management methods. Ann Sri Lanka Dept Agric 19:98–112

Rathnamalala RBSKK, Wijayagunasekara HNP, Hemachandra KS (2014) Predatory fauna associated with mealybugs and scales with emphasis *on Spalgis epius*. Ann Sri Lanka Dept Agric 16:183–193

Rutherford A (1914) *Xyleborus compactus* Eichh. A borer of tea and coffee. Trop Agric 42:220–222

Samarakoon SAMC, Priyadharshani TDC, Hemachandra KS, Sirisena UGAI (2017) Assessment of egg parasitoids in paddy and vegetable farming systems in Puliyankulama and Mahailluppullama area, Anuradhapura district of Sri Lanka. Proc. of 4th Ruhuna International Science and Technology Conference, University of Matara. 26th January 2017, 4

Sarathchandra SR, Perera DTB, Abeysekara ASK, Wickramasinghe S, Kahawatta UC, Nugaliyadde L (2016) Efficacy of Carbosulfan 200G/L SC and Diazinon 500G/L EW for the management of Paddy bug (*Leptocorisa oratorius*) in Sri Lanka. Ann Sri Lanka Dept Agric 18:173–176

Senadhira MAP (1967) A bibliographical guide to the animal parasites and arthropod pests of Ceylon. Ceylon J Sci Biol Sci 6:130–216

Senanayake NS, Somachandra KP, Hadji TKAI, Bandara AUK (2014) Effect of organic and inorganic mulches on aphid population and yield of tomato. Ann Sri Lanka Dept Agric 16:135–141

Senaratne KADW (1984) Report of the entomology division. Technical Report. In: TRI annual report. Tea Research Institute of Sri Lanka, Thalawakele, pp 34–48

Shalika SVN, & Hemachandra KS (2008) Insect diversity and abundance in vegetation in conventional and pesticide free mixed vegetable ecosystems. In: Proceedings of Annual Symposium of Sri Lanka Association of Advancement of Science, 64, 264

Singhamuni SAA, Hemachandra KS (2013) Study of the interactions among Cruciferaceae crops, cabbage feeding Lepidopterans and their egg parasitoids. Trop Agric Res 25(1):120–126

Singhamuni SAA, Hemachandra KS, Sirisena UGAI (2015a) Potential for mass rearing of the egg parasitoids, *Trichogramma chilonis* and *Tricogramma achaeae* (Hymenoptera: Trichogrammatidae) on *Corcyra cephalonica* eggs. Trop Agric Res 26(1):1–12

Singhamuni SAA, Jayasuriya MIUF, Hemachandra KS, Sirisena UGAI (2015b) Evaluation of the potential of *Trichogramma chilonis* Ishii (Hymenoptera: Trichogrammatidae) as a bio-control agent for *Trichoplusia ni*, cabbage semi-looper. Trop Agric Res 26(2):223–236

Sirisena UGAI, Hemachandra KS, Wijayagunasekara HNP (2013a) Invasive mealybugs of Sri Lanka: recent outbreaks and their threat on agriculture. Crop Life 7:105–110

Sirisena UGAI, Watson GW, Hemachandra KS, Wijayagunasekara HNP (2013b) Mealybugs (Hemiptera: Pseudococcidae) species on economically important fruit crops in Sri Lanka. Trop Agric Res 25(1):69–82

Sirisena UGAI, Watson GW, Hemachandra KS, Wijayagunasekara HNP (2016) Diversity of soft scale insects (Hemiptera, Sternorrhyncha, Cocoidea) in Sri Lanka. In: Chakravarthy AK, Sridhara S (eds) Arthropod diversity and conservation in the tropics and sub-tropics. Springer, Singapoore, pp 285–296

Sivapalan P (1971) Report of the entomology division. Technical Report. TRI Annual Report, 99–104

Sivapalan P, Vitarana SI (1975) Assessment of the Pest pheromone activity in virgin female moth of the tea tortrix *Homona cofferia* (Lepidoptera:Tortricidae) and its utilization in monitoring the moth population in tea fields. Tea Quart 45:24–28

Stockdale FA (1920) Two insect pests of tea in Ceylon. Trop Agric 55:276–279

Subasinghe SMC, Amarasena J (1988) Temporal distribution and chemical control of the maize stem-borer, *Chilo partellus* Swinhoe (Lepi: Pyrallidae). Trop Agric 144:123–134

Sumanapala AP (2017) A field guide to the dragonflies and damselflies of Sri Lanka. Ceylon Tea Services PLC, Colombo

Thevasagayam EE, Canagaslingham LSC (1960) Some observations of insect pests of dhal (*Cajanus cajan*) and their control. Trop Agric 116:287–298

Thevasagayam EE, Canagaslingham LSC (1961) Major insect pests of brinjal (*Solanum melongena*) and their control. Trop Agric 117:105–114

Tillekaratne K, Dharmaratne CMS, Edirisinghe JP (2008) Thrips species on common vegetables crops and their damage. In: Proceeding of the Peradeniya university international research sessions, vol 13. University of Peradeniya, Peradeniya, pp 294–295

Tillekaratne K, Edirisinghe JP, Gunetilleke CVS, Karunaratne WAIP (2011) Survey of thrips in Sri Lanka: a checklist of thrips species, their distribution and host plants. Ceylon J Sci Biol Sci 40(2):89–108

Uvarov BP (1930) Weather and climate in their relation to insects. Trop Agric 74:282–294

Vijayaratnam S, Balasingham B, Balakrishnan P, Kulendran S (2008) Effect of poultry manure on chilli yield and incidence of thrips in calcic red yellow latasols. Ann Sri Lanka Dept Agric 10:255–261

Vitarana SI (2000) 75 years of research in tea entomology. In: Modder WWD (ed) Twentieth century tea research in Sri Lanka. Tea Research Institute of Sri Lanka, Thalawakele, pp 111–160

Wahundeniya I (1993) Control of leaf eating caterpillars in cabbage using an insect growth regulator. Trop Agric 149:81

Wahundeniya I (2001) Role of insecticides in the control of leaf miner, *Leriomyza sativa,* damage in tomato. Ann Sri Lanka Dept Agric 3:289–301

Wahundeniya KB, Wahundeniya I, Ariyaratne I, De Silva YKK (2005) Control of pest and diseases of commercial leafy vegetables by farmers in the western province. Ann Sri Lanka Dept Agric 7:401

Walgama RS (2012) Ecology and integrated pest management of *Xyleborus fornicatus* (Coleoptera:Scolytidae) in Sri Lanka. Journal of Integrated Pest Management 3(4):1–8

Warshamana IK, Wijesinghe PRA, Perera GTS, Fernando TNP, Senarathne SMACU, Weligamage SS, Bandara HMR, Swaminathan T, Nugaliyadde L (2016) Efficacy of ethyl Formate (16.3% W/W) as a fumigant for the control of mealybugs in pineapple and pests of stored rice and maize. Ann Sri Lanka Dept Agric 18:165–168

Warshamana LK, Rajapaksha RWPM, Wanigasuriya SC (2012) The potential for introducing pests through imported consignments of agriculture commodities. Ann Sri Lanka Dept Agric 14:173–185

Wijayaratne LKW, Fernando MD, Palipane KB (2009) Control of insect pests under ware-house conditions using smoke generated from partial combustion of rice (paddy) husk. J Natl Sci Found 37(2):125–134

Wijerathna MAP, Edirisinghe JP (1995) Preliminary observations on graminaceous aphids (Homoptera:Aphidae) of the Peradeniya University park. Ceylon J Sci Biol Sci 24:34–41

Wijeratne PM (2000) Identification of insect pests of crops in Sri Lanka. Department of Agriculture, Peradeniya

Wijesekara A (2002) Synopsys of the Agromyzidae (Diptera) of Sri Lanka. Ceylon J Sci Biol Sci 29:41–62

Wijesekara A, Wijesinghe DP (2003) History of insect collection and review of insect diversity in Sri Lanka. Ceylon J Sci Biol Sci 31:43–59

Wijesekara GAW, Abeytunga KGW (1983) Seed treatment with insecticides for the control of *Ophiomyia* (*Melanagromyza*) *phaseoli* (Tryon) on cowpea in Sri Lanka. Trop Agric 139:43–50

Animal Feed Production in Sri Lanka: Past Present and Future

12

S. Premarathne and K. Samarasinghe

Abstract

Feed production for ruminants and nonruminants is quite different and therefore discussed under two separate sections in this chapter. There had been no habit of cultivating forage in Sri Lanka, thus ruminants are largely fed on naturally grown grasses available in different types of natural grasslands, roadsides, marginal lands, paddy bunds, home gardens, fallow paddy fields, and coconut lands. The quantity as well as quality of forage available in these sources are not adequate. However, with the introduction of European breeds of dairy cattle to medium/large-scale farms in the recent past, there is a growing trend in cultivating improved varieties of fodders such as Napior, fodder maize, and fodder sorghum. Production of grass silage is becoming popular but hay production is almost absent. Commercial feed production in Sri Lanka is limited to concentrate feeds which caters mainly to poultry subsector. Only a small quantity (4%) of the compounded feeds produced are fed to cattle, pigs, goats, shrimp, and others. Compounded feed industry heavily depends on imported feed raw materials which makes the feed costly. Therefore, there had been an increasing trend of producing feeds by farmers themselves although there are sufficient number of large-scale feed mills established in the country. The total compounded feed production has been doubled during the past decade, exceeding 1 million MT during the year 2016. The main raw materials produced locally include maize, rice polish, and coconut poonac, but the supply is insufficient. The animal feed production in Sri Lanka is totally handled by the private sector, and the government does the regulatory work. As the poultry subsector is growing rapidly and continuously, the demand for compounded feeds will continue to increase. However, availability of concentrate feed raw materials as well as forage will be limited in the future. Therefore, use of alternate feed resources and increasing feed efficiency will be a must to fill the gap.

S. Premarathne · K. Samarasinghe (✉)
Department of Animal Science, Faculty of Agriculture, University of Peradniya, Peradeniya, Sri Lanka

B. Marambe et al. (eds.), *Agricultural Research for Sustainable Food Systems in Sri Lanka*, https://doi.org/10.1007/978-981-15-2152-2_12

Keywords
Forage · Compounded feeds · Natural grasslands · Production systems · Poultry

12.1 Introduction

Feed is the most important component in any type of animal production system, especially for commercial purposes. While ruminants depend mainly on roughage feeds, properly prepared concentrates are necessary for the feeding of nonruminants. Therefore, the animal feed production can be considered under two subtopics: roughage feed production and concentrate feed production. About 70% of the agricultural land is solely devoted for crop agriculture, whereas in the remaining 30% of the land, both crops and animals are found. In the animal production sector, emphasis is given to dairy cattle farming and poultry (chicken) production, and there is no beef cattle industry in the country. The contribution of the livestock sector to the GDP is only about 0.8% (CBSL 2018), which is mainly from the dairy and poultry subsectors. The types of feeds, feed production systems, and feeding systems play different but important roles in the ruminant and nonruminant animal production sectors, and thus these two areas are discussed under two separate sections in this chapter.

12.2 Ruminant Animal Feed Production in Sri Lanka

Grasslands contribute to the livelihood of over 800 million people (White et al. 2000, cited by Reynolds et al. 1999) including many poor small holders in different parts of the world. Grazing livestock is the main source of livelihood of most of the rural farmers in Asia, including Sri Lanka where the agricultural population is 87% of the rural population although the income may be different with the involvement of other activities. Farm families, either cultivating food crops or engaged in plantation crops, keep one or several species of animals which are integral parts of their farm. Cattle and buffaloes are raised mainly for milk and draft, whereas goats are reared for meat production. Ruminant livestock feed is mainly derived from natural grasslands, roadsides, marginal lands, fallow paddy fields, and crop residues for a long period of time. However, the quantity as well as quality of forage obtained from communal grasslands, roadsides, and marginal lands is poor leading to low production by ruminants.

12.2.1 Types of Animals and Feeding Practices Adopted in Early Stages

In Sri Lanka, natural grasslands have spread over an area of 12,000 km^2 (Pemadasa 1990). Using grasslands for livestock through traditional means goes back to several centuries. However, as in many other countries, grasslands have deteriorated due to mis-management under the existing socioeconomic position in Asia. Nevertheless, grasslands remain important for livestock production and environmental stability. There is a continuing need to maintain a broad spectrum of production and conservation interests in order to ensure the effective and suitable management of the grassland resource (Premaratne et al. 2003).

It is common knowledge that the economic viability of grasslands as grazing grounds depends partly on their productivity and quality of constituent forage species. Further, annual rainfall and its seasonal variation, soil fertility, species composition, stocking rate, and anthropogenic and other biotic factors also play a prominent role (Murphy 1975). With an increase in human population, the traditional feeding, breeding, and surviving habitats for livestock have been acutely restricted in the country due to large-scale development activities pertaining to human life. Grassland types of Sri Lanka for livestock farming are given in Table 12.1.

Cattle and buffaloes are reared mainly under extensive systems using indigenous breeds and continuous grazing. Production of these animals is low due to poor genetic potential of the breeds and poor feeding and management practices adopted by the farmers. Buffaloes were mainly used for the preparation of paddy fields and threshing of paddy in the farm. One or two animals in a herd were milked just for the consumption for the farm family because there was no demand for milk from the community due to everyone owning animals.

12.2.2 Types of Animals and Feeding Practices Adopted at Present

Tables 12.2 and 12.3 present the animal production systems used in different parts of Sri Lanka with the main features of those systems.

In addition to natural grasses, pasture under coconut, forage available in home gardens, roadside Grass, forage available in marginal lands/fallow paddy fields, fodder trees/fodder legumes, crop residues, cultivated forage such as maize/corn (*Zea mays*) and sorghum (*Sorghum bicolor*) in private and large government farms, and total mixed rations in large farms are used at present.

12.2.2.1 Pasture under Coconut

Cattle and buffalo grazing on natural pasture under coconut is a common scenario in Sri Lanka. Coconut-livestock farming system provides a steady income throughout the year with less risk than intercrops. The microclimate under coconut will help rearing of superior crossbred animals, thereby increasing the milk production of animals. Livestock grazing on weeds would also reduce the cost of weeding in crop agriculture. In mixed farming systems, animals and poultry provide milk, meat, and eggs and thereby support achieving self-sufficiency in food supply for the farm family and recycling the nutrients within the system, thus reducing pollution and improving soil fertility and land productivity. Animals can perform numerous functions in smallholder systems. Rearing livestock in subsistence farming is particularly important when cropping risks are high. Livestock serve as a buffer when crop

Table 12.1 Grassland types of Sri Lanka for livestock farming

Grassland type	Rain fall (annual) (mm)	Elevation (m)	Distributed location	Dominant forage species found	Potential for livestock rearing	Stocking rate, number of cows (ha)	Important remarks
Dry Patana	1750–2250	500–2000	Uva Basin	*Arundinella* spp., *Pollinia* spp., *Ischaemum* spp., *Themeda tremula*, *Andropogan* spp.	Moderate	Low	The burning of grass just before the rain results in the soil being exposed and consequently getting eroded. This probably accounts for the absence of any trees.
Wet Patana	2500–400	>2000	Horton Plains Elk Plain Moon Plain Bopaththalawa	*Chrysopogan zeylanicum*	Poor	Medium	Grasses are tufted, coarse and wiry, scattered trees are prominent
Lowland Savanna	1450–1750	300–400	Bibile Monaragala	*Panicum* spp., *Themeda tremula*, *Desmodium* spp., *Temeda triandra*	High	Low	The grass cover is much taller. Fire-tolerant species are rather sporadic.

(continued)

Table 12.1 (continued)

Grassland type	Rain fall (annual) (mm)	Elevation (m)	Distributed location	Dominant forage species found	Potential for livestock rearing	Stocking rate, number of cows (ha)	Important remarks
Upland Savanna	1500–2000	400–500	Wellawaya Pethyagoda	*Panicum* spp., *Themeda tremula*, *Desmodium* spp., *Heteropogan triticeus* *Mimosa pudica*	Moderate	Medium	The grass cover is much taller. Soil is eroded and denuded as a result of frequent destructions of vegetation
Villu	1500–2000	0–200	Polonnaruwa Manampitiya Thamankaduwa Maduru oya	*Cynodon dactylon*, *Stenotaphrum secundatum*, *Bothriocloa glabra*	High	Low	Wet grasslands found in the flood plains of the rivers in the dry zone. Grasses are more succulent. Soil is rich in nutrients
Damana	1250–1750	0–100	Damana Ampara Inginiyagala	*Imperata cylindrica*	High	Low	Origin seems to be the results of forest clearing followed by repeated fire.
Talawa	2000–2500	0–200	Kalutara Galle Matara	*Cynodon dactylon*	High	Low	Arise as results of forest felling and chena cultivation in wet zone

Adopted from Pemadasa (1983) and Premaratne et al. (2003)

Table 12.2 Animal production systems used in different parts of Sri Lanka

Production systems	Elevation (m)	Rain fall (mm)	Temperature (°C)	Popular management system	Animal Types	Average daily milk production per cow (L)
Hill country	>450	>2000	10–32	Intensive	Pure exotic animals[a] and crosses	6–8
Mid-country	>450	>2000	10–32	Semi-intensive	Pure exotic animals and crosses, zebu crosses	4–5
Coconut triangle	<450	1500–2000	21–38	Tethered	Crosses of exotic breeds, zebu types, crosses of indigenous animals and buffaloes	3–3.5
Low country dry zone	<450	1000–1750	21–38	Extensive	Indigenous cattle, zebu cattle[b] and crosses, buffaloes	1–1.5
Low country wet zone	<450	1875–2500	24–35	Tethered	Crosses of exotic breeds and zebu type and indigenous animals and buffaloes[c]	3–3.5

Adapted from: Bandara (2007)
[a]Exotic breeds of cattle: Jersey, Friesian, and Ayrshire; [b]Zebu breeds of cattle: Sahiwal and Sindhi.
[c]Buffalo breeds: Murrah, Surthi, and Nili Ravi and indigenous buffaloes

yields do not meet family needs and can act as a "savings account," with offspring as the "interest." Diversification in livestock rearing extends the risk-reduction strategies of farmers beyond multiple cropping and thus increases the economic stability of the farming system. Spreading risk by practicing both crop and livestock production may lead to lower productivity within each sector than in specialized farms, but total land productivity may increase.

12.2.2.2 Home Gardens

Ruminant production in home gardens is based on the semi-intensive management system with tethered grazing of natural feed resources or improved forage under coconut and other perennials with improved breeds of animals (Zemmelink et al. 1999). Yields of grasses from homesteads and highland gardens are low as these areas are intensively cropped with trees and other perennial crops. This farming system provides regular supply of organic matter to the soil owing to daily excretion of feces and urine in large (cattle and buffalo) and small (sheep and goat) ruminants.

Table 12.3 Main features of ruminant production systems in Sri Lanka

Zone features	Dry zone	Coconut triangle	Mid-country	Up-Country Vegetable farmers and estate laborers	Wet zone and urban
Location	Dry zone districts in the North Central, North and East Provinces, and parts of Central, South, and North Western Provinces	Intermediate and wet zone areas of the North Western Province, and Gampaha district of the Western Province	Wet zone areas in the Central Province—Kandy and Matale districts	Nuwaraeliya district in the Central Province and Badulla district in the Uva Province	Districts in the Western, Southern, and Sabaragamuwa Provinces and Cities
Animal Types[a]	Indigenous cattle, zebu cattle and crosses, buffaloes	Crosses of exotic breeds, zebu types, crosses of indigenous animals and buffaloes	Pure exotic animals and crosses, and zebu crosses	Pure exotic animals and crosses	Crosses of exotic breeds and zebu type and indigenous animals and buffaloes
Dominant forage species	Natural grasses, *Panicum maximum*, *Brachiaria brizantha*, *Brachiaria ruziziensis*, *Napier*, Pusa Giant Napier, *Gliricidia sepium*, *Leucaena leucocephala*	Natural grasses, *Panicum maximum*, *Brachiaria brizantha*, *Brachiaria ruziziensis*, *Brachiaria miliformis*, *Brachiaria mutica*, *Digitaria decumbense*, *Gliricidia sepium*, *Leucaena leucocephala*	Natural grasses, *Panicum maximum*, *Brachiaria brizantha*, *Brachiaria ruziziensis*, CO-3, CO-4, *Napier*, *Setaria spacelata*, Pusa Giant Napier, *Gliricidia sepium*	Natural grasses, *Panicum maximum*, *Lolium perenne*, *Pennisetum clandestinum* (Kikiyu)	Natural grasses, *Panicum maximum*, *Brachiaria brizantha*, *Brachiaria ruziziensis*, CO-3, CO-4, Napier, Pusa Giant Napier, *Gliricidia sepium*, *Leucaena leucocephala*
Husbandry	Free gazing, or nomadic-type large herds, or sedentary small and medium herds	Medium-sized herds, limited grazing or tethered grazing under coconut palms	Small herds, some tethering, stall feeding	Small herds, zero grazing	Limited grazing, medium-sized herds or small herds, zero grazing

(continued)

Table 12.3 (continued)

Zone features	Dry zone	Coconut triangle	Mid-country	Up-Country Vegetable farmers and estate laborers	Wet zone and urban
Concentrate feeding	None/very little	None/small quantities to milking cows and calves; mainly rice bran, rice polish, and coconut poonac	Medium quantities to milking cows and calves; mainly rice bran, rice polish, and coconut poonac	Medium-high quantities to milking cows and calves; rice bran, rice polish, coconut poonac, and compounded feed	Medium quantities to milking cows and calves; mainly rice bran, rice polish, and coconut poonac
Herd size	Few to 10–25	5 cows and followers	2–3 cows	1–2 cows	2–3 cows
Average yield	2.1 L/cow/day, Total 300–400 L/cow over 180–200 day lactation	3–4 L/cow/day, Total 500–800 lit./cow over 200-day lactation	4–5 L/day, Total 1200–1600 L/cow over the lactation	6–8 lit/cow, Total 1800–2500 L/cow over the lactation	3 L/cow/day, Total 1000 L–1200 L/cow

Adopted from: Ranaweera and Attapattu (2006)

[a]Exotic breeds of cattle: Jersey, Friesian, and Ayrshire; [b]Zebu breeds of cattle: Sahiwal and Sindhi. [c]Buffalo breeds: Murrah, Surthi, and Nili Ravi and indigenous buffaloes

12.2.2.3 Roadside Grass

Use of roadside grass to feed ruminants is a common practice in developing countries including Sri Lanka. In general, animals are allowed to graze along the road sides, but cut and feed system is also common in hilly areas. Guinea "A" (*Panicum maximum*) is a native grass of Africa, which was introduced to Sri Lanka in the 1960s and is well-adapted to a wide range of soil types (Premaratne and Premalal 2006). Presently, species of Guinea grass are widespread across Sri Lanka. Harvesting of the grass at early growth stages would make it a high-quality, high-yielding forage suitable for animal feeding. It has become a freely available fodder grass for livestock farmers in Sri Lanka.

12.2.2.4 Marginal Lands/Fallow Paddy Fields

Most of the rural farmers use the weed cover or the forage found in marginal tea or rubber lands around the world. It is common to see animals grazing in paddy fields when the land is not being cropped with rice. However, grazing is not possible if the land is used to grow vegetables and other annual crops, though removing of grass (and weeds) is still possible. Rice fields are an important source of greenfeed, in the form of grass and ratoon crop, and grass growing on the field bunds throughout the growing period, except when bunds are being re-plastered. Paddy farmers also encourage others to harvest grasses growing in bunds for easy maintenance of bunds. As the fields are regularly fertilized, this feed is also of relatively high quality.

12.2.2.5 Fodder Trees/Fodder Legumes

Forage tree legumes such as *Gliricidia sepium*, *Erithrina* spp., *Leucaena leucocephala*, *Caliandra callothirus*, and *Acasia* spp. are being cultivated as a source of mulch, nitrogenous fertilizer, shade tree, fuel wood, agro-forestry tree, supportive crop for intercrops like black pepper, and as an animal feed in many homesteads. Rapid growth rate, the ability to grow as a live fence, and use of sticks to generate biofuel are added advantages of these fodder-tree legumes. Use of this organic manure will help to cut down the cost of fertilizer and to increase the physical properties of soil compared to addition of mineral fertilizer. However, the access to feed largely depends on the availability of household labor and this in turn depends on the size and demographic structure of the households, and the availability of alternative employment opportunities. In general, male farmers are responsible for harvesting fodder from trees in the Asian region.

Forage tree legumes are considered as high-quality forage for livestock. The special feature with the legumes is their ability to fix atmospheric nitrogen into plant proteins, and it is the largest and cheapest source of nitrogen for crops. In many parts of the world, forage tree legumes are cultivated as field borders, fence lines, or in-home gardens. These are harvested under cut and carry system and used as a supplement with low-quality roughage such as crop residues. Establishment period of legumes can be more than 1 year and generally legume trees are uncut until they reach a height of 1–1.5 m. These are considered as important sources of nutrients such as proteins, minerals, and fiber. However, the optimum dietary level of the leguminous forage on dry matter basis is 30–40% as these are high in proteins and also contain anti-nutritive factors such as condensed tannin, mimosine, and lignin.

12.2.2.6 Crop Residues

The availability of fibrous crop residues in Sri Lanka is over 3.2 million mt of dry matter. These estimates have been calculated using the grain to residue ratio applicable to each crop such as rice straw, maize stover, and sugarcane tops. The quality of crop residues is poor, but it can be fed to ruminants during the dry period when feed is scarce.

12.2.2.7 Cultivated Forage, Maize, and Sorghum in Private and Large Government Farms

Most of the large farms maintain high-producing European breeds or/and European × Indian crossbred animals and, farm-grown fodder for animal production, which are high in feeding value and palatability. Further, cereals such as maize and sorghum are also cultivated in an economically viable manner where land scarcity is a major constraint to dairy production. They are designed to increase the production of fodder, thus minimizing labor wages and increasing land productivity. High-quality, high-yielding, and disease-resistant fodder cultivars are selected for this purpose. Low-cost fodder harvesters, 4-wheel tractors, and simple irrigation systems are being used to minimize the production cost. According to Sarmini and Premaratne (2017), maize and sorghum as fodder grasses have performed better

Table 12.4 Forage recommendation for different systems

Common name	Scientific name	Cut and carry system	Grazing system	Cooler climatic areas
Fodder/pasture grasses				
Hybrid Napier—CO-3	*Pennisetum purpureum X, P. americanum*	X		
Hybrid Napier—CO-4	*Pennisetum purpureum X, P. americanum*	X		
Guinea-VRI 435	*Panicum maximum*	X		
Guinea-TD 58	*Panicum maximum*	X		
Guinea-Makueni	*Panicum maximum*	**X**		
Setaria	*Setaria sphacelata*	X		
Signal grass	*Brachiari brizantha* or *B. decumbens*		x	
Ruzi grass	*B. ruziziensis*		x	
Rye grass	*Lollium perenne*		x	x
Kikiyu grass	*Pennisetum clandestinum*		x	x
Gamba grass	*Andropogan gayanus*	X		
Fodder sorghum (sugar graze, jumbo, BMR6)	*Sorghum bicolor*	X		x
Fodder maize (Pacific, local varieties))	*Zea mays*	X		x
Fodder legumes				
Gliricidia	*Gliricidia sepium*	X		
Ipil-Ipil	*Leucaena leucocephala*	**X**		
Desmanthus	*Desmanthus virgatus*	**X**		
Sesbania	*Sesbania sesban*	**X**		
Erythrina	*Erythrina orientalis*	**X**		
Caliandra	*Caliandra calothrysus*	**X**		x
Pasture legumes				
White clover	*Trifolium repens*		x	x
Red clover	*Trifolium pratens*		x	x
Stylo-fine stem	*Stylosanthus guyanensis*		x	
Stylo-Verano	*Stylosanthus hamate*		x	
Greenleaf Desmodium	*Desmodium intortum*		x	
Silverleaf Desmodium	*Desmodium uncinatum*		x	

than CO-3 in terms of plant growth, yield, nutrient composition, and cost effectiveness in the northern region of Sri Lanka. Recommended forage for cut and carry system, grazing system, and cooler climatic areas is given in Table 12.4.

12.2.3 Forage Conservation as Silage/Hay

Most of the small- and large-scale farmers preserve excess forage as hay or silage for later use. Maize (corn) silage and CO-3 or CO-4 silages are common in Sri Lanka at the moment. Small-scale farmers produce barrel silage, whereas tower or bunker silos are maintained in large farms. It is important to use crops high in soluble carbohydrates such as maize, sorghum, or *Pakchong* 1 (Super Napier) to make silage so that high-quality silage can be produced at a low cost. There are a few farmers who produce silage for sale in Mahawali areas of Sri Lanka and in the Northern province using maize, sorghum, and palmyrah (*Borassus flabellifer*) leaves. According to Sarmini et al. (2017), a mixture of 10% palmyrah leaves with maize or sorghum can be recommended for making high-quality silage in the Northern region of Sri Lanka.

Hay making is not a common practice of small-scale farmers in Sri Lanka at present. Hay making should be improved in the country because transportation of hay is cheaper and easier when compared to silage.

12.2.4 Preparation of Leguminous Block

Pasture and fodder production during the rainy season is comparatively high in the dry zone and coconut triangle of Sri Lanka. Therefore, excess feed can be harvested and preserved for the dry season to maintain a uniform production throughout the year. Forage can be preserved in the form of leaf meals and feed blocks. Leguminous leaf meals are an alternative feed/protein source for livestock during dry seasons. Leaf meals can be pressed into blocks/briquettes with/without incorporating other concentrate feed ingredients such as coconut poonac, rice polish/bran, and molasses so that keeping quality can be increased and a market value can be obtained.

Gliricidia is widely used as a source of protein for ruminants, especially during dry spell. It could be fed as a fresh diet or in dried form as leaf meals. According to Somasiri et al. (2010a), Gliricidia leaf meal can be formed into blocks to improve the keeping quality using a briquette machine applying hydraulic pressure. Furthermore, these blocks can be improved nutritionally by mixing leaf meal with other feed ingredients such as coconut poonac and rice bran, and could be stored up to 3 months without deteriorating the quality, if packed properly. According to Somasiri et al. (2010a), 75% Gliricidia +25% coconut poonac and 75% Gliricidia +12.5% coconut poonac +12.5% rice bran were the best two recipes with respect to pressing ability, free fatty acid content, cost per leaf meal block, and requirement of unit of fresh leaves for the preparation of a unit of dried leaf meal. Transportation of these blocks is easier compared to silage, and producers could earn a living with this cottage industry. Preparation of leguminous block can be popularized in the country especially in the North, North Central, and Western Provinces as fodder legumes are freely grown in these regions. They can also be used as an effective feed for livestock, especially for dairy cows during the drought period (Somasiri et al. 2010b).

12.2.5 Nonconventional Feed Resources

The availability of these resources depends on the climatic factors as well as the varieties used in the field. Estimation of these resources is also incomplete due to the nonavailability of information. Even though these feed resources are available in the country, collection of the ingredients is an issue at present. Therefore, a mechanism should be designed to collect these materials from areas where they are available and to distribute throughout the country.

12.2.6 Concentrates and Brewery Wastes

Concentrates such as coconut (*Cocos nucifera*) poonac, rice (*O. sativa*) polish, rice bran, broken rice, maize, bran from pulses, agro-industrial by-products such as molasses, minerals, and other ingredients, formulated feeds, and brewery wastes are fed to ruminants according to the animal's requirements and production. Even though these feeds are expensive compared to forage, animal production can be increased and maintained at a higher level especially for European breeds/crosses under intensive management systems. Amount of concentrates given to ruminants can be reduced when high-quality forage is provided to animals. The two main ingredients that are used in ruminant feeding are rice bran and coconut poonac. At an extraction rate of 4%, the potential availability of rice bran is estimated to be 112,000 MT per year, whereas the availability of coconut poonac varies between 10,000 and 20,000 MT per year (Ibrahim and Siriwardene 2001). Brewers dried grains (BDG) is a by-product of the beer industry, and the brewery waste is a nonconventional feed ingredient that can also be used as an energy source in livestock feed. It contains a wide variety of essential nutrients needed for livestock, and therefore can be included as a cheaper feed source in poultry, pigs, and cattle ration (Samarasinghe 2007).

12.2.7 Formulation of Balanced Rations

This involves the selection and allocation of feed ingredients in such a way that nutrient requirements of animals are fulfilled while the cost of ration is kept low. Traditionally, farmers have used least cost ration formulations for this purpose; however, income maximization may not always be attained by this method. Use of trial and error method or a simple linear program using a computer could overcome this problem.

12.2.8 Total Mixed Rations (TMR)

The total mixed rations (TMRs) are being used effectively at commercial-scale private and government farms for European cattle breeds/crosses under intensive management systems. The TMR, or complete ration, is a practice of weighing and blending all feedstuffs into a nutritionally balanced ration so that it ensures every bite consumed by an animal is the same in a feeding system. Lammers et al. (2013) reported that the use of TMR improved feed efficiency. It provides adequate nourishment to meet the needs of ruminants, thus helping them to achieve maximum performance. These TMRs are formulated in the farm itself with the available feed ingredients according to the nutrient requirements of the animal. However, the use of TMRs by small-scale farmers are still not common due to the lack of knowledge and subsistence level of farming.

The TMR system has many potential benefits, and it warrants consideration by any farmer anticipating a change in his/her animal herd. Furthermore, TMR provides greater accuracy in the feed formulation and feeding because an animal will not consume significantly more or less forage or concentrate in a properly mixed TMR. Thus, Snowdon (1991) explained that feeding TMR would prevent livestock from expressing a preference for one type of forage over another. A greater variety of ingredients will allow more flexibility in formulating the ration for various production groups (Lammers et al. 2013).

The main problem of TMR is the inability to feed cows as individuals. The TMR can be developed for different animal groups. If not grouped according to production, cows in late lactation are likely to get too fat (Lammers et al. 2013). Further, Lammers et al. (2013) reported that the TMR system might not be economical for farms with a small herd size. The process of mixing ration required accuracy because over-mixing could cause serious health problems, and under-mixing would lead to less efficient feed utilization by the cows. Accurate measurements also require additional costs and maintenance. Some farmers do not choose to adopt TMR as it is not well-suited for their housing and feeding facilities. Moreover, if the TMR diet was not balanced correctly or mixed properly, the performance could become worse (Lammers et al. 2013).

12.2.9 Future Demands for Ruminant Animal Feeds and Way Forward

Green forage/fodders are the cheapest source of feed in any ruminant production system. Therefore, high-quality fresh fodder which is high in digestibility and other nutrients such as protein, minerals, vitamins, and energy should be made available for farmers to provide their animals with nutritious forage throughout the year. To achieve this target, forage should be grown as a crop similar to that of other cash crops. Currently ruminant production in Sri Lanka is largely a subsistence-level enterprise and use of farm-grown fodder is a new concept. Therefore, it is important to initiate intensive fodder cultivation, maize, and sorghum cultivation for animal feed and fodder marketing to produce high-quality green fodder aiming at commercializing ruminant production in the country. Further to imported varieties, local maize varieties such as *Badra* can be grown successfully as an animal feed because the cost of seeds is cheaper compared to imported varieties.

It is important to advise farmers to cultivate their own fodder blocks in the household using waste water from the kitchen or dairy washings. A number of high-yielding, high-quality forage species have been developed in India and other tropical countries. Use of these forage species together with organic manure available in the farm could increase the availability of forage for livestock. These fodder initiatives can be translated to smallholders to their benefits. Production of high-quality forage could reduce the greenhouse gas production and emissions by ruminants and thereby reducing global warming.

Furthermore, conservation of forage should be practiced by farmers so that animal feed as hay or silage will be available throughout the year. Commercialization of hay and baled silage production can overcome feed scarcity issues in the urban areas. Production of bag silage for sale using maize or sorghum should be encouraged for crop farmers. Commercial production of leaf meal blocks/leguminous blocks by farmers or cooperative societies should be promoted and encouraged to improve the digestibility of the ruminant diet and upgrade the livelihood of people. The challenge for the future is to develop technologies that will make it possible to use the resources available for direct feeding or for incorporation in commercial production of feed mixtures.

12.3 Feed Production for Nonruminants

12.3.1 History

12.3.1.1 Types of Animals and Feeding Practices Adopted in Early Years

Crop and livestock mixed farming has been the tradition of Sri Lankan farmers. Until recent years, there were no farmers rearing only livestock. In ancient Sri Lanka, both ruminants and nonruminants were reared extensively at subsistence level in small scale. Rearing chicken, the most prominent poultry species in the country, and pigs in backyards has a long history. Until the year 1955, there was no commercial poultry production, and the entire poultry sector was represented by 1.1 million village chickens reared in backyards. There was no demand for compounded feeds as the birds found their own feeds from backyards. Rearing indigenous pigs on scavenging also did not demand any commercial feed. Eggs and meat were imported to the country to fill the gap in the demand.

12.3.1.2 Development of Poultry Farming as a Commercial Industry

In mid-1950s, the Government of Sri Lanka identified the potential to develop the poultry sector in the country aiming at supplying animal-based protein food, mainly eggs, to people in the country to combat protein malnutrition. In order to develop the poultry sector as an industry, the Government established the Central Poultry Breeding Station and Hatchery in 1955 at Karandagolla, Kundasale, in the Central Province of Sri Lanka, and started importation of White Leghorn breeders, the most popular high-laying breed in the world at that time. The idea was to produce high-producing day-old layer chicks to be distributed among farmers. Farmers were supplied with limited number of day-old chicks (100–200 chicks per farmer at a time) for commercial farming instead of subsistence-level farming. At the same time, the deep litter housing system was introduced replacing backyard farming.

As a result of the government intervention, the poultry production in the country was increased by 85% during the first decade since 1955, and the importation of table eggs was terminated in 1963. The next focus was to establish the broiler industry to supply chicken meat at an affordable price. The first batch of broiler parents were imported to Sri Lanka in 1965, and the farmers were encouraged to start commercial broiler farms by issuing day-old broiler chicks through the Central Poultry Breeding Station at Kundasale. Within a year, the broiler meat consumption was increased from zero to 14.5% of the total chicken meat consumption. Since early 1980s, the poultry sector, especially the broiler subsector, expanded rapidly becoming the most developed subsector of the livestock sector in the country. Consequently, the production and availability of table eggs and broiler meat increased steadily as illustrated in Fig. 12.1.

In the year 2016, there were 180,635 farms rearing cattle, 48,925 farms rearing goats, 6539 farms rearing swine, and 42,703 farms rearing commercial chicken (broiler 8722 and layer 33,981) in Sri Lanka (DAPH 2017). The chicken population was increased up to 21 million birds in 2016 with an annual average growth of 12.7% from 2012 to 2016. Such a fast growth was not observed in any other livestock subsector in the country. Today all commercial chicks (both layers and boilers) as well as their parents are produced within the country using imported grandparents of high genetic potential. The scale of operation is converting from small to medium and large, and housing from conventional open sheds to automated closed houses. These developments demanded expansions in other supportive industries such as feeds and drugs, house construction, and marketing.

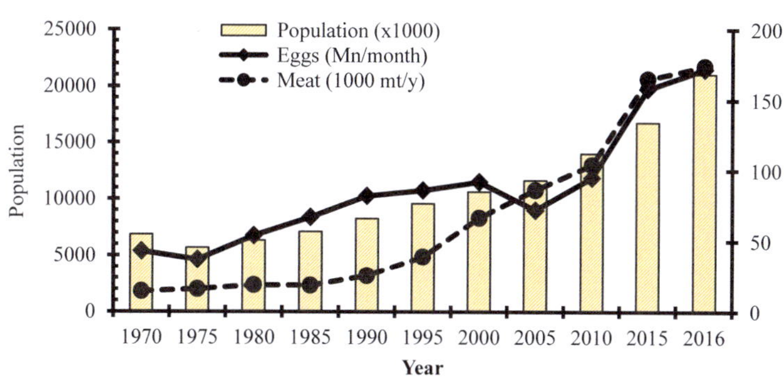

Fig. 12.1 Chicken population and annual production of table eggs and broiler meat in Sri Lanka during 1970–2016

12.3.2 Importance of the Feed Industry

Asian countries like Sri Lanka did not have a culture of cultivating and maintaining pasture lands to feed ruminants until recently. Sri Lanka still does not have an "organized feed industry" catering to ruminant animals except for a limited amount of concentrates produced for dairy cows. In the case of monogastric animals, it is necessary to feed the animals with concentrates. Even though the pig production is not expanded and developed, the poultry industry, which is the most developed sector in Sri Lanka, consists of flocks of modern commercial strains. Unlike village chicken, these strains require extremely high-quality feeds for a satisfactory performance as they are so sensitive for the plane of nutrition. A modern broiler chick under local conditions grows to about 1.8 kg body weight in about 35 days, thus having an average weight gain of about 51 g/day. In the case of layers, a laying rate of about 95% is achieved during the peak production while having an average laying rate of about 80–85%. Such a high performance can be achieved only under proper feeding management conditions. Deficiency of a single nutrient in feed can cause a significant reduction in the performance of birds. Therefore, the feeds provided to poultry should be properly balanced for all nutrients and carefully produced. This can be achieved only if there is an organized feed industry.

12.3.3 Establishment of the Feed Industry and Production of Compounded Animal Feeds

After establishing the poultry subsector as a commercial industry in 1955, a demand was created for compounded concentrate feeds. As the quantities required were small at the beginning, imported feeds served the purpose. Compounded feeds were imported and marketed in the country by three main private sector companies. The government-owned Ceylon Oils and Fats Corporation was producing a limited amount of compounded feeds using locally available feed raw materials, but it could not successfully cater to the increasing demand. In line with the open economic policy introduced to the country in 1978, the government encouraged the private sector to invest in local industries by offering various tax benefits. Consequently, the Government of Sri Lanka started a business partnership with Prima (Limited) of Singapore in 1982 to commence feed milling in large scale, thus establishing a feed industry in Sri Lanka. Ceylon Grain Elevators (Limited), a subsidiary of Prima (Limited) of Singapore, started compounded feed production in 1983 under the brand name "Prima®." According to the partnership agreement, the Government-owned Ceylon Oils and Fats Corporation was closed and its milling facility located at Seeduwa in the Western Province was handed over to Prima (Limited) of Singapore in 1982. The Ceylon Grain Elevators (Limited) was having the monopoly of importing maize for animal feed and was manufacturing feed under duty-free concessions for about 15 years since then.

With the increase in demand for poultry feeds owing to expanding poultry industry, two other multinational companies came into feed milling in Sri Lanka. Gold

Coin Feedmills, an Asian multinational company based in Singapore, started animal feed milling in Sri Lanka in 1993. Their subsidiary company, Coin Feedmills (Lanka) Limited, started to produce animal feeds at their factory located in Mattakkuliya in the Western Province under the brand name “Gold Coin®.” During the same year, Nutrena (Pvt) Ltd., a subsidiary company of Cargills Inc., USA, established a modern feed mill at Ekala in the Western Province and started feed milling under the brand name “Nutrena®,” thus becoming the third large-scale feed miller in the country. The Nutrena feed mill was however sold to the then CIC Group of Companies in 2002, and feed milling was continued by their subsidiary company CIC Feeds (Pvt) Ltd. under the brand name “CIC Feeds®.” During the past few years, there were five additional large-scale feed mills established in Sri Lanka as shown in Table 12.5.

It is interesting to note that about 96% of the compounded animal feeds produced in Sri Lanka is for poultry, while all other types of compounded feeds (dairy, pig, goat, shrimp, horse, other) amount to only 4% (Table 12.6). This is because the poultry sector has been the fastest growing animal production sector in the country. Compounded feed production has increased annually during the past few decades (Table 12.6), and sudden peaks in the production can be seen in 1992, 2000, and then after 2010 (Fig. 12.2). Upward trend in the production of compounded animal feeds in Sri Lanka showed a similar trend as the growth of the poultry sector. The country has produced over 1 million mt of poultry feeds in 2016 showing a growth of about 100% during 2010–2016 (Fig. 12.2). Establishment of new large-scale feed mills (Table 12.5) and expansion of broiler production through large closed-type houses introduced by large-scale producers could be attributable to the remarkable increase in the feed production seen after the year 2000. Today, there are 13 large-scale broiler producers having closed-type houses and contributing to about 30% of the broiler production in the country (Samarakoon 2018 Personal communication).

Table 12.5 Large-scale compounded feed millers in Sri Lanka

Year of establishment	Name of the feed miller	Maximum capacity MT/moth
1983	Ceylon Grain Elevators	20,000
1993	Gold Coin Feedmills (Lanka) Ltd	10,000
1993	Neutrena Feed Mills Pvt Ltd	10,000
2002	CIC Feeds Pvt Ltd (successor of Nutrena Feed Mills)	10,000
2013	New Hope Lanka Limited	20,000
	Pussella Farms	20,000
	Cosmo Feeds	20,000
	Crown Feeds (Pvt) Ltd.	20,000
2016	Fortune Agro Industries (Pvt) Ltd.	30,000

Table 12.6 Production of concentrate feeds by commercial feed millers and farmers in Sri Lanka (MT/year) (Source: DAPH 2017)

	Poultry feeds			Other types of feeds
Year	Total	Commercial	Self-mixed	Commercial
2009	479,785	272,563	207,222 (43%)	
2010	484,511	285,510	199,000 (41%)	13,740 (3%)
2011	594,253	386,253	208,000 (35%)	41,000 (7%)
2012	661,339	363,737	351,500 (53%)	23,580 (3%)
2013	769,090	384,872	384,220 (50%)	103,370 (12%)
2014	812,624	426,628	385,996 (48%)	34,550 (4%)
2015	898,210	501,810	396,390 (44%)	35,150 (4%)
2016	1,033,850	620,310	413,540 (40%)	44,800 (4%)

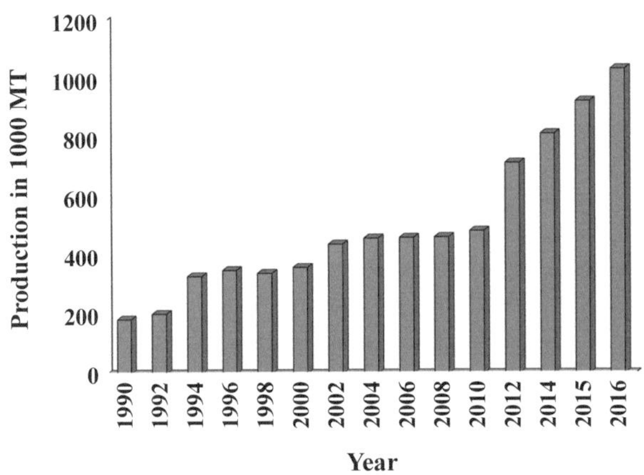

Fig. 12.2 Annual production of poultry feed in Sri Lanka 1990–2016 (source: FAO 2018)

12.3.4 Production of Self-Mixed Poultry Feeds

Commercial feeds are relatively costly as most of the feed raw materials are imported. Therefore, there had been a growing tendency for poultry farmers to mix their own feeds. A field study conducted about 16 years ago, when the self-mixed feeds amounted to about 10% of the total poultry feed production, revealed that more than 40% of layer farmers in Kurunegala District in the North Western province, where the highest poultry population is found, formulate and use their own feed requirements (Jayaweera and Samarasinghe 2002). Table 12.6 depicts that the amount of self-mixed feed has gradually increased to about 53% of the total poultry feed production from 2000 to 2012. The latest data shows that the self-mixed feed amount is in the decline and reduced to 40% in 2016 (DAPH 2017). The study conducted by Jayaweera and Samarasinghe (2002) found that self-mixed feeds are not up to the standards in terms of nutrient levels and hygienic conditions. Most of the feeds formulated by farmers were 8% deficient in sulfur amino acids, 20% lower in energy/protein ratio, and 40% higher in total ash. They contained excessive amounts

(60% more than the requirement) of calcium and phosphorus. Inferior quality of self-mixed feeds obviously reduces the performance of birds. Around 50% of the farmers interviewed in the study admitted that they get a low egg production from their flocks. However, the farmers were satisfied with the self-mixed feeds as the feed was about 18% cheaper than the commercial feed, which has led to increase their profits by 10%.

12.3.5 Availability of Concentrate Feed Raw Materials in Sri Lanka

Similar to other Asian countries, the two main feed raw materials used in producing compounded feeds in Sri Lanka are maize and soybean meal. In a traditional feed formula, the content of maize is about 40–50%. Locally produced rice polish/rice bran and coconut poonac had been in use in the feed industry in Sri Lanka from the beginning. Even though the country was successful in establishing modern feed mills to produce animal feeds, until recently, sufficient attention has not been paid to produce feed raw materials locally. As a result, the local compounded feed industry has been heavily dependent on imported raw materials, thus making the feed expensive. Dependence on imported feed raw materials may be sustainable as long as the economy is strong, and the producing countries are willing to export them to Sri Lanka. During the year 2005, Sri Lankan feed millers found it difficult to import cereals for feed manufacturing as the producing countries in the region seemed to have no surpluses to export. Importing cereals from western countries was obviously not economical. As a consequence, a significant amount of maize was replaced by locally produced rice, which was banned by the government later on, in order to stabilize the rice prices in the local market. Both the Government of Sri Lanka and the feed millers finally realized the importance of producing feed raw materials locally, which of course was a benefit gained of this situation. A Presidential Task Force was thus established to promote cultivation of maize and soya bean with private sector investment.

12.3.5.1 Availability of Maize

The main consumer of maize in Sri Lanka is the animal feed millers. Maize is also demanded for human consumption as tender green cobs and dried grains for "*Thriposha*" production, a processed food for mothers and babies. It is estimated that about 10–15% of the *Maha* season crop (September–February) and 50% of the *Yala* season crop (March–August) are harvested as tender green cobs for direct human consumption. Therefore, only a part of the local supply of maize is available for compounded feed manufacturing. Until the year 2005, the local maize production was limited to about 20% of the requirement and therefore the balance was imported (Table 12.7).

After the establishment of the Presidential Task Force for maize, the local production of maize started to increase annually making more local maize available for feed production. There was an increase in local maize production in 2006 contributing about 47,530 MT to the national needs (approximately 36% of the requirement; Table 12.7).

Table 12.7 Estimated demand for feed production and availability of maize in Sri Lanka (Source: FAO 2018)

Year	Total availability (MT)	Local production (MT)	Imports (MT)	Local production (% of total)
2003	166,348	29,650	136,698	18
2004	188,638	35,200	153,438	19
2005	188,729	41,800	146,929	22
2006	131,573	47,530	84,043	36
2007	135,198	56,440	78,758	42
2008	219,095	135,900	83,195	62
2009	157,804	129,770	28,034	82
2010	172,899	161,690	11,209	94
2011	146,041	137,797	8244	94
2012	219,680	202,150	17,530	92
2013	225,723	209,042	16,681	93
2014	329,282	241,144	88,138	73
2015	327,035	258,398	68,637	79
2016	822,860	779,803	43,057	95

In spite of increasing compounded feed production, which in turn increased the demand for maize, the country became almost self-sufficient in maize during 2012. Until the year 2016, the locally produced maize could look after the total requirement for animal feed production. However, since the rate of annual growth in the compounded feed production was higher than the annual expansion of the maize production, a gap between the local production and the demand for maize has gradually developed since 2014. Extreme weather conditions and unavailability of agrochemicals have also affected local maize production during the recent past. Therefore, the country had to import small quantities of maize from time to time when there was a shortage in the local production during the last few years.

12.3.5.2 Availability of Soybean Meal

Soybean meal is the most widely used protein source in animal feeds in Sri Lanka as in other countries in the world. Almost the entire amount of soybean meal required for the feed industry is imported at present. From the limited data available, it appears that Sri Lanka has used locally produced soybean meal until late 1970s when soybeans were produced and used for oil extraction by the former Ceylon Oils and Fats Corporation. The importation of soybean meal for animal feed industry started in 1979 and boosted up in mid-1980s as a result of the expanding poultry industry (Fig. 12.3). Thereafter, the importation of soybean meal was annually increased at a higher rate (Table 12.8 and Fig. 12.3).

Owing to the absence of local production of soybean meal to cater to the needs, the feed millers had to depend totally on imports. The limited amounts of soybeans produced in Sri Lanka was totally used for human food industry and not available for animal feeding. However, until late 1970s, soybean was a major rainfed crop grown in highlands during the wet season and used for oil extraction, thus

Table 12.8 Availability of Soybeans and soybean meal (source: FAO 2018)

Year	Local production (MT)	Imports (MT)	Import of soybean meal (MT)
2008	3030	2445	119,522
2009	3790	1682	104, 759
2010	7520	1632	114,677
2011	3840	0	159,470
2012	1670	101	148,517
2013	13,445	1122	152,235
2014	6790	18	161,352
2015	6578	7293	177,201
2016	6714	7146	194,983

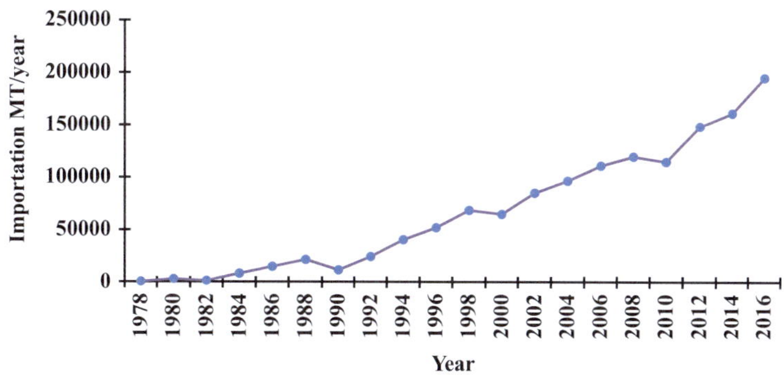

Fig. 12.3 Annual importation of Soybean meal to Sri Lanka during 1978–2016 (Source: FAO 2018)

supplying some soybean meal for the feed industry. In mid 1970s, the local soybean production rapidly declined due to the lower demand caused by the closure of the government-owned major oil extraction facility (former Ceylon Oils and Fats Corporation). However, currently there is a move to cultivate soybeans locally, and facilities for oil extraction are already established by the private sector.

12.3.5.3 Production of Coconut Poonac

Coconut poonac (coconut oil meal, copra meal, or coconut oil cake) is the only oil cake which is produced locally in considerable amounts and used for animal feed in Sri Lanka. As coconut is one of the main three plantation crops grown in the country, a large quantity of coconuts are produced annually. While a part of the nuts is consumed fresh in cooking, the other part is processed to produce either desiccated coconut, coconut milk powder, or copra which is used for oil extraction. From the coconut oil industry, poonac is obtained as a by-product which is used in animal feeds as a protein supplement.

Assuming 40% as the average extraction rate of coconut oil, the estimated amount of coconut poonac produced in the country during 2000–2014 (Table 12.9) shows that the poonac production is subjected to high fluctuations and in a declining

Table 12.9 Estimated production of coconut poonac and rice polish/rice bran

	Annual production (MT)			
Year	Coconut oil[a]	Coconut poonac[b]	Rice (Paddy)[a]	Rice polish[c]
2000	44,407	29,309	2,859,900	285,990
2002	30,100	19,866	2,859,480	285,948
2004	14,500	9570	2,628,000	262,800
2006	37,892	25,068	3,342,000	334,200
2008	59,018	38,952	3,875,000	389,520
2010	65,133	42,988	4,300,820	430,082
2012	55,190	36,425	3,845,941	384,594
2014	45,268	29,877	3,381,000	338,100
2016	–	–	4,117,336	411,737

[a]Source: FAO (2018)
[b]Estimated assuming 40% extraction rate
[c]Estimated assuming 10% extraction rate

trend since 2010. The quality of coconut poonac varies marginally depending on the method of oil extraction, solvent extraction, or expeller extraction. Solvent-extracted poonac contains relatively low fat and little more protein, whereas expeller poonac contains low protein (20%) and more fat (approx. 12%). Sri Lanka mostly produces expeller poonac having a low quality. Though it contains more gross energy due to high fat content, the metabolizable energy value for monogastric animals is rather low (about 1700 kcal/kg) due to the high fiber content (12–14%) in poonac. Expeller coconut poonac also has a poor keeping quality. Coconut fat is also prone to rapid rancidity resulting in poor feeding value. The feed millers are thus reluctant to incorporate coconut poonac in pig and poultry feeds. Locally produced coconut poonac is therefore used mainly for the production of dairy cattle feed.

12.3.5.4 Production of Rice Polish/Rice Bran

Paddy is the main cereal crop grown in Sri Lanka for centuries as it provides the main staple food, rice. When rough rice is milled to produce edible rice, valuable by-products are obtained which are extensively used in the animal feed industry. Rice polish/bran is a major constituent in compounded animal feeds while a small amount of broken rice is also used depending on the availability.

Rice is processed in large- and small-scale millers, currently by using roller type mills. In both situations, rough rice may be processed either raw or after steaming or parboiling, thus producing raw rice, steamed rice, or parboiled rice, respectively. Roller type rice mills are efficient as the amount of broken rice produced is low. At present, there is almost no broken rice reaching the feed industry from local rice mills, but feed millers use to import broken rice whenever it is cost-effective. However, a large quantity of rice bran/polish is produced as a valuable by-product, which is efficiently used by feed millers. Traditionally, the bran produced from the milling of raw rice is referred to as rice polish, while the other is named as rice bran. Usually the extraction rate of rice bran/polish is about 10% of rough rice milled. Table 12.9 depicts the estimated production of rice bran/polish assuming that the total harvest of paddy is milled to produce edible rice. A fraction of rice bran/polish produced in the country is exported, not due to a surplus but for economic reasons. Accordingly, the actual availability of rice bran/polish in the country would be lower than the estimated values.

12.3.5.5 Other Feed Raw Materials

There are various other types of feed raw materials used for the production of compounded feeds. The amount of type and quantity of these raw materials depend on the availability and price, and the competitiveness with maize and soya bean. Other cereals such as rice and wheat are used as alternatives to maize whenever maize is not available or expensive. A considerable amount of wheat bran is produced in Sri Lanka as wheat milling is done using imported grains to produce wheat flour. Poultry by-product meal produced locally is also available to some of the feed millers. Imported raw materials such as fish meal, meat and bone meal, gingerly poonac, and maize gluten meal are used in varying quantities depending on their availability and price. All vitamin mineral supplements, amino acid supplements, calcium and phosphorus supplements (mainly calcium carbonate and di-calcium phosphate), and various types of nonnutritive additives required for feed milling are imported.

12.3.6 Roles Played by the Private and Public Sectors in Animal Feed Production

Compounded animal feed production in Sri Lanka is totally handled by the private sector. As already depicted in Table 12.5, there are eight large-scale feed manufacturers in the country who are having modern feed milling facilities. Even the needs of feeds by the government-own livestock farms are supplied by the private feed millers. It was the private sector who started large-scale feed milling in Sri Lanka. Feed raw materials are either directly imported by the feed millers themselves or purchased through local raw material suppliers. Quality control of the raw materials as well as the finished products is done at the feed mill using their own laboratories and trial farms. In addition to the large-scale feed manufacturers, there are two medium-scale feed millers and a number of small-scale feed producers including self-feed producers. The commercial feed millers have established their own marketing network through wholesale dealers and retail dealers.

The role of the government in animal feed production is limited to a regulatory work. The commercial feed production and raw material importation are regulated by Animal Feed Act No. 15 of 1986, which was amended in 2016 as Animal Feed (Amendment) Act, No. 15 of 2016. The Registrar of Animal Feeds of the Department of Animal Production and Health (DAPH) is the licensing authority for the purposes of the Animal Feed Act. According to the Act, all feed manufacturers, sellers, and importers should have a license from the Feed Registrar to perform their activities. The Act also has given provision to establish a Feed Advisory Committee comprising of experts from different sectors to advise the line Ministry on issues related to animal feeds. The feed millers are also benefitted from tax relaxations on importation of feed raw materials.

12.3.7 Future Demands for Compounded Animal Feeds and Way Forward

Alexandrators and Bruisma (2012) estimated that the global demand for animal-sourced food will increase by 70% by the year 2050, and the main contributor will be the poultry sector. In Sri Lanka, during the past decade, the poultry sector has shown an annual growth of 8.3%. If this trend continues to the next decade, there will be 51 million birds demanding 3.2 million MT of feeds by 2027. When the requirement for other types of animals, especially dairy cattle, is considered, the estimated feed requirement would still be higher. The supply of feed raw materials to produce this amount of feed will be a big challenge to the country.

When the increasing human population and land limitation for agriculture are considered, the possibility for increasing feed crop production is remote. It will also be difficult to import feed from other countries as they also face the same problem. The global usage of maize for animal feed has reduced from 63% of the total production in 2005 to 54% in 2013 (FAO 2018), indicating that non-feed usage of maize is getting increased and a limited amount is left for animal feeding. Though the global production of soya beans has increased from 145 million MT in 2005 to 179 million MT in 2013, the usage of soya bean for animal feeding remained unchanged as 18% (FAO 2018). These two examples clearly show that the availability of the supply of animal feed raw materials will not be sufficient enough to meet the future feed requirement. Thus, the shortage of feed raw materials will be the main constraint to livestock and poultry production in future.

Two alternatives can be suggested to overcome the feed problem, i.e., (1) use of alternate feed resources (nontraditional feeds and by-products), and (2) increase feed utilization efficiency through proper feeding management along with precise feed formulation, comfortable environment, genetic improvement, health management, and the use of appropriate feed additives. The second alternative is commercially more viable as there can be many practical problems in using by-products and other feed resources.

References

Alexandrators N, Bruisma J (2012) World agriculture towards 2030/2050: the 2012 revision. Food and Agriculture Organization of the United Nations, Rome

Bandara BHWMUS (2007) The current status of smallholder dairy system in Sri Lanka. Available at: http://www.ilri.org/infoserv/webpub/fulldocs/south_south/ch07.htm

Central Bank of Sri Lanka (2018) Annual report. Central Bank of Sri Lanka, Colombo

DAPH (2017) Livestock statistics. Department of Animal Production and Health, Getambe

FAO (2018) FAOSTAT, statistical database of FAO. Food and Agriculture Organization of the United Nations, Rome. http://FAO.org

Ibrahim MNM & Siriwardene JAdeS (2001) Feeding systems and ration formulation. Part 1. Nutrition of the Buffalo. Water Buffalo in Asia. Publication of the SAREC/NARESA Buffalo Information Dissemination Programme, National Science Foundation, Colombo, Sri Lanka

Jayaweera MDN and Samarasinghe K (2002) Investigation of present status of small-scale self-mixing poultry feed manufacturers in the North-western province of Sri Lanka. Directed study report, Board of Study in Animal Science, PGIA, University of Peradeniya

Lammers BP, Heinrichs AJ and Ishler VA (2013). Use of total mixed rations (TMR) for dairy cows. http://extension.psu.edu/animals/dairy/health/nutrition/nutrition-and-feeding/dietformulation-and-evaluation/use-of-total-mixed-rations-tmr-for-dairy-cows/view, Accessed 20 October 2016

Murphy PG (1975) Net primary productivity in tropical terrestrial ecosystems. In: Leith H, Wittaker RH (eds) Primary production of the biosphere. Springer, New York

Pemadasa MA (1983) Grasslands. In: Fernando CH (ed) Ecology and biogeography in Sri Lanka. Junk Publishers, pp 99–131

Pemadasa MA (1990) Tropical grasslands of Sri Lanka and India. J Biogeogr 17:395–400

Premaratne S and Premalal GGC (2006) Country pasture/forage resource profile of Sri Lanka. http://www.fao.org/ag/AGP/AGPC/doc/counprof/Srilanka/Srilanka.html

Premaratne S, Premalal GGC, Jayawardena VP (2003) Sustainable management of grassland resource for ruminant livestock production in Sri Lanka. Trop Agric Res Ext 6:60–65

Ranaweera, N.F.C. & Attapattu (2006) In: Ranaweera N.F.C. (2009) Sri Lanka: Opportunities for dairy sector growth, Smallholder dairy development: Lessons learned in Asia, Rome: Food and Agriculture Organization of the United Nations. http://www.fao.org/docrep/011/i0588e/I0588E08.htm, Accessed 29 May 2014

Reynolds SG, Batello C, Bass ST (1999) Perspectives on rangeland development. In: Proceedings of the VI International Rangeland Congress, pp. 160–165. Townswille

Samarakoone (2018) Personal communications

Samarasinghe K (2007) Feeds and feed formulation for poultry in Sri Lanka. Department of Animal Science, University of Peradeniya, Sri Lanka

Sarmini M, Premaratne S (2017) Yield and nutritional quality potential of three fodder grasses in the northern region of Sri Lanka. Trop Agric Res 28(2):175–182

Sarmini M, Premaratne S, Premalal GGC, Mahipala MBPK (2017) Ensiling property and nutritional quality of Palmyrah (*Borassus flabellifer*) leaves incorporated corn (*Zea mays*) or sorghum (*Sorghum bicolor* (L.) moench) silage. Trop Agric Res 29(1):25–31

Snowdon, M. (1991) Total Mixed Rations for Dairy Cattle. Department of Agriculture, Aquaculture and Fisheries. https://www2.gnb.ca/content/gnb/en/departments/10/agriculture/content/livestock/cattle/rations.html, Accessed 20 January 2016

Somasiri SC, Premaratne S, Gunathilake HAJ, Abeysoma HA, Dematawewa CMB, Satsara JHMN (2010b) Effect of Gliricidia (*Gliricidia sepium*) leaf meal blocks on intake, live weight gain and Milk yield of dairy cows. Trop Agric Res 22(1):76–83

Somasiri SC, Premaratne S, Gunathilake HAJ, Abeysoma HA, Satsara JHMN (2010a) Development of a leguminous leaf meal block as an animal feed. Trop Agric Res 21(4):412–420

White R, Murray S, Rohweder M (2000) Pilot analysis of global ecosystems (PAGE): Grassland ecosystems. World Resource Institute (WRI), Washington, DC, p 1000

Zemmelink G, Premaratne S, Ibrahim MNM, Leegwater PH (1999) Feeding of dairy cattle in the Forest-garden farms of Kandy, Sri Lanka. Trop Anim Health Prod 31(5):307–319

Livestock and Poultry to Assure Sustainability in the Food System

13

G. L. L. P. Silva and C. M. B. Dematawewa

Abstract

Livestock and poultry are two integral parts in the food production systems in many countries in the globe, including Sri Lanka. Hence, assuring the sustainable utilization of resources and systems associated with those animal components are critically important. The important role that livestock and poultry perform in the production environment, the evolution of diverse farming systems that have helped sustainable utilization of the available resources by the animal industry, trends experienced in resources with especial reference to genetic resources, and changes in animal populations and production have been described in this chapter focusing on the Sri Lankan context, in comparison with dynamics experienced in regional and global scenarios. Institutional setup and related changes have been presented in the context of diverse production environment in supporting the understanding of the governing environment of the diverse systems. Influence of the trends and dynamics of the industry in bringing in sustainability to food systems of the country has been discussed with emphasis on resource diversity, genetic richness and mixed production environment. The approaches in managing the livestock systems for sustainability and associated challenges and strategies with some suggestions for research orientations are presented.

Keywords

Animal-sourced food · Productivity · Genetic improvement

G. L. L. P. Silva (✉) · C. M. B. Dematawewa
Department of Animal Science, Faculty of Agriculture, University of Peradeniya, Peradeniya, Sri Lanka

B. Marambe et al. (eds.), *Agricultural Research for Sustainable Food Systems in Sri Lanka*, https://doi.org/10.1007/978-981-15-2152-2_13

13.1 Introduction

Sustainability of an agricultural system is inherently associated with integrity of social, environmental and economic aspects. Food and Agriculture Organization (FAO) of the United Nations (UN) recognizes four main criteria for sustainability assessments of food and agriculture systems in the world, namely, environmental integrity, economic resilience, social well-being and good governance (FAO 2013). As a nation striving to achieve Sustainable Development Goals (SDGs), Sri Lanka has set one of the targets for SDG-2 to ensure sustainable food production systems and implement resilient agricultural practices that increase productivity and production, help maintain ecosystems, strengthen capacity for adaptation to climate change, extreme weather, drought, flood and other disasters and that progressively improve land and soil quality within the next 12 years (http://www.statistics.gov.lk/sdg/index.php/sdg/target/2).

Animal component occupies a distinct role in sustainable food production systems. Animal products are an excellent source of high-quality protein, fat, vitamins and minerals such as zinc, iron, selenium as well as calcium and phosphorus (Heywood 2013). The FAO estimates that livestock and poultry contribute to around 12.9% of global calories and 27.9% of protein supply directly through provision of meat, milk, eggs, offal, etc. (FAO 2011), in addition to the heavily undervalued contributions to the ecosystems that ensure sustainability of those production systems. Animal-sourced foods (ASF) are becoming increasingly important as those tend to occupy a growing proportion of the global food plate regardless of the individual variations unique to the status of development of individual countries and their dietary restrictions. Meals that include a few or no servings of different food groups such as fruits and vegetables, pulses, dairy products, fish and meat are considered by food scientists as lacking in both balance and variety (Jayawardena et al. 2012). Consumers of many countries in the developing world are undergoing a rapid nutritional transition at present seeking variety in their food plate. Consequently, a dramatic increase in consumption of ASF can be seen over the past few decades where Sri Lanka is no exception.

During a time span of 25 years from 1985 to 2009, the per capita calorie supply from ASF has shown a 38.3% increase compared to about 8.5% increase of vegetable- and cereal-based food in Sri Lanka (Wijesekara 2015). Based on the similar trends in consumption transition in the world, FAO (2010) predicted that the demand for ASF would increase to the heights that would pressurize the global production of meat to be doubled by 2050 compared to the status of production in 1990. By the year 2020, the world would be passing the half-way milestone of that prediction. Parallel to the demand for increase of production, FAO (2010) has also not forgotten to warn that the current phase of industrial livestock production practices would not be sufficient to achieve the targeted sustainability goals. The reason behind this warning is due to that livestock is currently the single largest user of land in the world, accounting for 70% of all agricultural land and 30% of total land surface. Fortunately, the world scenario is not directly reflected in Sri Lanka, where only 8.5% (0.56 million ha) of the total land area is occupied by livestock (mixed or sole) and 99% of those are smallholdings, i.e., less than 2 ha (Perera and Jayasuriya 2008). This warrants a country-specific strategy for sustainable food and production systems in Sri Lanka.

This chapter describes the role play of livestock and poultry in the sustainable food systems in relation to achieving national targets with special reference to trends and dynamics of smallholder animal production systems, and animal genetic resources. The specific focus is on the genetic resources of domesticated avian and mammalian species that contribute to food production and agriculture in Sri Lanka.

13.2 Livestock Systems in Evolution

Animal farming is considered as a traditional income-generating activity in Sri Lanka for centuries. The sustenance of this livelihood function has been ensured by its integrity with socio-economic and cultural aspects of the society and the system where it is operating. While satisfying the nutritional needs of the communities in Sri Lanka by providing high-quality protein through the main produces such as milk, meat and eggs, the livestock systems have been playing a critical role offering non-food-based services including draft power for agriculture and transport, and subsidiary products including hides, skins and manure. Their contribution to manage the high risk and vulnerability associated with crop farming, especially under the unpredictable climatic and environmental conditions, has being invaluable. Evolving in a country with an agricultural economy and a long history of agriculture-based civilization, farm animal management has been playing a multifaceted role in the socio-economic and cultural context of Sri Lanka. Currently, livestock is an integral component of the agriculture sector of the country where smallholder operations are predominant.

Many of the traditional livestock and poultry systems are not inherently complex owing to the short operational distance where a considerable proportion of production, harvesting, processing, marketing, consumption, and disposal take place within the system. Especially, the inputs needed and outputs generated need not travel much beyond the boundaries of the systems. A major concern around the world is that the livestock production could have a significant impact on the environment as it uses considerable amount of water, land and feed resources. However, a lesser known fact is that it also provides non-food services contributing to the food production systems through improvement of environmental health. The traditional production systems of the country (Ibrahim et al. 1999) provide a platform to utilize livestock and poultry for optimizing the resource utilization such as through the use of marginal land areas, scavenged feeding, and recycling of nutrients, which are not possible with sole crop production system (Ibrahim and Schiere 2002). Traditional livestock systems also add to the balance of energy and protein (food balance) available for human consumption. Many traditional livestock systems have evolved through practices such as grazing, scavenging and recycling. Being largely based on environmental norms and inputs, livestock component in traditional system competes the least for human food and poses the least disturbances on health of environment, making the system of operation sustainable in the long run.

With the sustainability being its forte, the livestock systems all over the world have evolved into diverse production systems accommodating a variety of animal genetic resources playing multifaceted roles (Abeygunawardena et al. 2012). This process is clearly evident in the evolution of dairy system of Sri Lanka. Despite the fact that crop cultivation is dominant in the agriculture system of the country, a large number of livestock production systems has evolved based on the diversity of resource availability of the system. For example, the traditional dairy system is purely based on the crop agriculture farming pattern where cattle and buffaloes fill the gaps in different dimensions of the system such as time, land and other input utilization (Silva et al. 2011). Goat and sheep production systems also have traditionally evolved with the crop-based production environment (Silva et al. 2009; Vijitha and Silva 2013). Thus, many of the ruminant production systems are crop-based, and livestock-dependent processes appears as a component occupying 17.9% of the system (Ranaweera 2007). Pig production of the country that has been predominant in the coastal area of the country in the past, however, has recently spread towards North Western areas of the country with a commercial orientation (Dematawewa et al. 2009; Silva et al. 2016a). The backyard poultry system, though showing little dependency on crop farming, is the most widespread and viable operation in the rural as well as peri-urban areas of the country (Silva et al. 2016b; Bett et al. 2014a).

The backyard poultry or village chicken production system is one of the key operations in the rural food systems in Sri Lanka contributing to the nutritional, livelihood and social needs of many rural communities in the country (Thilini et al. 2016; Abeykoon et al. 2014; Wijayasena et al. 2014; Rajapaksha et al. 2014; Abeykoon et al. 2013). Even within one agro-climatic zone in the country, many livestock production systems have been evolved based on different factors within which farming is in operation, especially the agro-climatic zone, resource availability and socio-economic conditions along with cultural affiliations (Table 13.1). According to the classification suggested by Teufel et al. (2010), out of the total agriculture production systems in Sri Lanka, livestock-only system is represented by 0.08% and it is one of the lowest representations in the region.

13.2.1 Animal Diversity and Livestock Systems

Existence of a wide diversity of animals, both within and among species, has invariably been allowed in different production systems. The boundaries of these production systems are mainly demarcated by the agro-climatic factors, resource base and the socio-economic conditions. In particular, the animal genetic resources for food and agriculture provide a biological capital on which livestock production systems are built. As in the other countries of the region, Sri Lanka is blessed with a high diversity in animal genetic resources (Pushpakumara and Silva 2008) that has gradually built up over the centuries of agriculture-based civilization of the country

Table 13.1 Different livestock and poultry production systems across the agro-climatic zones in Sri Lanka (Source: Ibrahim et al. 1999; Silva et al. 2010)

Agro-climatic zone	System of rearing[a]
Low-country dry zone	– Rainfed rice-based cattle farming system – Smallholder buffalo system – Vegetable-based intensive Jaffna dairy system – Irrigated rice-based dairy system – Large-scale intensive livestock system (state/private owned) – Crop-based semi-intensive Jaffna sheep system – Crop-based nomadic Jaffna sheep system – Semi-intensive small-scale poultry system – Backyard poultry farming system – Buy-back poultry system
Low-country intermediate zone	– Rainfed rice-based cattle farming system – Smallholder buffalo system – Irrigated rice-based dairy system – Large-scale intensive livestock system (state/private owned) – Rainfed highland crop-based goat farming system – Coconut-based goat farming system – Plantation-based semi-intensive sheep farming system – Peri-urban pig farming system – Intensive commercial-scale poultry system – Semi-intensive small-scale poultry system – Backyard poultry farming system – Buy-back poultry system
Low-country wet zone	– Off-farm income-based Peri-urban dairy farming system – Smallholder buffalo system – Coconut-based dairy farming system – Coconut-based goat farming system – Plantation-based semi-intensive sheep farming system – Crop-based pig farming system – Peri-urban pig farming system – Backyard pig farming system – Intensive commercial-scale poultry system
Mid-country intermediate zone	– Off-farm income-based peri-urban dairy farming system – Large-scale intensive livestock system (state/private owned) – Irrigated rice-based dairy system
Mid-country wet zone	– Tea-estate-based cattle farming system – Kandyan homegarden system – Off-farm income-based peri-urban dairy farming system
Up-country intermediate zone	– Tea-estate-based cattle farming system – Kandyan homegarden system – Off-farm income-based Peri-urban dairy farming system – Estate-based goat farming system
Up-country wet zone	– Tea-estate-based cattle farming system – Vegetable-based cattle farming system – Large-scale intensive livestock system (state/private owned) – Estate-based goat farming system

[a]Identification of the system of rearing is indicated as it appears on the original reference

(Jalatge 1980; Chandrasiri 2002; Silva et al. 2010), through natural selection with appropriate adaptation to variety of environmental impacts. Diverse production systems play a critical role in creating opportunity for utilization of those resources.

Recognizing the full range of production and management systems and identifying the germplasm that is optimal for each system are essential for sustainable utilization of resources as well as the sustainability of the system (Notter 1999). Economic output, which is mainly expected from domestic animals, is largely contributed by commercial breeds in leading production systems in the country. However, majority of the systems in rural areas, which depend on local livestock and poultry, often provides more diverse ecosystem services along with their crucial role in food security, nutrition and health. The production systems involved in grazing or scavenging systems (extensive system of rearing) located especially in relatively harsh climate conditions permit the use of resources within the system, but are not readily suitable for crop production (Ibrahim and Schiere 2002). By making use of diverse genetic resources, different production systems explore the adaptive capacities of those species and breeds, which can thrive under harsh environmental conditions. Therefore, both highly specialized and underutilized genetic resources have unique roles to play in the respective farming systems facilitating the sustainable utilization of different genetic resources. Table 13.2 depicts the diverse animal genetic resources utilized in major livestock production systems.

13.2.2 Diversity of Animal Genetic Resources

The genetic diversity of a farm animal species directly corresponds to the characteristics of production conditions (Notter 1999). As the farm animal management has been playing a significant role in different social and economic strata of Sri Lanka, the production conditions have also been diversified according to the need of the strata. Variation of genetic resources has to be invariably high corresponding to the diverse production conditions. On the contrary, genetic diversity is relatively low when the production conditions are highly standardized, as generally found in developed countries (Notter 1999). In the attempts to improve the production of ASF for the country, Sri Lanka has followed various strategies in the past such as improving the genetic potential of economic traits of farm animals, especially milk production (Buvanendran 1967; Nadarajah 1968; Wijeratne 1970; Balachandran 1972; Jalatge 1980; Chandrasiri 2002; Dematawewa and Silva 2000). The results of such attempts and also the genetic variability preserved in genotypes of indigenous farm animals, most of which are blends of different genotypes brought into the country during kings or colonial era of the country (Chandrasiri 2002), have resulted in a highly diverse within-species genetic variability in farm animals.

In general, farm animal genetic resources in the country could be categorized into two main groups, namely, the locally adapted (including indigenous categories) and continually imported livestock.

Table 13.2 Different animal genetic resources and livestock production systems in Sri Lanka (Source: Silva et al. 2010)

Species	Farming system	Breed/type
Cattle and Buffalo	Estate system	Jersey, Friesian, Ayrshire and their crosses
	Kandyan homegarden system (mid country smallholder system)	Jersey, Friesian, Ayrshire and their crosses
	Coconut-based dairy farming system – Cattle	Jersey crosses, Sahiwal crosses, Sindhi crosses, Indigenous
	– Buffalo	Indian crosses, Indigenous
	Peri-urban dairy system – Cattle	Jersey, Friesian, Ayrshire and their crosses
	– Buffalo	Murrah, Nili-Ravi, and their crosses
	Small holder buffalo system	Indigenous, Murrah, Nili-Ravi, and their crosses
	Jaffna Peninsula system	Jersey, Sahiwal, and their crosses
	Irrigated rice-based system	Indigenous, Zebu crosses Indigenous, Temperate crosses
Goat	Rainfed Highland system	Indigenous, Jamnapari crosses
	Coconut-based system	Indigenous, Jamnapari crosses
	Tea estate system	Saanen
Sheep	Crop-based semi-intensive system	Jaffna local sheep
Pig	Crop-based system	Large white, Landrace, Indigenous
	Livestock-based system	Large White, Landrace, Duroc and their crosses
	Back yard system	Indigenous, Crossbreds
Poultry	Intensive system	Commercial strains
	Semi-intensive system	Commercial strains, Indigenous chicken
	Scavenging system	Indigenous chicken
	Buy-back system	Commercial strains

13.2.2.1 Locally Adapted Livestock and Poultry

Livestock breed types that have been localized for more than 40 years and continued to perform under present context for more than seven generations are categorized as locally adapted breed types (Chandrasiri 2002). Genetic and phenotypic characterization of some of the locally adapted species have been attempted (Thilakaratne 1984; Silva et al. 2008, 2010, 2016a, 2017; Subalini et al. 2010, 2013; Bett et al. 2014b; Liyanage et al. 2015; Rathnakumara et al. 2015; Periasamy et al. 2017) as given in Table 13.3, and some more are yet to be identified as breeds by distinguishing them from other reported breeds elsewhere. The inherent ability of locally adapted breeds to withstand harsh environments and their tolerance against specific diseases and environmental stresses could never be undervalued in the local production systems. Some of these breed types have been evolved from their wild ancestors adapted in different geographical locations for centuries.

Table 13.3 Different livestock breeds introduced to Sri Lanka

Locally adapted categories	
Species	**Category and the description**
Cattle	Lankan cattle (*Batu-harak*) – A different category of cattle and described as an archaic cattle with little relation to the indo-Pakistani zebu Thamankaduwa cattle – A distinct category of cattle, which has been evolved with a blend of native cattle with several Indian zebu breeds, brought to the northern and eastern regions Cape cattle/Hatton Cattle - It is evident that the genotype of Hatton cattle consists with a mixture of different *B. taurus* genotype with lesser extent of native cattle. However this cattle type is now believed extinct
Buffalo	Lankan buffalo – A unique category that possesses swamp type phenotypes and river type genotype (50 chromosomes)
Goat	Lankan goat – A blend of different genotypes exist in the country for centuries, thus origin is not known Kottukachchiya goat – A breed originated from indiscriminate crossing of non-descript animals from South India with selected Lankan goats with little directional selection Sri Lankan Boer goat – A stabilized crossbred goat type developed by crossing imported German Boer with purebred Jamnapari, purebred Kottukachchiya and Jamnapari × Kottukachchiya crossbreds. Later disappeared as the crossbred was not stabilized and the genotype got diluted
Sheep	Jaffna Local Sheep is the only native breed of sheep and unique animal, which closely resembles the Suffolk breed of England. It descents a blend of genotypes of south Indian sheep breeds
Pigs	Village pig is the domestic descendent of the wild pig (*Sus scrofa affinis*) in Sri Lanka with a slow directional selection for production traits
Chicken	Rho-White – A crossbred of white (Leghorn lines adapted to local conditions) and brown lines (Rhode Island Red adapted to local conditions) produced for local industry. Later disappeared due to segregation of genotype and lack of farmer acceptance Village chicken - A very diverse and unique group of birds, which have close genetic relationship to Red Jungle Fowl (*Gallus gallus*) and Gray Jungle Gowl (*Gallus sonneratii*). Possesses a high genetic and phenotypic diversity, grouping diverse categories of birds such as Game or long legged, Necked-neck, Crown, Frizzled, etc. and are well adapted to the harsh environments and backyard system of rearing
Imported breeds	
Species	**Breeds introduced**
Cattle	*Bos Taurus* – Friesian, Jersey, Dairy Shorthorn, Holstein, Meuse Rhine Yssel (MRY), Ayrshire, Brown Swiss, Red Poll *Bos indicus* – Sindhi, Sahiwal, Tharpakar, Haryana, Khillari, Kangayam, Gir Synthetic breeds – Australian Milking Zebu (AMZ), Australian Friesian Sahiwal (AFS), Sunandini
Buffaloes	Murrah, Surti, Nili-Ravi, Mehsana
Goats	Jamnapari, Saanen, Beetal, Boer, German Fawn
Sheep	Polled Dorset, Bikaneri, Madras Red, South Down, Wiltshire Horn, Bannur
Pigs	Large White, Landrace, Duroc
Rabbits	New Zealand White, Californian White, Belgian Red, Vienna Blue, Flemish Giant, New Zealand Red

(continued)

Table 13.3 (continued)

Chicken	Plymouth Rock, Sussex, Cornish White, Leghorns, Rhode Island Red improved broiler strains (Hubbard varieties, Indian River (IR), Cobb 500, Ross 308, Arbor Acres) and layer strains (Hy Line, Bovan, Shaver)[a]
Turkeys[b]	Medium White, bronze, buff
Ducks[b]	Velovi, Petrock, Khaki Campbell, Vigova, Indian Runner, Pekin, Aylesbury, Orpington
Muscovy[b]	White variety, Black variety
Geese[b]	Toulouse, Emden, African, Chinese, Canada
Quails	Japanese quail

Compiled based on information from Chandrasiri (2002)

[a]Dr. Duminda Dissanayake; Senior Technical Manager, Poultry sector company [personal communication]

[b]Hevapokara (1995)

13.2.2.2 Continually Imported Breeds

The animal genetic resources belonging to this category are brought to the country for commercial purposes, and hence are in commercial oriented production systems. Often, they have been brought to the country for upgrading the national herds of livestock (Table 13.3).

13.2.3 Trends in Productivity

Existence of a dominant small-scale farming system in Sri Lankan agriculture sector brings in an array of consequences. Livestock production systems are well known to be increasingly vulnerable to climate change, globalization, increasing price of inputs and degradation of the natural resource base. Simultaneously, there is an increasing public pressure for livestock production systems that contribute to environmental health while providing provisional services required by the farming communities. Enhancement of production and productivity is required to fulfil the food demands of the ever-increasing population despite all those challenges. In the global context, livestock production target of doubling current production in 2050 (FAO 2017) has to be achieved while restricting or reversing its negative impact on the environment. This necessitates a drive towards multifunctional landscapes in animal production. Sri Lanka, with its diverse production systems, especially with dominating mixed-production practices, has the correct landscape to address those multifunctional aspects needed. Whether Sri Lanka has functional adequacy in the context of provisional services of animal industry, particularly with dominating small-scale mixed production system, is a question to be addressed and a challenge to be surmounted.

Sri Lankan livestock industry can be assessed with respect to the production of major ASFs. Table 13.4 depicts the national productivity dimensions of milk, egg and poultry meat, which are the major categories of ASFs utilized in Sri Lanka, in the context of world and regional productivity levels. Accordingly, there is an

Table 13.4 National, regional and global average production values for major categories of ASFs utilized in Sri Lanka (Source: FAO 2016)

Area	Annual average milk production (kg/animal)	Annual average egg production (kg/bird)	Annual average poultry meat production (kg/bird)
Sri Lanka[a]	435	13.65	1.15
Asia	775	9.93	1.40
Global	1034	10.41	1.71

[a]Total production of the country

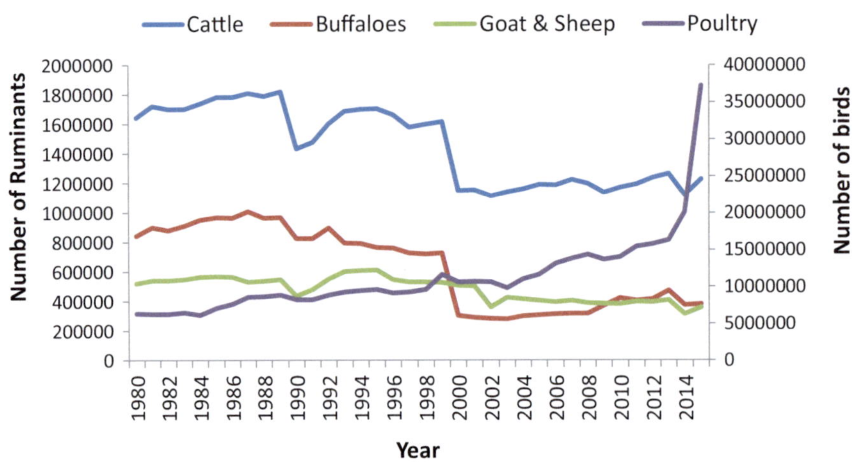

Fig. 13.1 Changes of farm animal populations in three major ruminant species and chicken during 1980-2015 (source: DCS – various issues)

obvious opportunity for improvement, which needs innovative interventions through the existing production environment, i.e. mixed smallholder production system. The recent milk production levels of the country appear to be about 3.1 and 2.2 L/milking cow/day for cattle and buffaloes, respectively (calculated based on DCS 2017a). Those two dairy sectors have shown 42.8% and 41.7% increases in milk production (L/cow/day) during the last decade (2008–2017). Moreover, the poultry sector egg productivity has shown a phenomenal growth of 77% during the same period (calculated based on the statistics available in DCS 2017a). Production statistics of those two sectors provide evidence that increase in productivity depends on the production environment and compatible intervention strategies.

13.2.3.1 Trends in Animal Populations

Farm animal populations of the country has shown a gradual decreasing trend over the past three and half decades, except for chicken population (Fig. 13.1) with the current national herd sizes of 1, 0.28, 0.28, 0.01, 0.09 and 21.2 millions of animals for cattle, buffalo, goat, sheep, pig and chicken, respectively (DCS 2017a). However, the populations of all ruminant categories have recorded a stable phase after the

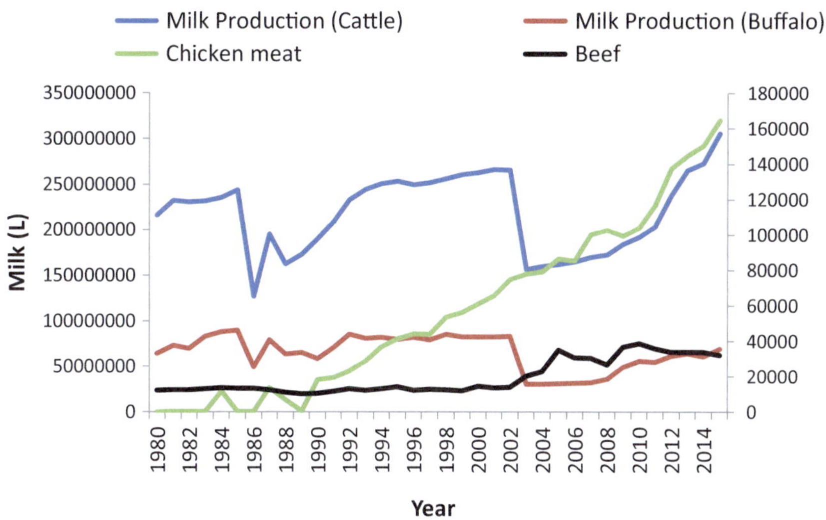

Fig. 13.2 Changes in volumes of production of different animal produces during 1980-2015 (source: DCS – various issues)

island-wide census of agriculture conducted in 2002 (the estimated numbers before year 2002 were corrected with the island-wide census). However, cattle population showed a minor increment in population due to the recent importation of cattle under mega dairy project. The poultry population growth was remarkable for the period (after last island-wide census to date, 2002–2017), which recorded 98.5% increase in population.

13.2.3.2 Trends in Animal Productions

All major animal products showed an increasing trend during the last three and half decades (Figs. 13.2 and 13.3) despite the decreasing trends in certain ruminant populations. However, there has been a distinct increase in bird numbers, especially during the last 5 years, parallel to the increase of broiler production of the country (Fig. 13.2). The population and production dynamics are more visible in the past decade probably owing to the change in the socio-economic status of the people in the country. These trends showed no deviation to the industry behaviour observed in the region as well as in the globe, and also in accordance with the predictions of Vasileska and Rechkoska (2012).

In the global arena, livestock and poultry contribute to around 12.9% of calories and 27.9% of protein directly through provision of meat, milk, eggs and offal (FAO 2011). Sri Lanka does not deviate from the world average (DCS 2016b), where out of the average total protein supply of 73.45 g/day of a Sri Lankan, 25 g of protein/day is supplied from direct consumption of ASF. The proportional contribution of different categories of food of animal origin to the total amount of protein supplementation is given in Fig. 13.4.

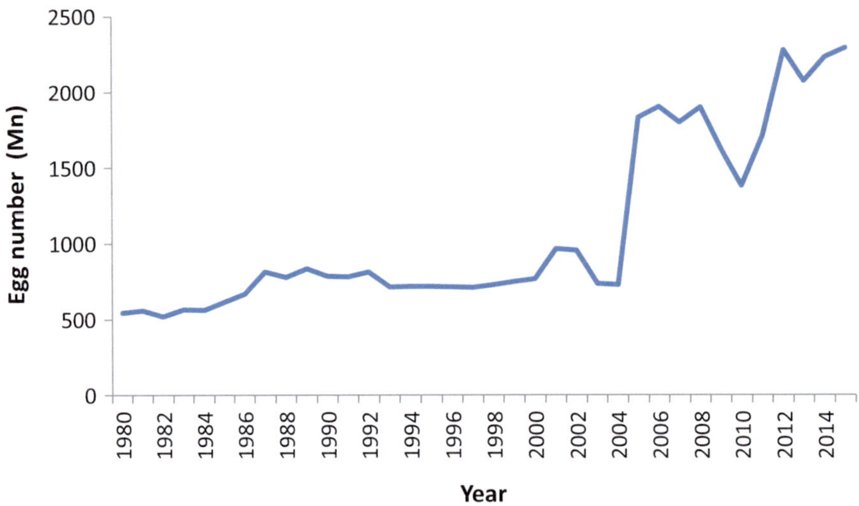

Fig. 13.3 Changes of egg production (indicated by number) during 1980–2015 (source: DCS – various issues)

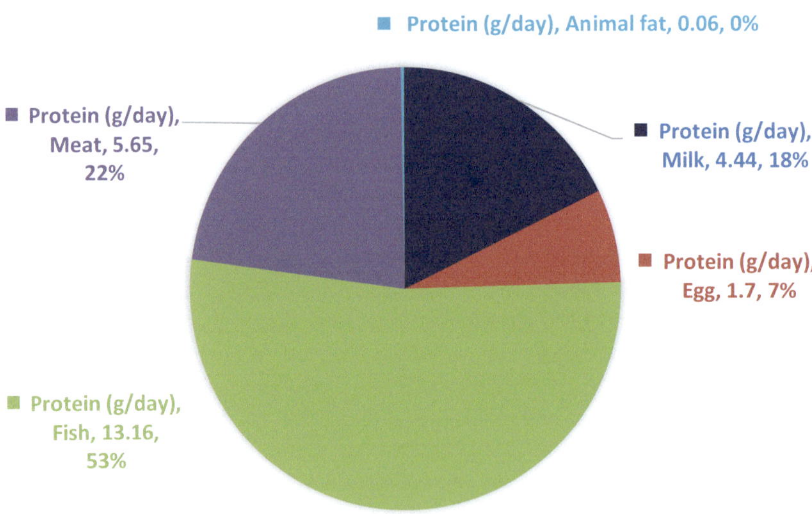

Fig. 13.4 Proportions of different food categories of animal origin in total animal protein basket of a Sri Lankan (source: DCS 2016b)

Improvement in productivity is vital to meet the demand while abating the adverse effects on the environmental health. This is clearly evident from the changes in volumes of animal production over the past few years. The rates of increase of the major animal products of the country are substantial during the last decade resulting in a phenomenal increase in the per-capita availability of those products (Table 13.5).

Table 13.5 Dynamics of animal produces and their per-capita availability during 1981-2017 (source: DCS 1981-2017)

Product type	Current annual production	Rate of increase per annum[a]	Current per capita availability	Rate of increase/year[a]
Milk	374 million L	17.2 million L	48.56 L	2.2 L
Chicken meat	164,500 MT	7925 MT	7.82 kg	0.32 kg
Egg	2294 million eggs	39 million eggs	109 eggs	4.3 eggs
Beef	32,180 MT	N/A	1.5 kg/year	N/A
Mutton	1353 MT	N/A	0.09 kg/year	N/A

N/A not available, *MT* metric tons
[a]Base year 2007

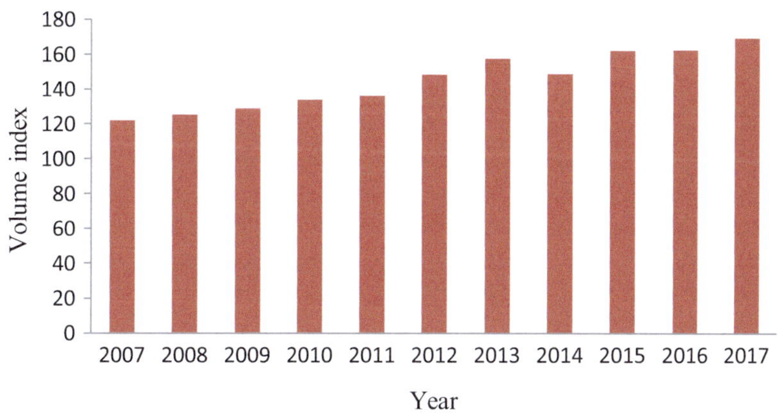

Fig. 13.5 Change of volume index of livestock products during 2007-2017 (assuming base year 2002 as 100%; source: DCS 2017b)

13.2.3.3 Volume of Production

Output of the livestock industry is a clear indicator of the status of production. Volume index of agriculture sector production of Sri Lanka shows that the livestock and livestock products have seen a steady increase over the past decade (Fig. 13.5). However, the increase of volume of production is not yet sufficient to fulfil the demands of the sector as far as certain commodities are concerned. Both in-house production and imports are necessary to satisfy the demand for those commodities. For example, Sri Lanka has spent around Rs. 36.8 billion (245 million USD) for processed animal products in the year 2016 (DAPH 2017), majority (81.5%) of which being spent on powdered milk. This indicates the opportunity that exists for expansion of the industry further to fulfil the demand of the country.

Thc present phase of increase in production and productivity is not sufficient to meet the growing demand. Dairy subsector in particular has shown a considerable rate of increase over the past decade (Table 13.5). However, the increase has not eased the importation bill of powdered milk for the same period. Despite the fact that dairy development is the country's priority in sector development plan, the current domestic milk production fulfils only around 40% of the requirement (DAPH 2016). Importantly, the pace of increase of demand for dairy products surpasses the pace in which the in-house production is growing.

13.2.4 Trends in Institutional Set-up

Historically Sri Lanka has had a dedicated Ministry responsible for Livestock Development, which is the main government organization responsible for policy planning, implementation, monitoring and resource mobilization. The main line agency that comes under the ministry is the central DAPH, which was established in 1978 as the main state organization responsible for livestock development in Sri Lanka. Before 1978, the subject of animal production and health was amalgamated with the Department of Agriculture (DoA) since its establishment in 1912, until the DAPH was established in 1978. Policy implementation in areas of animal health and disease control, animal production and breeding, related resource management, monitoring and evaluation, dissemination of information, implementation of laws and regulations, animal quarantine, welfare and disease surveillance are the main responsibilities of DAPH. Most of those activities are directed to the end users by the nine provincial governments through provincial DAPHs, which operate under provincial ministries since devolution of administrative power decentralization became effective from 1987. However, the devolution of responsibilities for delivery of many of the services in the livestock sector to the provincial administration and subsequent decentralization to the divisional level has introduced a cumbersome administrative structure reducing the efficiency of field-level functions. As described by Perera (1996), the lines of authority are complex and involve duality between the line agency and provincial administration. Thus, decentralization of power has created an extra burden to the limited human resources who are supposed to meet farmers' needs. Timely solutions are unreachable under the cumbersome bureaucracy constraining operational effectiveness.

National Livestock Development Board (NLDB) is a commercially oriented government organization, which is mandated to produce quality breeding materials, livestock and agricultural products to enhance the socio-economic standards of Sri Lankans, and comes within the purview of the Ministry responsible for the subject of animal production and health. Milk Industries of Lanka Company Limited (MILCO) that operates under the same Ministry, Mahaweli Livestock Enterprises Ltd. of the Mahaweli Authority of Sri Lanka (Ministry of Mahaweli Development and Environment), some cooperative societies and a handful of private organizations produce milk products, value-added milk products and other animal produces while providing various support services for dairy farmers. However, compared to the private organizations, government plays a prominent role in providing services to the dairy sector (Chandrasiri 2002; Ranaweera 2010).

The poultry industry, however, is successfully managed by the private organizations while the government involvement has been limited to monitoring of importation of parent/grandparent stocks and feed ingredients. Among different subsectors of the livestock industry, poultry is the only subsector that has achieved self-sufficiency and has been capable of maintaining the sustainability. Small ruminant industry of the country has not gained any prominence during the history of livestock industry, and hence has stagnated in terms of both animal numbers (except for random fluctuations of animal numbers) and production, despite its great potential to contribute to the rural economy. The swine industry is restricted to certain areas of the country and both the government institutions and the private sector involvements can be seen in the operations of the industry.

There were several attempts by various farmer societies and non-governmental organizations scattered all over the island to provide management and technical know-how to promote livestock production along with welfare schemes for members. In contrast, the private sector provides inputs to almost all the levels of the market chain as could be seen clearly in the poultry industry. Some private sector organizations are involved in production and marketing of poultry meat and eggs, hatchery operations, contract operations, manufacture and marketing of animal feed, etc. In such scenarios, large-scale private poultry farms belonging to multinational companies or large-scale national enterprises act as individual organizations in leading the market integration.

13.3 The Players Today

In spite of a widespread smallholder system that exists in Sri Lankan Agriculture sector, including the livestock subsector and its comparatively low contribution of 0.8% to the national GDP (CBSL 2017), there is a well-recognized fair share of contribution of the sector to the national food security. Its contribution to the food security is especially highlighted when the sustainability of food systems is considered and particularly in the context of empowering vulnerable communities. Given the developing status of the country with 4.1% of the population being very poor, a considerable portion of the population is on the brink of poverty (DCS 2016a) and about 26.9% of the population being employed in agriculture sector creates an opportunity for expansion of the livestock sector. It should not be an insurmountable challenge to assure sustainability in the national food system if the sector is driven in an efficient and appropriate resource utilization environment.

The role play of livestock and poultry in the variety of production systems is a crucial factor in sustainable food production operations, especially when the productivity of the smallholders is considered in the country. As Lin (2011) described for crop systems, the livestock systems also need to address the decision on choice of correct animal/breed, efficiency of input utilization and appropriate production operation for any animal food production system to become sustainable, and to produce safe and quality food. Identifying and balancing the key players acting today (Fig. 13.6) could be a directive for the future plans and growth of the sector.

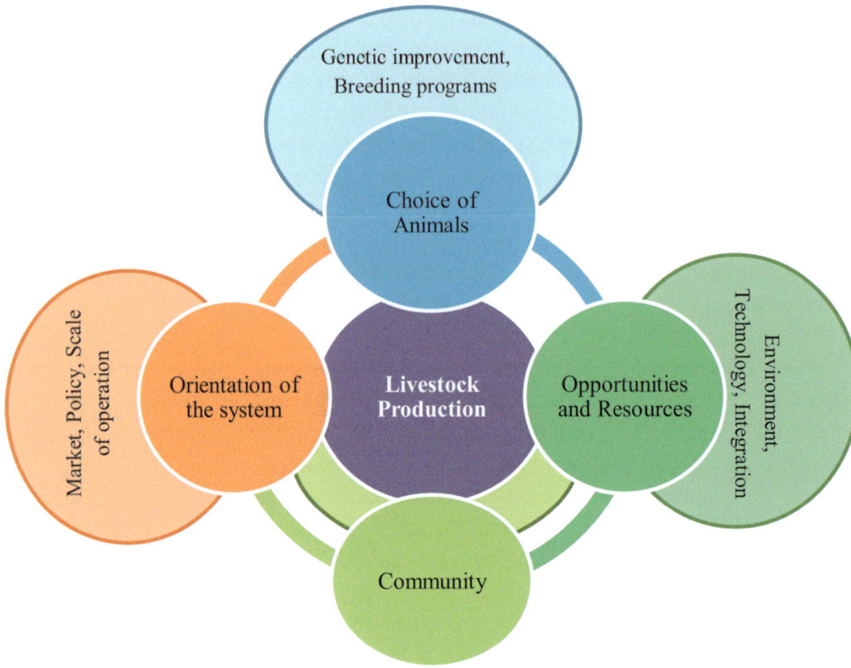

Fig. 13.6 Diagrammatic representation of the players of the livestock sector

13.3.1 The Role of Animal Diversity

This current dichotomy of animal industry in Sri Lanka consists of high input-high output intensified animal system and medium to low input animal production operations (Abeyratne 2007). Diverse animal resources are needed to cater to the needs of existing divisions of operations. The improved genotypes, which are more suitable for the high input-high output system, have received attention during the past 5–6 decades of existence of animal rearing as an industry. However, more widespread medium to low input production systems are catering to the highly diverse production environments which need diverse genetic resources (Abeygunawardena et al. 2012). There has been a lack of focus of the successive authorities in Sri Lanka towards improvement of particularly the low input systems. Any focus of improvement should be devoted equally to both ends of the dichotomous industry due to the following reasons.

- High input-high output system fulfils the national requirement of the country for the self-sufficiency in the ASF considering the changes in dietary patterns of growing and changing populations.
- Medium to low input system fulfils the environmental, social and cultural requirement while contributing to the national animal-based food requirements. This system balances the livestock sector against the claims of (1) impacts on biodiversity; (2) impacts of global change and (3) impacts over energy and water security.

13.3.2 Genetic Improvement Programmes

Herd improvement programmes and systematic dairy operations in Sri Lanka date back to the king's era and to colonial regimes (Chandrasiri 2002). However, the animal importation could be traced back to several occasions during the history of animal industry of Sri Lanka. In 1930s and 1940s, Sindhi cattle and Murrah buffalos were imported from India, Sahiwal cattle, Nili Ravi Buffalo and Beetal from Pakistan, and Surthi buffaloes and Jamnapari goats from India and Pakistan (Perera 1996), aiming for herd improvement programmes of the country. Temperate cattle types were introduced later in the breeding programmes. Table 13.6 summarizes the recorded live-animal importation events to date. Beside those, Sri Lanka continually imported semen from European (*Bos taurus*) as well as Zebu (*Bos indicus*) cattle breeds for upgrading programme of national cattle and buffalo herds. Tables 13.7 and 13.8 summarize the milestones achieved over the years in genetic improvement programmes in relation to institutional development and breeding technologies of the country.

Genetic improvement of commercial chicken has not been a priority in the state sector interventions except for the development of Rho-White chicken undertaken by the Central Poultry Research Station (CPRS) of DAPH in 1970s. This was to cater to the demand prevailed at that time for a locally adapted commercial breed during the transition period of backyard poultry production system to intensive type of commercially oriented poultry systems. Except for such attempt, the chicken subsector improvement is totally monitored by the private ventures with the importation of high-quality day old chicks and island-wide network of hatchery operations along with contract grower systems.

Animal genetic improvement programmes of the country have been supported by national as well as international agencies. The current status of genetic improvement and animal production are the results of such efforts. The implementation of improvement programmes has taken place in national, regional or divisional levels of the country (Table 13.9). However, the national targets of improvement may or may not have met in such occasions.

13.3.3 Resources, Environment and Animals

More than seven decades-long implementation of breeding and selection programmes with a wide range of objectives has converted the national farm animal herds into mixed genotypes with varying levels of genetic potentials. Therefore, the animal genetic resources in Sri Lanka are highly diverse and have a capacity of tackling the dichotomy of the production as described previously in this chapter. The commercially oriented system of production utilizes the improved genotypes with high inputs for which the national animal breeding policy guidelines (DAPH 2010) provide directives necessary for optimum utilization of such resources under varying environmental and production conditions. As discussed in detail on several occasions and correctly pointed out by Perera and Jayasuriya (2008), development of this sector will not only help to achieve food security, import substitution or even self-sufficiency in animal products, but also prevent the labour migration between

Table 13.6 History of live-animal importation to Sri Lanka (✓ indicates one occasion of importation during the given time period)

Species and breeds	Year 1936–40	1941–45	1946–50	1951–55	1956–60	1961–65	1966–70	1971–75	1976–80	1981–85	1986–90	1991–95	1996–00	2001–05[a]	2006–10[a]	2011–15	2016–to date
Cattle																	
Kerry	✓✓																
Ayrshire	✓					✓											
Scindhi	✓	✓	✓	✓													
Sahiwal	✓			✓						✓	✓✓	✓					
Kangayan	✓																
Khillari	✓			✓					✓								
Tarpakar				✓✓							✓						
Jersey						✓✓							✓			✓[b]	
Fresian						✓			✓	✓	✓		✓			✓[b]	
Hariyana							✓✓										
Shorthorn						✓											
Gir									✓								
Crossbred (JxF)																✓[b]✓[b]	✓[b]✓[b]✓
AFS													✓				
Breed unknown	✓		✓✓	✓		✓	✓	✓			✓	✓				✓✓	
Buffalo																	
Murrah	✓	✓		✓													
Surti									✓								
Nili Ravi										✓	✓✓	✓					
Breed unknown							✓										

Species and breeds	Year 1936–40	1941–45	1946–50	1951–55	1956–60	1961–65	1966–70	1971–75	1976–80	1981–85	1986–90	1991–95	1996–00	2001–05[a]	2006–10[a]	2011–15	2016–to date
Goat																	
Jamnapari	✓	✓		✓			✓[b]						✓				
Surti																	
Sannen										✓	✓						
Beetal											✓	✓					
Boer							✓			✓	✓						
Breed unknown	✓	✓	✓	✓		✓						✓✓✓	✓			✓	
Sheep																	
Bikenari							✓[b]										
Bannur									✓[b]								
Red Madras									✓[b]								
Breed unknown	✓																
Pigs																	
Large white										✓	✓		✓			✓	
Landrace										✓	✓		✓			✓	
Duroc																✓	
Breed unknown												✓	✓✓				
Rabbits																	
Breed unknown							✓					✓					

Sources: Wickramaratne (2000); Dr. L.W.N. Samaranayake, Director/Animal Breeding of the DAPH – personal communication
[a]Data not available
[b]Importation of >1000 animals

Table 13.7 Milestones in the history of institutional development aiming at livestock breeding and improvement (source: DAPH 2010)

Year	Activity
1935	Beginning of cattle breeding programme, in Karagoda-Uyangoda and Polonnaruwa farms
1936	Establishment of Ridiyagama farm for cattle and buffalo breeding
1941	Establishment of upcountry dairy farm for cattle breeding at Bopaththalawa
1950	Establishment of dry zone cattle breeding farms in Thamankaduwa, Kandakaduwa and Thrikonamaduwa
1950s	Establishment of Central Livestock Research Station (CLRS) in Polonnaruwa farm
1954	Establishment of Central Poultry Research Station (CPRS) in Kundasale
1961	Establishment of goat breeding station at Kottukachchiya
1961	Establishment of swine breeding station at Welisara
1962	Establishment of Artificial Insemination Center at Kundasale.
1970	Cattle crossbreeding programmes in Undugoda farm
1978	Establishment of separate Department of Animal Production and Health (DAPH) and handing over of 21 livestock farms managed by Department of Agriculture to DAPH
Late 1970s	Mahaweli Authority of Sri Lanka established farms for dairy and draught animal development
1983	Systematic goat breeding programme commenced under Sri Lankan German Goat Project
1990s	Handing over of state farms to National Livestock Development Board (NLDB) with a mandate to produce improved breeding materials for national requirement
1994	Formulation of Animal Breeding Policy Guidelines for Sri Lanka
2001	Two biggest dairy farms were privatized to involve the private sector in dairy development
2010	The National Livestock Breeding Policy Guidelines and Strategies for Sri Lanka was updated
2012	Open up of large-scale importation of dairy breeds for genetic improvement of the country

sectors for economic benefits. Further, fostering the rural mixed farming system with livestock as an instrument for social and economic advancement of the rural poor, through employment, sustainable income and improved nutrition, would help efficient utilization of available resources while contributing to the healthy environment overcoming the adverse effects of animal farming.

In order to perform the intended roles, the genotypes of animals need to match with the production environment and the input levels. Once high production efficiency is achieved, the farming exercise invariably becomes market oriented. Livestock industry of Sri Lanka, as in any given animal industry, needs a solid and country-specific breed structure for each animal species which are playing a role contributing to the industry. For example, as the dairy sector of Sri Lanka comprises several different breeds and their crosses, continuous and high-quality supply of breeding materials according to the recommendations for environmental and production norms have to be in place targeting the industry demand for breeding materials. The wide array of farming systems in the country (Table 13.2) needs to utilize the animals with diverse genetic potentials. Having appropriate animals in the

Table 13.8 Milestones passed in livestock breeding and improvement in terms of Breeding technologies

Year	Technology	Activity
1937	Artificial insemination (AI)	Commencing AI activities
1938		The first AI-calf '*Simon*' was born at the Meewatura farm, Peradeniya
1947		Establishment of Semen Collection Center in Gatambe
1950		National AI programme was started
1952		Establishment of the first semen supply centre
1954		Commencement of the buffalo semen collection
1960		Establishment of semen processing stations in Kundasale and Tinnaweli
1961		Swine AI was started on trial basis
1966		Commencement of deep frozen semen production
1970		AI service established at field level in Jaffna
1979		Beginning of private AI service
2001		Commencement of commercial application of swine AI at Horakelle NLDB farm
2004		Establishment of Swine Semen Production Center at Kotadeniyawa
2007	Embryo transfer	The first ET goat '*Peradeniya Kumari*' was born at the Experimental Farm of Faculty of Veterinary Medicine & Animal Science, University of Peradeniya
2007		The first ET calf (*Apeksha*) was born at Bopaththalawa NLDB farm

Based on DAPH (2010)

correct farming system is one of the first and foremost steps in establishing the sustainable utilization of available resources in the system.

13.4 Managing Livestock Systems for Sustainability

The concept of sustainability in a farming system identifies the combined activities that operate with low environmental impacts, yet ensuring food and nutrition security for present and future generations. Thus, sustainable systems are protective and respectful of diversity and environment (Hoffmann and Baumung 2013). Sustainability of the system, therefore, always relies primarily on the biodiversity preserved while reaching the required output in terms of provisional services. As described in detail by Marambe and Silva (2012), responsible utilization of biodiversity is one of the main characteristics of sustainable agriculture systems. When the livestock production system of Sri Lanka is considered, a large proportion of the land area is with livestock mixed farming category where livestock-only production occupies only 1% of the total agriculture land area of the country (Mapa et al. 2002). This indicates that the sustainability of the livestock systems in Sri Lanka relies mainly on the mixed (rainfed or irrigated) farming system.

Though intensification of livestock production is a widely accepted, modern production-oriented strategy for reaching self-sufficiency for the country, finding

Table 13.9 Description of the programmes implemented in Sri Lanka for genetic improvement of farm animals (sources: Gamage et al. 2010; DAPH 2010)

Year	Project	Activities	Extent of operation
1972	FAO/SIDA Animal Breeding Project	Establishment of breeding network, Extension campaigns Improvement of deep frozen semen production in Kundasale	All island
1979	Integrated Rural Development Project	Facilitating AI services	Implemented in Kurunagale, Matara, Nuwara Eliya, Matale, Puttalam, Badulla and Anuradhapura
1983-88	Sri Lanka/Asian Development Bank (SL-ADB) Livestock Development and Breeding Project	Improvement of the AI centre at Kundasale	Improved field AI service in Jaffna, Killinochchi, Kandy, Puttalam, Kegalle, Kurunegala, Galle, Kalutara and Matara
1983-89	Sri Lanka-German Goat Development Project (Federal Republic of Germany GTZ)	Technical corporation for establishing the nucleus herds and breed improvement	Implemented in Gampaha, Jaffna, Mannar, Vavunia, Mullaitivu, Trinco, Kegalle, Puttalam and Anuradhapura
1983-92	Livestock Planning Project (Federal Republic of Germany GTZ)	Planning of breeding programmes	Implementation in limited districts
1984-88	Sri Lanka/Swiss Livestock Development Project	Improvement of AI station, Polonnaruwa	Improved field AI in Polonnaruwa, Batticaloe, Trincomalee and Ampara
1986-91	SL-Netherlands Livestock Development Project	Facilitating AI services	Implemented in NLDB farms in many locations
1987-96	Sri Lanka Agriculture Research Project	Improvement of animal breeding activities	All island
1990-95	Minipe/Nagadeepa Irrigation Rehabilitation Project	Genetic improvement of existing herds through stud-bull services	Minipe and Nagadeepa project areas
1992-94	Smallholder Integrated Livestock Extension Project (SILEP)	Improve livestock breeding activities	n/a[b]
1992-94	Artificial Insemination Incentive Scheme[a]	Improvement of AI success rate	n/a[b]
1992-94	Heifer Calf Rearing Scheme I[a]	Improvement of breeding performance	All island
1998-2003	Livestock breeding project[a]	Improvement of AI	All island
2008	Heifer Calf Rearing Scheme II[a]	Improvement of early calf growth	All island
2009	Genetic, feeding and management improvement project for dairy cattle (GFMI)	Improvement of AI alone with progeny testing activities	Central province and Kurunegala district

(continued)

Table 13.9 (continued)

Year	Project	Activities	Extent of operation
2010	Dairy Cattle & Buffalo Development Project	Genetic improvement of dairy animals	n/a[b]
2012–16	Sri Lankan Dairy Development Project	Increase the milk production of NLDB farms while producing high-quality breeding materials to the farmers	National Livestock Development Board (NLDB) farms, and later extended to medium to large-scale entrepreneurs

[a]With local funds
[b]Infrmation are not available

strategies to foster the existing, naturally evolved production systems is the viable and appropriate option when the sustainability is considered. Frison et al. (2011) reported of the approaches used in intensification strategies containing the seeds of their own destruction, in the form of increased release of greenhouse gases, depletion of water supplies and degraded soils. Hence, building production systems that deliver results of intensification (provisional service) without hampering the natural input to the system is a key at present. This encompasses the idea of producing more output from the same area of land while reducing the negative environmental impacts and at the same time increasing contributions to natural capital and the flow of environment (Pretty et al. 2011). Thus, the production goal of the present-day livestock system while travelling the 'road to self-sufficiency' needs to focus on sustainable and intelligent management of production and consumption (Nellemann et al. 2009). Sri Lanka, by virtue of its inheritance, owns diverse production systems and domesticated animal genetic resources. Though it is a complex landscape of production, it paves the way for environmental and social justice for sustainability within the production systems. Traditionally evolved systems in the country already combine various mixtures of livestock species with crops, trees or plantations. In managing with the aim of sustainability, the challenge is to increase productivity of the traditional systems, which enables the country to enjoy high usable biomass production while conserving the natural resources, which in turn nurtures the whole production system. The theory behind this is a simple approach of balancing the flow of energy and materials within the ecosystem.

The approach to manage livestock systems in a sustainable manner leading to produce more from the same area of land is a widely discussed topic in the global arena (Heywood 2013). Table 13.10 presents a summary of the options derived based on the approaches proposed for crops.

Managing the livestock system in sustainable manner needs the realization of the fact that, ASF production is tied with the system and ultimately becomes a dependent of numerous ecosystem services that operate at various levels. External inputs need to be carefully manipulated in order to secure the delicate operation of such services at each level from lowest to highest and from microscopic to landscape scales.

Table 13.10 Approach to manage livestock systems for sustainability (source: Collette et al. 2011)

Criteria	Approach
Production intensification	Bring intensification along with socio-economic and environmental benefits
Genetic diversity	Exploit animal genetic diversity that are suitable for a range of agro-ecosystems and production practices
Good management practices	Minimize the environmental foot prints (soil, water, waste disposal)
Technology	Smart and precision technologies to minimize the water foot prints
Institution and policy	Formulate sectorial development policies and prepare institutions to encourage smallholders to adopt sustainable production intensification

13.4.1 Challenges and Strategies

The annual rates of increase of the three main ASF items in Sri Lanka, namely, chicken egg, chicken meat and dairy products are 11.2%, 5.7% and 9.5%, respectively, during the last decade (DCS 2017a). The current rate of increase of dairy products is not sufficient for the fulfilment of national demand, which indicates that the industry should grow at a much higher rate to achieve self-sufficiency. The major challenge is on increasing productivity while minimizing the environmental degradation.

Satisfying the differential demands of the production dichotomy of the livestock industry, i.e. high input intensified production and the low input production systems, is the second major challenge, which needs thorough investigations in finding solutions. Existence of the industry dichotomy is essential for the survival of the sustainability of the industry as well as the fulfilment of the demands for production and environment.

As identified by Pingali and McCullough (2010), meeting the growing demand for animal products while sustaining productive assets of natural resources is one of the major challenges the mankind has to face today. Resource competition is likely to increase, for example, through the decreasing availability of and competition for land and water (Hoffmann and Baumung 2013). Poor industry performance, industry-related resource degradation and community-related socio-economic issues are the adverse outcomes of the above two major processes. Finite resources (water, fossil fuels and land), processing facilities (hygiene, packaging and transport), production pressures for animal-sourced food items, and environmental events (drought, floods, global warming and disease outbreaks) have been threatening the sustainability of traditional livestock systems. Therefore, it is imperative to launch strategic research on animal production systems for maximization of available resources and their efficient and effective application, governance of animal welfare standards and environmental sustainability.

Various development strategies for livestock sector have been proposed in many occasions based on national-level findings as well as regional and individual studies (Ranaweera 2007; MLD 2008; Perera and Jayasuriya 2008; Subasinghe and Abeygunawardena 2011; Alahakoon et al. 2016; Subasinghe 2016). The strategies proposed to date directly or indirectly focus on the two major challenges mentioned above (Table 13.11).

Table 13.11 Challenges and strategies sustainable livestock production systems (Sources: Ranaweera 2007; MLD 2008; Perera and Jayasuriya 2008; Subasinghe and Abeygunawardena 2011; Hoffmann and Baumung 2013; Heywood 2013; UN-HLTF 2015; Alahakoon et al. 2016; Subasinghe 2016)

Major challenge	Strategies for …
Increase productivity while harnessing the environmental degradation	• Increase productivity for the resources used – Breeding strategies for efficient production – Nutritional strategies to enhance nutritional quality and availability – Market orientation of production system
	• Sustainable utilization of natural resources – Management strategies for enhancing productivity in unit area of land, units of input resources (water, feed etc.) – Utilization of available resources/underutilized resources – Use of technologies for regenerating and recycling the resources within the system
	• Creating awareness on ecosystem services – Identification facilitation needed for ecosystem services within the given system while maximizing the provisional services
	– Effective governance system with respect to the use and the protection of natural resources
	– Developing strategies for building up resilience
Satisfying the differential demands of dichotomy of the industry	• Identify the functional differences in the dichotomy – Inputs (including physical and human resources) need for commercial level high input operations such as breeding, nutrition, reproductive, health related aspects – Inputs (including physical and human resources) need for low- medium input operations such as breeding, nutrition, reproductive, health related aspects
	• Quantity and quality maintenance in each system – Trade-off between the quantity and quality in different systems – Identification of differences in strategies according to the level of operation and integration
	• Guaranteeing market for input as well as output – Policy oriented decisions – Institution level interventions – Unbiased service orientation (for both high input and low to medium input levels)
	• Identifying sustainable utilization of genetic resources – Understand the breeding objectives for different systems – Capacity building of communities for proper choice of breeding animals

13.4.2 Research Orientation

Livestock and poultry research has an important place in supporting development policies, strategies and programmes. Once developed based on strong scientific evidence, such policies, strategies and programmes should not be changed according to political agenda as long-term implementation of those strategies relies heavily on political will. Any new improvement to be done in national policies and strategies periodically also must be based on scientific evidence. As described by Subasinghe (2016), a national policy is the expression of government activity and it could be oriented towards the betterment of the sector. For a country like Sri Lanka in particular, more appropriate research orientation must be focused on seeking solutions to development-related problems of the sector (Subasinghe 2016) and also to find remedies to technical drawbacks in the existing system. Thus, the research orientation of livestock sector needs a two-level approach, (a) short-term research programmes to address the technical drawback of the current system, and (b) long-term research programmes based on the industry needs and the future projections of the industry.

13.5 Conclusions

Livestock and poultry industries in Sri Lanka are a rich ground with a significant potential for improvement. There is an immense opportunity in the livestock and poultry sectors for expansion to cater to the rapidly growing consumer demand, and assure food and nutritional security of the nation. The threats such as competition for limited resources, especially feed, and lack of futuristic and strategic approach in the present phase of growth of the industry hinder the opportune advantages the industry could have clutched. In addition, the industry is also facing serious challenges of absorbing the pressures for increasing productivity while abating the environmental degradation. The industrial setup with an arena of dominating mixed production system could be used as a driving force and strength if governed by favourable and appropriate policy environment, in bringing in solutions to many of the treats and challenges that the industry is facing today. Thus, the industry dichotomy could be better utilized with supportive policy orientation driven by short-term and long-term research, planning and strategies for the sustainability of livestock food system in the country.

References

Abeygunawardena HA, Rathnayaka D, Jayathilaka WMAP (2012) Characteristics of cattle farming systems in Sri Lanka. J Natl Sci Found 25(1). https://doi.org/10.4038/jnsfsr.v25i1.5016

Abeykoon MNDF, Weerahewa J, Silva GLLP (2013) Determinants of market participation of indigenous poultry farmers: a case study in Anuradhapura District in Sri Lanka. Trop Agric Res 24(4):347–361

Abeykoon MNDF, Weerahewa J, Weligamage P, Silva GLLP (2014) Willingness to pay for chicks of different indigenous chicken types: an application of experimental auctions. Trop Agric Res 26(1):162–174

Abeyratne AS (2007) A review of the livestock industry of Sri Lanka: past performance and future trends. A.S. Abeyratne, Kandy

Alahakoon AU, Jo C, Jayasena DD (2016) An overview of meat industry in Sri Lanka: a comprehensive review. Korean J Food Sci Anim Res 36(2):137–144

Balachandran K (1972) Artificial insemination in cattle – some misconceptions in Ceylon. Anim Prod Health Bull 6(3):109–115

Bett RC, Bhuiyan AKFH, Khan MS, Silva GLLP, Thuy LT, Islam F, Abeykoon MND, Nguyen TH, Sadef S, Mwai O, Ibrahim MNM (2014b) Phenotypic variation of native chicken populations in the south and South East Asia. Int J Poult Sci 13(8):449–460. ISSN 1682-8356

Bett RC, Bhuiyan AKFH, Khan MS, Silva GLLP, Thuy LT, Sarker SC, Abeykoon MND, Nguyen TH, Sadef S, Kariuki E, Baltenweck I, Poole J, Mwai O, Ibrahim MNM (2014a) Indigenous chicken production in the south and South East Asia. Livest Res Rural Dev 26(12). http://www.lrrd.org/lrrd26/12/bett26229.html

Buvanendran V (1967) Acclimatisation of temperate cattle in natural selection. Ceylon Vet J XV(1):1–4

CBSL (2017) Annual report. Central Bank of Sri Lanka, Colombo

Chandrasiri ADN (2002) The state of genetic diversity. In: Chandrasiri ADN (ed) Country report on the state of animal genetic resources of Sri Lanka. FAO, Rome. Available at http://www.fao.org/docrep/pdf/010/a1250e/annexes/CountryReports/SriLanka.pdf. Accessed 28 April 2018

Collette L, Hodgkin T, Kassam A, Kenmore P, Lipper L, Nolte C, Stamoulis K, Steduto P (2011) Save and grow: a policy makers' guide to sustainable intensification of smallholder crop production. Food and Agriculture Organization of the United Nations, Rome

DAPH (2010) The National Livestock Breeding Policy; guidelines and strategies for Sri Lanka. Department of Animal Production and Health, Gatambe

DAPH (2016) Annual report 2016. Department of Animal Production and Health, Gatambe, Peradeniya

DAPH (2017) News bulletin. Department of Animal Production and Health, Gatambe, Peradeniya. Available at <http://www.daph.gov.lk/web/images/content_image/news_bulletins/livestock_statistical/import_data_2016.pdf>

DCS (2016a) Poverty indicators; Department of Census and Statistics Household Income and Expenditure Survey – 2016. Ministry of National Policies and Economic Affairs

DCS (2016b) Statistical abstracts – 2016. Department of Census and Statistics, Colombo

DCS (2017a) National livestock statistics. Department of Census and Statistics, Colombo. Available at http://www.statistics.gov.lk/agriculture/Livestock/Livestock Statistics.html

DCS (2017b) Agriculture and environment statistics division. Department of Census and Statistics, Colombo. Available at http://www.statistics.gov.lk/agriculture/indices/VolumeIndex.html

Dematawewa CMB, Silva GLLP (2000) Genetic improvement of buffaloes in Asia with special reference to Sri Lanka. In: Siriwardena JAS, Subasinghe DHA, Horadagoda NU, Abeygunawardena H (eds) Proceedings of the Third Asian Buffalo Congress on Changing Role of the Buffalo in the New Millennium in Asia, Kandy, Sri Lanka, pp 129–142

Dematawewa CMB, Silva GLLP, Premasundara AS (2009) Swine industry of Sri Lanka. Department of Animal Production and Health, Gatambe, Peradeniya. (ISBN 978-955-8004-02-9), p 135

FAO (2010) The state of food and agriculture 2009: livestock in the balance. Food and Agriculture Organization of the United Nations, Rome

FAO (2011) World livestock 2011 – livestock in food security. Food and Agriculture Organization of the United Nations, Rome

FAO (2013) Sustainability assessment of food and agriculture systems (SAFA), guidelines (version 3). Food and Agriculture Organization of the United Nations, Rome

FAO (2016) FAOSTAT, Food and Agriculture Organization: available at http://www.fao.org/faostat/en/?#data/

FAO (2017) The future of food and agriculture – trends and challenges. Food and Agriculture Organization of the United Nations, Rome

Frison EA, Cherfas J, Hodgkin T (2011) Agricultural biodiversity is essential for a sustainable improvement in food and nutrition security. Sustainability 3:238–253. https://doi.org/10.3390/su3010238.

Gamage DVSS, Silva P, Marambe B (2010) Capacity building for conservation and utilization of far animal genetic resources of Sri Lanka (chapter 5). In: Silva P (ed) Indigenous animal genetic resources of Sri Lanka – status, potential and opportunities. GEF-UNEP-ILRI –FAnGr Asia Project (ISBN 978-955-589-120-2)

Hevapokara A (1995) Sathvavarga. Department of Animal Production and Health, Peradeniya

Heywood VH (2013) Overview of agricultural biodiversity and its contribution to nutrition and health. In: Fanzo J, Hunter D, Borelli T, Mattei F (eds) Diversifying food and diets: using agricultural biodiversity to improve nutrition and health. Routledge, Oxon, pp 35–67. OX14 4RN. ISBN: 978-0-203-12726-1(ebk)

Hoffmann I, Baumung R (2013) The role of livestock and livestock diversity in sustainable diets. In: Fanzo J, Hunter D, Borelli T, Mattei F (eds) Diversifying food and diets: using agricultural biodiversity to improvenutrition and health. Routledge, Oxon, pp 68–87. OX14 4RN. ISBN: 978-0-203-12726-1(ebk)

Ibrahim MNM, & Schiere JB (2002) Introduction to agriculture system studies. Faculty of Agriculture, University of Peradeniya, Sri Lanka. ISBN 9555-97074-1-8.

Ibrahim MNM, Staal SJ, Daniel SLA, Thorpe W (1999) Appraisal of the Sri Lanka dairy sector, volume II: main report, Department of Animal Science, University of Peradeniya, International Livestock Research Institute, Ministry of Livestock Development and Estate Infrastructure, Colombo, Sri Lanka

Jalatge EFA (1980) Animal genetic resources in Sri Lanka. Proceedings of SABRAO Workshop on Animal Genetic Resources in Asia and Oceania. *Nikken Shiryo* No. 47. Tropical Agriculture Research Center, Japan

Jayawardena R, Byrne NM, Soares MJ, Katulanda P, Hills AP (2012) Food consumption of Sri Lankan adults: an appraisal of serving characteristics. Public Health Nutr 16(4):653–658. https://doi.org/10.1017/S1368980012003011

Lin BB (2011) Resilience in agriculture through crop diversification: adaptive management for environmental change. BioScience 61(3):183–193. https://doi.org/10.1525/bio.2011.61.3.4

Liyanage RP, Dematawewa CMB, Silva GLLP (2015) Comparative study on morphological and morphometric features of village chicken in Sri Lanka. Trop Agric Res 26(2):262–273

Mapa RB, Kumaragamage D, Gunarathne WDL, Dassanayake AR (2002) Land use in Sri Lanka: past, present and the future. Proceedings of 17th World Congress of Soil Science held in Thailand, Volume 31. pp. 1–7.

Marambe B, Silva P (2012) Sustainability management in agriculture – a systems approach. chapter 33. In: Madu CN, Kuei C-H (eds) Handbook of sustainability management. World Scientific Publishers, Singapore. ISBN: 978-981-4354-81-3 [ISBN: 978-981-4354-82-0 (ebk)], pp 687–712. https://doi.org/10.1142/9789814354820_0033

MLD (2008) Discussion paper on public-private joint program for development of the dairy industry in Sri Lanka. Ministry of Livestock Development, Colombo

Nadarajah R (1968) Limitations to artificial insemination service in Ceylon. Anim Prod Health Bull 2(4):99–107

Nellemann C, MacDevette M, Manders T, Eickhout B, Svihus B, Prins AG, Kaltenborn BP (eds) (2009) The environmental food crisis – the environment's role in averting future food crises. A UNEP rapid response assessment. United Nations Environment Programme, GRID-Arendal, www.grida.no. BirkelandTrykkeri AS, Norway. February, 2009. ISBN: 978-82-7701-054-0. Available at <https://gridarendal-website-live.s3.amazonaws.com/production/documents/:s_document/221/original/FoodCrisis_lores.pdf?1486728701>

Notter DR (1999) The importance of genetic diversity in livestock populations of the future. J Anim Sci 77:61–69

Perera ANF (1996) Genetic improvement of cattle, buffaloes, and small ruminants in Sri Lanka. Country report submitted to the workshop on Genetic Improvement of Livestock in Asia. Conducted by Food and Agriculture Organization of United Nations. Held in Kandalama, Sri Lanka

Perera BMAO, Jayasuriya MCN (2008) The dairy industry in Sri Lanka: current status and future directions for a greater role in national development. J Natl Sci Found Sri Lanka 36(Special issue):115–126

Periasamy K, Vahidi SMF, Silva P, Faruque MO, Naqvi AN, Muladno, Cao JH, Zhao SH, Le Thi T, Pichler R, Podesta MG, Shamsuddin M, Boettcher P, Garcia JF, Han J-L, Marsan PA, Diallo A, Viljoen GJ (2017) Mapping molecular diversity of indigenous goat genetic resources of Asia. Small Rumin Res 148:2–10. http://dx.doi.org/10.1016/j.smallrumres.2016.12.035

Pingali P, McCullough E (2010) Drivers of change in global agriculture and livestock systems. In: Steinfeld H, Mooney HA, Schneider F, Neville LE (eds) Livestock in a changing landscape Vol 1: drivers, consequences, and responses. Island Press, Washington, pp 5–10

Pretty J, Toulmin C, Williams S (2011) Sustainable intensification, increasing productivity in African food and agricultural systems. Int J Agric Sustain 9(1):5–24

Pushpakumara DKNG, & Silva P (2008) Agriculture biodiversity in Sri Lanka (*in Sinhala*). Biodiversity Secretariat, Ministry of Environment and Natural Resources, Sri Lanka

Rajapaksha T, Abeykoon MNDF, Silva GLLP (2014) Potential of expanding the capacity of indigenous poultry farmers in Sri Lanka: a case study in Thirappane area. In: Proceedings of the Peradeniya university international research sessions (iPURSE). University of Peradeniya, Peradeniya, p 700

Ranaweera NFC (2007) Improved market access and smallholder dairy farmer participation for sustainable dairy development: Lessons learned – Sri Lanka. Food and Agriculture Organization, Regional Office for Asia and the Pacific, Bangkok, Thailand

Ranaweera NFC (2010) Sri Lanka: opportunities for dairy sector growth. In: Morgan N (ed) Smallholder dairy development: lessons learned in Asia, Animal Production And Health Commission for Asia and the Pacific. Food and Agriculture Organization of United Nations, Rome. http://www.fao.org/docrep/011/i0588e/i0588e08.htm, Accessed 11 June 2018

Rathnakumara WMTD, Abeygunawardena WW, Deepani MLNAR, Amarasinghe AAAWK Lokugalappatti LGS, Samaranayake LWN, Rathnayake DRTG, Alexander PABD, Rabel RAC (2015) Phenotypic Characterization of Jaffna Local Sheep, an Indigenous Livestock Breed of Sri Lanka. 67th Annual Scientific Sessions of the Sri Lanka Veterinary Association, p.14. https://www.researchgate.net/publication/300106791_Phenotypic_Characterization_of_Jaffna_Local_Sheep_an_Indigenous_Livestock_Breed_of_Sri_Lanka. Accessed 11 Jun 2018

Silva GLLP, Thuy LT, Abeykoon ND, Hanh NTH, Bett RC, Okeyo M, Ibrahim MNM (2016a) Comparative study of indigenous pig production in Vietnam and Sri Lanka. Int J Livest Prod 7(10):83–93. https://doi.org/10.5897/IJLP2016.0306

Silva P, Dematawewa CMB, Chandrasiri N (2010) Farm animal genetic resources (Chapter 1). In: Silva P (ed) Indigenous animal genetic resources of Sri Lanka – status, potential and opportunities. GEF-UNEP-ILRI –FanGr Asia Project (ISBN 978-955-589-120-2)

Silva P, Dematawewa CMB, Goonawardena A, Chandrasiri ADN, Pushpakumara A (2009) Production systems and characteristics of indigenous small ruminants in Sri Lanka. Paper presented at the International Symposium on Sustainable Improvement of Animal Production and Health, FAO/IAEA Joint Program, held in 9–11 June, Austria, Vienna, pp 96–97

Silva P, Dematawewa CMB, Kurukulasuriya M, Utsunomiya YT, Garcia JF, Pichler R, Thiruvenkadan AK, Ramasamy S, Han J-L, Periasamy K (2017) Genetic diversity analysis of major Sri Lankan goat populations using microsatellite and mitochondrial DNA D-loop variations. Small Rumin Res 148:2–10. https://doi.org/10.1016/j.smallrumres.2016.12.030

Silva P, Guan X, Ho-shing O, Jones J, Xu J, Hui D, Notter D, Smith E (2008) Mitochondrial DNA-based analysis of genetic variation and relatedness among Sri Lankan indigenous chickens and the Ceylon jungle fowl (*Gallus lafayetti*). Anim Genet. https://doi.org/10.1111/j.1365-2052.2008.01783.x

Silva P, Jianlin H, Hanotte O, Chandrasiri ADN, Herath HMSP (2011) Indigenous cattle in Sri Lanka: production systems and genetic diversity. In: Odongo NE, Garcia M, Viljeon GJ (eds) Sustainable improvement of animal production and health. Food and Agriculture Organization of the United Nations, Rome, pp 45–48

Silva P, Liyanage RP, Senadheera S, Dematawewa CMB (2016b) Monograph on indigenous chicken in Sri Lanka. UNEP-GEF-ILRI FAnGR Asia Project, University of Peradeniya, Sri Lanka

Subalini E, Silva GLLP, Dematawewa CMB (2010) Phenotypic characterization and production performance of village pigs in Sri Lanka. Trop Agric Res 21(2):198–208

Subalini E, Silva GLLP, Dematawewa CMB (2013) Molecular characterization of native swine types in Sri Lanka. Trop Agric Res 24(2):175–185

Subasinghe DHA (2016) Small ruminant industry in Sri Lanka – part III. Sri Lanka Vet J 63(1):41–48

Subasinghe DHA, Abeygunawardena H (2011) Dairy industry in Sri Lanka - part I. Sri Lanka Vet J 63(1):29–34

Teufel N, Markemann A, Kaufmann B, Valle Zárate A, Otte J (2010) Livestock production systems in South Asia and the greater Mekong sub-region: a quantitative description of livestock production in Bangladesh, Cambodia, India, Lao PDR, Nepal, Pakistan, Sri Lanka, Thailand, and Viet Nam. Pro-poor Livestock Policy Initiative (PPLPI) working Paper No 48. Food and Agriculture Organization – Animal Production and Health Division, Rome, Italy

Thilakaratne N (1984) Report on the FAO/UNEP pilot project on animal genetic resources data bank, Veterinary Research Institute, Sri Lanka

Thilini HWL, Korala Gedara P, Silva GLLP (2016) Impact of intervention on socio-economic status of village chicken rearing farmers: a case in Karuwalagaswewa and Thirappane Veterinary Divisions. Proceedings of the Peradeniya University International Research Sessions (iPURSE), Peradeniya, Sri Lanka, November 4–5, 2016, vol 20, p 11

UN-HLTF (2015) Advisory notes by the HLTF working group to respond to the 5 'Zero Hunger Challenge' element. The United Nations Secretary General's High Level Task Force on Global Food and Nutrition Security. Available at http://www.un.org/en/issues/food/taskforce/pdf/HLTF%20-%20ZHC%20Advisory%20Notes.pdf, and accessed on 28/4/2018

Vasileska A, Rechkoska G (2012) Global and regional food consumption patterns and trends. Procedia Soc Behav Sci 44:363–369

Vijitha T, Silva GLLP (2013) Evaluation of the present status of Jaffna local sheep. In: Rathnathunga NVI (ed) Proceedings of University Research Sessions of University of Peradeniya. University of Peradeniya, Peradeniya, p 95. (ISSN: 1391-4111, ISBN: 955-589-092-7)

Wickramaratne SHG (2000) Impact of livestock technology in Sri Lanka. The technical report submitted to ILRI for the SLP project titled 'Increase of livestock productivity in mixed crop livestock farming system', Animal Breeding division, Department of Animal Production and Health, Perdeniya, Sri Lanka

Wijayasena AMPSS, De Alwis DJ Silva GLLP, Abeykoon MNDF (2014) Cost benefit Analysis of village chicken production in Puttlam district of Sri Lanka: a case study. Proceedings of the Peradeniya University International Research Sessions (iPURSE), 18, 224. July 4–5, Peradeniya, Sri Lanka, July 4–5

Wijeratne WVS (1970) Crossbreeding Sinhala cattle with Jersey and Friesian in Ceylon. Anim Prod 12:473–483

Wijesekara G (2015) Changing patterns of food consumption in Sri Lanka: 1985–2009. ASARC Working Paper. Australian National University, Australia. [Available at https://crawford.anu.edu.au/acde/asarc/pdf/papers/2015/WP2015_02.pdf, accessed 11 April 2018]

Fisheries Sector Contribution for Sustainable Food System: Past, Present, and Future

14

Saman Athauda and Nimal Chandraratna

Abstract

At global scale, the capture fisheries and aquaculture provide 3 billion people with almost 20% of their average per capita intake of animal protein, and a further 1.3 billion people with about 15% of their per capita intake. This share can exceed 50% in some countries. Fish are extremely important and fish farming has developed rapidly over the last 30 years: total dietary protein from fish is between 50% and 60% in Sri Lanka. Fish provides a similarly significant proportion of protein in the human diets in most small island nations in the world. The fisheries sector of Sri Lanka consists of three main subsectors, namely, coastal, offshore and deep sea, and inland and aquaculture. These three subsectors employ around 250,000 active fishers and another 100,000 in support services in Sri Lanka. Sri Lankan fisheries are managed by two core legislative instruments, namely, Fisheries and Aquatic Resources Act No. 2 of 1996 and Fisheries Act No. 59 of 1979, along with several regulations to assist in the implementation of these acts. The industry has seen a change of fortune in the recent past with a boost with the reinstatement of GSP Plus from the USA and lifting of export banning of fish to European Union (EU) in the recent past. Sri Lankan tuna, shrimps, and crabs have quality of unique taste and texture. Meanwhile, the Sri Lankan government agencies are also seeking to position the country's marine and inland fishing industries as major export earning sectors as well main source of economics and food security considering its nutritional impact on nations consuming plant-based diets. Thus, governance needs to ensure that fisheries and aquaculture adapt to the impacts of climate change and improve the resilience of

S. Athauda (✉)
Department of Animal Science, Faculty of Agriculture, University of Peradeniya, Peradeniya, Sri Lanka
e-mail: sbathauda@pdn.ac.lk

N. Chandraratna
National Aquaculture Development Authority (NAQDA), Pelawatte, Battaramulla, Sri Lanka

B. Marambe et al. (eds.), *Agricultural Research for Sustainable Food Systems in Sri Lanka*, https://doi.org/10.1007/978-981-15-2152-2_14

food production systems where one of the solutions is integration. Availability of these natural aquatic resources provides immense opportunities to achieve high economic growth with guaranteed food and nutritional security for its population in near future by blue economic revolution.

Keywords
Fisheries · Aquaculture · Sustainable food system · Integration · Sri Lanka

14.1 Introduction

14.1.1 World Fisheries Resources and Captures: Current Context

Fisheries and aquaculture remain important sources of food, nutrition, income, and livelihoods for hundreds of millions of people around the world. Starting from 1950, global catches rose until 1988. With capture fishery production relatively static since the late 1980s, aquaculture has been responsible for the impressive growth in the supply of fish for human consumption. Whereas aquaculture provided only 7% of fish for human consumption in 1974, this share had increased to 26% in 1994 and 39% in 2004. Global total capture production in 2014 was 93.4 million mt, of which 81.5 million mt was from marine waters and 11.9 million mt from inland waters (SOFIA 2016). China is the main fish producer and the largest exporter of fish and fishery products. For marine fisheries production, China remained the major producer, followed by Indonesia, the United States of America, and the Russian Federation.

The Northwest Pacific remained the most productive area for marine capture fisheries, followed by the Western Central Pacific, the Northeast Atlantic, and the Eastern Indian Ocean. With the exception of the Northeast Atlantic, these areas have shown increases in catches compared with the average for the decade 2003–2012. The situation in the Mediterranean and Black Sea is alarming, as catches have dropped by one-third since 2007, mainly attributable to reduced landings of small pelagics such as anchovy and sardine, but with most species groups being affected. World catches in inland waters were about 11.9 million mt in 2014, continuing a positive trend that has resulted in a 37% increase in the last decade. Sixteen countries have an annual inland water catches exceeding 200,000 mt, and they represent 80% of the world total with China leading in both global marine and freshwater fish production (SOFIA 2018).

Production from aquaculture in 2014 amounted to 73.8 million mt, while China accounted for more than 60% of (45.5 million mt). Other major producers were India, Viet Nam, Bangladesh, and Egypt. In addition, 27.3 million mt of aquatic plants (US$ 5.6 billion) were cultured. Aquatic plant farming, overwhelmingly of seaweeds, has been growing rapidly and is now practiced in about 50 countries. As indicated in Fig. 14.1 the world per capita apparent fish consumption increased from an average of 9.9 kg in the 1960s to 14.4 kg in the 1990s and 19.7 kg in 2013, with preliminary estimates for 2014 and 2015 pointing towards further growth beyond 20 kg (SOFIA 2016). In year 2016 it was reported as 20.3 kg (SOFIA 2018).

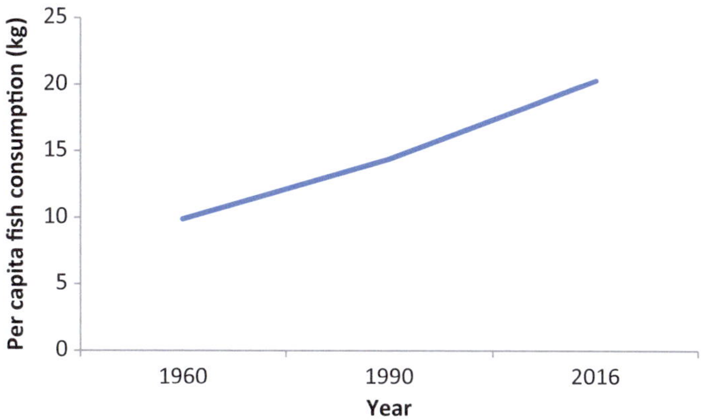

Fig. 14.1 Trend in global per capita fish consumption since 1960–2016 (Source: SOFIA 2016 and SOFIA 2018)

14.1.2 Historical Development of Fisheries Sector in Sri Lanka

The fisheries sector of Sri Lanka consists of three main subsectors, namely, (a) coastal, (b) offshore and deep sea, and (c) inland and aquaculture. These three subsectors employ around 250,000 active fishers and another 100,000 in support services in Sri Lanka. This workforce represents a population of some 1 million people. At present there are some organizations dedicated for the development of fisheries industry in Sri Lanka. The National Aquatic Resources Research and Development Agency (NARA) looks after research and development, and the National Aquaculture Development Authority (NAQDA) focuses its activities on the development and management of all freshwater aquatic resources in the country and also promotes development of aquaculture and sea farming. The Ceylon Fisheries Cooperation maintains cold storage facilities and carries out production and sale of fishery by-products, while the Ceylon Fisheries Harbors' Cooperation maintains fisheries infrastructure facilities such as proper landing facilities through construction, maintenance, and management of harbors and anchorages. The Department of Fisheries deals with management, regulation, conservation, and development of fisheries and aquatic resources and the Cey-Nor Foundation Limited carries out building, manufacturing, and selling of fishing crafts, engines, and gear, and the operation of workshops for repairing of fishing crafts.

Fishing is the largest extractive use of wildlife in the world, and aquaculture is the most rapidly growing food production sector globally with an annual growth rate of over 6% since 2002. The recent trends in marine fisheries worldwide are overcapacity of fleets both near shore and offshore resulting in dramatic and widespread declines in catches per unit efforts, the movement of the bulk of the world's fishing capacity from the developed to developing countries, declining catches, economic hardships for small artisanal fishers, and reduction in fish for poor consumers.

Sri Lanka has a well-established fishery industry. The Exclusive Economic Zone (EEZ) covers 517,000 km^2, of which some 27,800 km^2 form a continental shelf that is referred to as the coastal subsector. The balance beyond the continental shelf and out to the 200 nm EEZ boundary is considered the offshore and deep-sea subsector. The total fish production increased marginally over the years even though the offshore fish catch reported an increase while the coastal fish catch declined, due to high fishing pressures which lead to reduce the catch per unit effort (CPUE) and some unfavorable weather conditions prevail in coastal. The total fish production reported in Sri Lanka in 2016 was 530,920 mt, of which 17,593 mt was exported.

14.1.3 Current Situation of Fisheries Resources and Capture in Sri Lanka

Sri Lanka has exclusive economic rights for fishing in the ocean area of 517,000 km^2 and 21,500 km^2 of territorial water in marine water resources along the coastal line of 1770 km. This is in addition to the inland water bodies 262,000 ha of freshwater and 158,000 ha of brackish-water resources, which makes fishery to be one of the promising industries in the country. More importantly it contains a variety of coastal habitats that include estuaries and lagoons, mangroves, sea grass beds, salt marshes, coral reefs, and large extents of beaches and dunes that are vital for ecological functioning and maintenance of coastal biodiversity.

About 610 species of coastal fish have been reported from Sri Lankan waters, of which the more common species caught are *Sardinella* spp., *Amblygaster* spp., *Rastrelliger* spp., *Auxis thazard*, *Anchova commersoni*, and *Hirundichthys coromandelensis*. Most of these species live near the surface or high in the water column (pelagic species). These small pelagics account for about 40% of the coastal fish catch. Species such as *Lethrinus* spp., *Trichiurus* spp., *Caranx* spp., species of skates and rays, *Cynoglossus* spp., *Johnius* spp., and *Tolithus* spp. are bottom dwellers (demersal species). In addition, there are various mid-water species.

Katsuwonus pelamis and *Thunnus albacares* dominate the large pelagic catches from offshore and deep-sea fisheries. These are migratory fish species and therefore fall under stocks shared with other countries. Other important species are *Scomberomorus commerson*, *Platypterus* spp., *Tetrapturus angustirostris*, *T. audax*, *Makaira nigricans*, *M. indica*, *Xiphias gladius*, and *Coryphaena hippurus*. Moreover, about 60 species of sharks (De Silva 1984–1985) live in the oceanic waters of Sri Lanka. Some of the more common shark species are *Carcharhinus falciformis*, *C. longimanus*, *C. melanopterus*, *Alopias pelagicus*, *Sphyrna zygaena*, and *S. lewini*.

About 215 demersal species (Sivasubramaniam and Maldeniya 1985) have been reported from the oceanic waters around Sri Lanka. The commercially important, larger *Lutjanus* species are *L. lentjan*, *L. nebulous*, and *Pristipomoides* spp.; *Epinephelus* spp. are the demersal ones (Rajasuriya 2014). Although indigenous freshwater fish species like *Labeo dussumieri* and *Puntius sarana* are found in inland fish catches, their commercial importance is quite low. Introduced fish species such as tilapias (*Oreochromis mossambicus* and *O. niloticus*) dominates the inland fish landings. Attempts have also been made to introduce Indian and Chinese carp species into reservoirs in the past and their stocking is continuing with culture-based fisheries practices conducted by NAQDA. There is close to 2.4 million direct and indirect employments being generated in fisheries sector. The key stakeholders are fishermen, breeders, processors, and logistics, cold chain, packing, and other service suppliers.

Sri Lankan fisheries are managed by two core legislative instruments, namely, Fisheries and Aquatic Resources Act. No. 2 of 1996 and the Fisheries (Regulations for Foreign Fishing Boats) Act No. 59 of 1979, along with several regulations to assist in the implementation of these acts. Yet many aspects including a three-decade long war, poaching and lack of coastal and ocean security, as well as poor understanding of sustainable fishing methods have seen the downfall of a promising industry abundant of resources. Over the years, the successive governments of Sri Lanka and international agencies have acknowledged that this enormous resource for fish farming or aquaculture has not been exploited in a sustainable way because, among other things, most of the marine fishing is focusing on coastal areas in which harvesting goes at near MSY (Maximum Sustainable Yield) level, but not the offshore and deep sea where there are still underutilization of fisheries resources. As per inland fishing, changing policies by changing governments ensure politics play a bigger role than rational, sound, and pragmatic policies for the fisheries sector.

The year 2017, however, showed a series of changes in the local and global environment, which can make a positive impact on the Sri Lankan fishery industry in the long run. The industry, which has seen a change of fortune since the end of the civil war and the revival of the fishing industry in the Eastern and Northern regions of Sri Lanka, received a power boost with the reinstatement of GSP Plus from the USA and lifting of export banning of fish to European Union (EU) in the recent past, which can boost the exports of fish and fishery products. Meanwhile, the Sri Lankan government agencies are also seeking to position the country's marine and inland fishing industries as major export earning sectors by establishing aquaculture and marine industry parks in the Eastern and Southern coasts of the country. Two aquaculture industrial parks are to be established in the Mannar and Batticaloa districts while the establishment of another aquaculture park is being proposed in Hambanthota district.

In assistance of the Sri Lankan government's attempts to provide a quick boost to the local fishing industry, the South Korean government, too, has extended a soft loan facility to develop four multi-purpose harbors in Maadagal, Chalai, Udappuwa, and the island of Delft in North and North-Western provinces in Sri Lanka (EDB 2017a). Moreover, two technical training colleges on fishing and aquaculture were established to enhance the technical capacity of the Sri Lankan fishing community in developing sustainable and energy-efficient practices in fishing.

Sustainability in the local fishing industry is the main concern that has been addressed aggressively at various levels. While Sri Lanka's first Marine Finfish Hatchery was inaugurated in Tharmapuram in Batticaloa (Eastern province) to facilitate the development of marine finfish farming and produce sufficient amounts of marine fish to meet the demands of the national and international markets without exhausting the natural supplies, the Sri Lankan parliament has unanimously amended the Fisheries and Aquatic Resources Act No.2 of 1996 recently in July 2017 to ban bottom-trawling, a practice which has threatened the future of the aquatic resources in Sri Lankan oceans (EDB 2017a). Under the fisheries sector development strategy (EDB 2017b), a modern and technically improved Vessel Monitoring System (VMS) will be established in order to curtail IUU fishing; this will be able to prevent national fishing vessels from drifting to other countries' territorial waters.

With the promising developments in technology training, infrastructure development, and sustainable harvesting practices, the fishing industry in Sri Lanka is in for a better future in the years to come.

14.1.4 Situation on Current Fish Utilization and Marketing

The share of world fish production utilized for direct human consumption has increased significantly in recent decades, up from 67% in the 1960s to 87% or more than 146 million mt in 2014 (SOFIA 2016). During the same year, The remaining 21 million mt was destined for nonfood products, of which 76% was reduced to fishmeal and fish oil, which are still considered as the most nutritious and digestible ingredients for farmed fish feeds.

Developing economies, whose exports represented just 37% of world trade in 1976, saw their share rise to 54% of total fishery export value and 60% of the quantity (live weight) by 2014 (SOFIA 2016). Fishery trade represents a significant source of foreign currency earnings for many developing countries, in addition to its important role in income generation, employment, food security, and nutrition.

Sri Lanka is ranked among the first 50 countries in the world exports with total share of 0.2% in the world export market (EDB 2017b). Sri Lanka has been exporting fish and fishery products to Europe, America, and Asia markets over the years. Based on the coastal areas surrounding the country, the fishing and fisheries processing industries in Sri Lanka attract many workers who have been unemployed and displaced by the war, tsunami, and other social and economic factors. After the 30 years' war is over, liberalizing the huge area of land and coast in the North and East are available for Aquaculture and sea farming of various varieties of fish and also to develop processing factories.

There are 52 government-approved fishing and processing facilities spread across the country, out of which 32 have been approved by the European Union as per their safety (EDB 2017a), sustainable fishing, and safe food processing practices; Sri Lankan fishing industry can expect a better future in exporting seafood and seafood-based products to the EU. Further, improvement of the country's long line fishing fleet used for tuna and billfish fishing, two types of fish, has a larger market in EU and Eastern-Asia.

Sri Lankan tuna, shrimps, and crabs have quality of unique taste and texture. Most of the seafood processing plants are situated in the Western Province in the Colombo and Gampaha districts due to its easy access to the airport. There are few associations that support the seafood sector by addressing the sector-related issues through coordination with the government institutions. There are more than 75 medium- and large-scale companies engaged in exporting seafoods but only 32 companies have EU-Approved processing plants (EDB 2017b).

In addition to the exports, Sri Lanka has to import a substantial amount of dried fish, sprats, and canned fish annually in order to carter the excess domestic demand owing to a low level of domestic production of dry fish and sprats in the country.

14.1.5 Trends in Future Products and Markets

Potential for the future expansion of the fisheries sector rests with more concentration on value addition (especially focusing on exports) and convenient food production. Further, organic aquaculture production, especially organic shrimps, targeting high-end markets, also has the potential for expansion by using the available reservoirs for freshwater fish culture. Commercial culture of oysters, mussels, and seaweed farming are lucrative areas to look into.

14.2 Fish Production in Sri Lanka

14.2.1 Inland Capture Fishery

Sri Lanka has an extensive freshwater and brackish-water resource to sustain viable fishing activities. These include irrigation reservoirs, seasonal tanks, and brackish-water resources such as estuaries, lagoons, or marshes. The total inland and aquaculture fish production in 2016 was 73,930 mt and has contributed 14% to the total fish production of the country in the same year (Annual Performance Report 2016). Anuradhapura (18%), Puttalam (14%), and Ampara (13%) districts are dominant inland fish producing districts in the country. Capture fish production is mainly done with exotic fish species, which can breed naturally under Sri Lankan conditions, such as tilapia, common carp, and other indigenous fish species, i.e., Hiri kanaya (*Labeo* spp.), Lula (snakehead), and catfishes (butter catfish; stinging catfish, walking catfish). In addition to finfish being caught, freshwater prawn is also captured from aquatic resources where they are abundant. Capture fish production in inland fisheries by major species is presented in Table 14.1.

14.2.2 Inland Culture Fishery

Inland culture-based fishery has been performing since the introduction of exotic tilapia in the 1950s. At present Chinese major carps (bighead, silver, grass carp) and Indian major carps (catla, rohu, mrigal) are the major group of fish stocked in both perennial and seasonal inland aquatic resources. In addition, Genetically Improved Farm Tilapia (GIFT) has also been stocked in perennial reservoirs

Table 14.1 Inland capture fisheries by major species (MT)

Species	2012	2013	2014	2015	2016
Tilapia	39,950	39,070	46,610	40,504	43,836
Carps/Mrigal	3570	3450	3920	2847	3363
Catla/Rohu	12,460	8980	11,020	9117	7772
Hiri kanaya	670	590	580	358	230
Lula	1170	2040	2230	1582	1849
Freshwater prawn	290	540	460	374	705

Table 14.2 Annual fish production by sub sectors (mt)

Sector	2012	2013	2014	2015	2016
Marine	417,220	445,930	459,300	452,890	456,990
Coastal	257,540	267,980	278,850	369,020	274,160
Offshore/deep sea	159,680	177,950	180,450	183,870	182,830
Inland & Aquaculture	68,950	66,910	75,750	67,300	73,930

periodically to increase their population, while in seasonal tanks, it is done on annual basis as the culture-based fisheries is a cheaper source of animal protein for rural fork. Stocking of fingerlings continuously into inland water bodies has contributed significantly for the development of inland fisheries as well as the increase of fish production. Major, medium, and minor reservoirs as well as seasonal tanks were mainly targeted for stocking of fingerlings and of them medium reservoirs were major and about 32% of fingerlings have being released into them in 2016 (Annual Report NAQDA-2017).

14.2.3 Marine Capture Fishery

As stated earlier in this chapter, fish from the marine fisheries is the dominant subsector in terms of production, employment, and fishing fleet strength (Table 14.2). There are totally 15 fisheries administrative districts of the country. Of them, Tangalle and Galle districts are dominant and have contributed over 26% to the total marine fish production of the country in 2016, followed by Kalutara (10%), Puttlam (9%), Negombo (7%), Matara (7%), Chilaw (7%), and Jaffna (7%) districts that have also contributed a considerable proportion to the total marine fish production of the country. Skipjack tuna (Balaya) and yellow fin tuna (Kelawalla) are dominant species that have contributed 10.4% and 8.7%, respectively, to the total marine fish production in 2016. However, production of Skipjack tuna has decreased by 11.7%, compared to the previous year. Both tunas are classified under the world top 10 species been captured in marine fisheries.

14.2.4 Aquaculture

Global aquaculture production has increased steadily in the recent years as nearly 46% of the world's fish produced for human consumption comes from farmed sources (SOFIA 2016). It is the fastest growing food production sector and has become an important element of economic growth and poverty reduction plans in many countries. As Sri Lanka has vast water bodies and suitable environmental conditions, the country has an enormous potential to increase the production of finfish and shellfish through the sustainable development of aquaculture. The contribution from inland and aquaculture subsector is low. However, since 1980, fish culture in seasonal village tanks, marine shrimp culture in coastal earthen ponds, and live ornamental fish exports have reached commercial dimensions. Presently, tilapia and catla are dominant in inland and aquaculture fish production which was contributed 59.3% to the total fish inland production in 2016 (Annual Report NAQDA 2017). Among others, the cultured shrimp contributes 8.2% to the total and was mainly focused on export market. In addition to aquatic fauna that has been cultured, culture of aquatic fauna in all three types of water resources (freshwater, brackish water, and marine water) is now being popularized in certain areas of the country owing higher foreign exchange earnings by exporting them and their products for various commercial purposes. The future plans to develop country's aquaculture industry is mainly focused on the high-value aquatic resources by setting up a fishery mega zone to culture sea cucumber, sea bass, GIFT (tilapia), and seaweed on a commercial scale. Cage culture, mollusk, and seaweed culture are yet to be developed. Bivalve farming will be developed by which shellfish varieties such as oysters, mussels, clams, and cockles will be farmed for human consumption and exportation.

14.3 Fish as a Food

14.3.1 Protein Source

Fish provided more than 3.1 billion people with almost 20% of their average per capita intake of animal protein (Fish and human Nutrition, FAO). Further to being a rich source of easily digested, the high-quality fish proteins contain all essential amino acids. Even small quantities of fish can have a significant positive nutritional impact on nations consuming plant-based diets.

Fish provides a good source of high-quality protein and contains many vitamins and minerals all of which are vital for the healthy functioning of the body (Table 14.3). As an animal protein source, fish contributes to about 17% at the global level, but exceeding 50% of total animal protein intake in some small island developing states, as well as in Bangladesh, Cambodia, the Gambia, Ghana, Indonesia, Sierra Leone, and Sri Lanka (FAO 2014). Sri Lanka has a dietary concentration on one staple commodity (rice), which is deficient in essential amino acids such as lysine and methionine. Hence, fish are a vital food source that

Table 14.3 Nutritional aspects of fish compared to other animal origin food source per 100 g

Source	Calories	Protein (g)	Carbohydrates (g)	Fat (g)
Fish	110–140	20–25	0	1–5
Chicken (breast)	160	28	0	7
Beef (steak)	210	25	0	7
Lamb	250	30	0	7

Table 14.4 Nutritional composition of fish (100 g serving pack) (Source: Jag Pal et al. 2018)

Protein	Fat	Calories	Vitamin A	Vitamin B_{12}	Iron	Zinc	Calcium
20–25 g	1–5 g	110–140	2503 μg	0.50–14 μg	0.34–19 mg	0.6–4.7 mg	8.6–1900 mg

contains all essential amino acids. Though Sri Lanka is surrounded by the Indian Ocean, the per capita fish consumption in Sri Lanka is yet below the world average of 20 kg (FAO 2017). For better nutrition and food security, the challenge is to increase fish harvest in an affordable manner by which fish consumption can be increased to improve nutritional status. The freshwater fishes have traditionally been obtained from the inland capture fisheries resources like rivers, reservoirs, and wetlands. Since last two decades, freshwater aquaculture sector has expanded significantly along with development of culture-based fisheries, producing large quantity of freshwater fishes as affordable animal protein source for rural communities.

14.3.2 Vitamin Supplier

Fish provides essential fats (e.g., long-chain omega-3 fatty acids), vitamins (A, D, and B), and minerals (including calcium, iodine, zinc, iron, and selenium), particularly if eaten whole (Table 14.4). Whitefish such as seer (*Thora*) contains little fat (usually less than 1%) whereas oily fish such as sardines contain between 10 and 25%. However, as a result of its high fat content, oily fishes contain a range of fat-soluble vitamins (A, D, E, and K) and essential fatty acids.

14.3.3 Micro and Macro Nutrient Supplier

The nutrient composition of fish varies widely across species, particularly for micronutrients. Iron content ranges from 0.34 to 19 mg/100 g raw edible parts, zinc from 0.6 to 4.7 mg/100 g, calcium from 8.6 to 1900 mg/100 g, vitamin A from 0 to 2503 μg RAE/100 g, and vitamin B_{12} from 0.50 to 14 μg/100 g (Table 14.4).

14.3.4 Health Benefits of Fish

Fish is one of the best sources to provide essential fatty acids: omega-3, omega-6, EPA, and DHA. Fish is usually high in unsaturated fats and provides health benefits in protection against cardiovascular diseases. It also aids fetal and infant development of the human brain and nervous system. With its valuable nutritional properties, it can also play a major role in correcting unbalanced diets and, through substitution, in countering obesity in human beings.

Catfishes and snakehead have widely been caught naturally for food for hundreds of years in Asia, considering them excellent to eat due to its taste and high level of vitamin D, while some research has found that tilapia may be far less nutritious than generally believed as omega-3 fatty acid content is often far lower than that of other commonly eaten fish species. However, it contains high level of omega-6 fatty acid being a tropical fish species (Jag Pal et al. 2018).

14.4 Integrated Fish Farming Systems

Climate change is a fundamental threat to global food security, sustainable development, and poverty eradication. Thus, governance needs to ensure that fisheries and aquaculture adapt to the impacts of climate change and improve the resilience of food production systems where one of the solutions is integration. Integration amalgamates various food production systems in order to maximize resource utilization and minimize the waste and by-products.

There is a need to strengthen aquatic ecosystem governance to deal with the increasing use of water space and resources. Aquaponics is a symbiotic integration of two mature food production disciplines: (i) aquaculture, the practice of fish farming and (ii) hydroponics, the cultivation of plants in water without soil. Aquaponics combines the two within a closed recirculating system. In addition, available other farming systems can be incorporated with aquaculture practices as discussed below.

14.4.1 Integration with Livestock

The advantages of integration are obvious. Integrated fish farming is a process of farming where fish is produced in combination with other farm products and livestock. The system links each of the involved subsystems in it, such as fish, crops, and livestock, in such a way that the waste or by-products from one subsystem can be used as an input for the next system. An integrated agriculture system can ensure the maximum utilization of all resources, such as land, water, and feed, and also minimizes waste.

As far as fish production is concerned, livestock serves as the major purpose of providing organic manure for the fish ponds, thereby reducing the cost and need for providing compounded fish feeds and chemical fertilizers. By reducing the cost of fertilizers and feedstuffs, the overall cost of fish production is reduced and profits increased. Aquatic resources (i.e., pond or tank) available in livestock farms are mainly used for irrigation and wallowing purposes. These ecosystems can thus be utilized as the place for stocking fish. The pond ecosystem transforms the inert nutrients of the manure to digestible, protein-rich, live food for the fish. Therefore, to make livestock-fish integration efficient, suitable fish species (plantivores, detritivores, and herbivores) are needed. They may not be available in the local fish fauna but in the exotic fish introduced to Sri Lanka.

14.4.2 Integration with Poultry

Some extent of poultry fish integration can be seen in Sri Lanka, mainly with duck (water fowl). In this system duck house is built on the pond to allow manures to fall directly into the pond or it is located on the dike and manure is washed in daily. As ducks are raised on the pond surface, they drop their nutrient-rich manure food in the pond; the fish gather protein-rich natural food from the pond ecosystem or may consume directly the feed spilled by the ducks. In order to prevent ducks from eating small fish, larger size of fingerlings are stocked in the pond. The recommended fish species are tilapia or tilapia cum carp at the stocking density of 2000 fish/ha.

14.4.3 Integration with Paddy (Rice)

Rice-fish farming system is site and socioculturally specific. It is important to note that fish is secondary to rice, which is still the main component of the system. However, in Sri Lanka, rice farming is carried out at an intensive level and the fertilizer subsidy program would enable farmers to make more money from rice than fish. Further, factors such as shorter growing season, intensive mechanization and greater use of pesticides and herbicides, and low productivity of the system also limits rice-fish integration to be popularized in Sri Lanka. Table fish production under rice-fish integration is not productive as paddy varieties cultivated in Sri Lanka have a short cultivation cycle. The period for fish grow-out cycles will be limited to 2 or 3 months and the fish that are produced will be small. The small-size freshwater fish are not preferred by Sri Lankans, and thus rice-fish integration should not be intended for table fish production. A system has to be developed to culture high-priced ornamental fish species (i.e., guppy, platy, goldfish), which have a good export market.

14.4.4 Integration with Plantation Crops

Most of the tea estate in Sri Lanka possess water source that is connected to the cooling system of the power-generated engine used before electricity supply received. These water bodies are now being used for stocking of food fish to be cultured as an animal protein source for estate people. In addition, available water bodies in coconut plantations for irrigation purpose and for wallowing of grazing buffaloes have also be used to stock table fish in estates belonging to government and private sector. An aquaponic system can easily be set up at home where we can raise fish and grow plants at the same time. A recirculating water system from your fish tank would easily provide nutrients for your plants and also clean the water of your fish tank.

14.5 Economics and Social Contribution of Fishery

14.5.1 Economics Contribution

14.5.1.1 Fish and Fish Products Export

While coastal fisheries still dominate the overall fish output of the country, production from offshore and deep-sea fishing has been increasing rapidly since the early 1990s. This trend is largely in response to higher demand for tropical fish, particularly tuna species, from markets in industrial economies such as Japan, the EU, and the USA. The increase of offshore/deep-sea fish production by 2016 (Table 14.2) has gone hand in hand with a number of improvements throughout the entire chain such as more reliable and larger vessels with cold storage, modern navigation instruments and fishing gear, fish processing plants (for frozen fish), and laboratory and quality testing facilities. Tuna has rapidly become Sri Lanka's main fish export, overtaking cultured shrimp which dominated fish exports over the last two decades. Sri Lanka's fish products currently compete well in export market in terms of both price and quality, and most of the fish product categories (HS 06).

Export earnings have shown steady growth during recent years and now account for approximately 2% of the total GDP. Sri Lankan fish exports include fresh and frozen fish (tuna, swordfish, shark, seer, etc.). Tuna accounts for 51.5% of the total fish and fishery products exports of Sri Lanka. Export contribution of European market is 16.7%, Japanese market 3.7%, the USA 36.6%, and other states 43% (Annual Performance Report 2016). In order to expand the export market opportunities, discussions have been held with regard to exporting yellow fin tuna head to the Korean market. Crustaceans (i.e., prawns, lobsters, and crabs), sea cucumber and shark fins. The European Union is the main international buyer for Sri Lankan fisheries exports—36%, followed by the USA and Japan, accounting for 27% each of total fish exports by Sri Lanka (Annual Report CBSL 2017). Further, the rapidly growing opportunities for fish exports have yet to be fully exploited in light with lifting of banning of exportation of fish and fishery products to EU countries and GSP tax relief offered by the USA to Sri Lanka recently.

Meanwhile, fish exports increased by 41.8% in volume terms and 48.5% in rupee terms, in 2017 compared to 2016, following the removal of the ban on fish exports to the EU in 2016. Further, fish imports to Sri Lanka declined by 8.4% in volume terms and 3.4% in rupee terms in 2017 compared to that of 2016.

14.5.1.2 Fish Supply and Availability

A critical element of the supply chain is the broad base of numerous small-scale fishers who operate in small motorized, traditional craft and relatively large, multi-day boats that are able to exploit offshore and deep-sea fisheries and carry out fish production. Of the two marine fishing categories, the major share of production, about 60%, in 2016 (Table 14.2) comes from coastal fishing along the entire coastline where fishing operations are at near MSY level. Fishers harvest a variety of species that reflect the spatial variations of the fishing grounds. The output of coastal fishing is marketed through diverse channels to different, mainly domestic end, markets that include urban and rural retail fish outlets, small mobile vendors, supermarket chains, and the state-owned Ceylon Fisheries Cooperation (CFC) outlets. Urban wholesale markets, such as St. Johns Fish Market in Colombo and Kandy wholesale fish market, play an important role in distributing coastal fishery outputs. While coastal fisheries still dominate the overall fish output of the country, production from offshore and deep-sea fishing has been increasing rapidly since the early 1990s (Table 14.2). The per capita availability of fish is 15.3 kg. Currently there are 12 harbors on the coastline with the main fishery harbor situated in Colombo, and 37 anchorages and 710 fish landing sites (*thotupola*), which are also inadequate, poorly managed, and often lack even basic facilities and are to be improved in order to minimize postharvest losses and for better distribution of fish being harvested.

In progressive markets, some actors who are successfully performing multiple roles at different levels of the supply chain are contributing to the momentum of the system. For instance, coastal operators venturing into deep-sea fishing have moved from being local market suppliers to suppliers for the export market, along with the government subsidiary program introduced recently on purchasing of trawlers. In spite of these important signs of progress, Sri Lanka's fisheries sector has performed less than satisfactorily with domestic supply unable to provide the quantities of fish products required at affordable prices to meet the nutritional needs of the country's population.

14.5.1.3 Annual Fish Consumption per Capita

The Sri Lankan fishery sector serves both the domestic and export markets, with the domestic component attracting 75% of the marine fish production. Recently, the Ceylon Fisheries Corporation (CFC), which has established 130 retail fish outlets located island-wide, has extended its support to small fishermen by assisting them in marketing and providing price support under a guaranteed price scheme. The per capita consumption as at present is 17.3 kg, which is expected to increase up to 22 kg in near future. Daily basis per capita fish consumption has increased by 47.2 per day from January to September 2016 (Annual Performance Report 2016). Canned fish and dry fish also contribute immensely to the per capita fish consumption in the country.

14.5.1.4 Contribution to GDP

Fisheries sector contribution to national GDP was 1.3% for the last two consecutive years (CBSL 2017). There was a marginal increase in fish production in 2017 (0.1%) compared to 2016. However, marine fish production, which accounted for 84.6% of the total fish production in 2017, declined by 1.7% to 449,440 mt, while inland fish production increased by 10.7% to 81,870 mt. Prolonged dry weather in the deep sea and substantial postharvest losses had an adverse impact on marine fishing last year.

14.5.1.5 Other Indirect Contributions of the Fisheries Sector

Fifty-two government-approved fish processing facilities spread across the country have also provided employment opportunities for the people. The number of functioning ice factories in 2013 was 90 and has been increased up to 123 by 2016 (Annual Performance Report 2016). Ice production has remained at 3949 mt at the end of September 2016. Further, feed manufacturing, marketing, and transportation will also enhance with the development of aquaculture industry in the country.

14.5.2 Social Contribution

14.5.2.1 Contribution to Livelihood Development

Many projects have been launched by the Ministry of Fisheries during 2016 aiming the expansion of new employment opportunities for enhancing the life standards of fishers. Community-based sea cucumber farming as an alternative livelihood method in Northern and North-Western provinces, identification of suitable locations considering the environmental factors in chank shell farming, breeding edible fish with community participation in the abandoned clay pits, development of sea moss and seed cultivation have also been commenced.

14.5.2.2 Financial Strength

Total export value of food fish and fishery products accounts for Sri Lanka Rs. (SLR) 26,802 million in 2016 while ornamental fish exportation alone contributed another SLR. 2382 million to the country's economy.

14.5.2.3 Social Strength

Fishing fleet of Sri Lanka marine fishery comprises 15,022 small traditional craft (48% motorized) used in the lagoons and coastal areas, 8334 fiber glass speed boats with kerosene out board motors (18–24 ft. and 25–40 hp. motors), 1550 (21–24 ft.) day boats, and 1700 larger multi-day boats (32–52 ft.). The number of total fishers directly involved in marine fishery is over 2,70,000 and its created over 6,50,000 employments to the sector (Annual Performance Report 2016). Further, job opportunities have been created in freshwater fish fingerling production by the mini hatcheries, private ponds, and cages, which have accounted for 44.85% share of the total fish seed production in the country (Annual Report NAQDA 2017). Fish as a commodity has also created more employment in marketing, processing, packing, transportation, and input provision to the industry. Hence, the whole fishery sector created over 2 million jobs to the nation.

14.6 Conclusion

Recent actions taken by global community for the sustainable use of fisheries resources are commendable. The use of GIS and remote sensing has expanded the suite of modeling approaches to include higher-resolution and more reliable predictors of yield, including direct measures of primary production and hydrologic regime. For example, researchers have established relationships between chlorophyll concentrations as a measure of freshwater primary production and fishery yields worldwide. Further, progress has been made worldwide in managing by-catch and reducing discards. The FAO–GEF project Sustainable Management of By-catch in Latin America and Caribbean Trawl Fisheries (REBYC-II LAC) (2015–2019) have aimed to reduce food loss and support sustainable livelihoods by improving by-catch management and minimizing discards and sea-bed damage, thereby turning bottom trawl fisheries into responsible fisheries.

Measures that have been taken to control fishing operations within the exclusive economic zones are now considered much stronger. Steps being taken to combat illegal, unreported and unregulated IUU fishing prevent further build-up of fishing overcapacity and/or reduce it, and implement plans for the protection and conservation of marine fauna and flora.

In tropical and subtropical fisheries, gillnets and trammel nets are among the main gear types used in fisheries. A food loss and waste reduction project initiated by the FAO and focusing on the harvesting stage of the fish supply chain has been started with gillnet and trammel net fishing operations, the results of which should be of wider interest (FAO 2015a, b).

Among the varied research projects implemented in the recent past by NARA for encouraging fish export industry are, informing fishers by providing prediction on tuna fishing grounds, introduction of fishing gear for octopus harvesting, and controlling white spot syndrome of brackish-water prawn. Fish aggregation devices have been located in sea by NARA. Further, diverse strategies are being adopted by NAQDA to promote ornamental fisheries exports such as the introduction of new species of ornamental fish, fish breeding, prevention of fish diseases, provision of training to the farmers on high-quality fish feeds, technology development, diagnosing fish diseases, and assisting fishers to obtain loan facilities and organization of international exhibitions on ornamental fisheries.

Research recently conducted were on the Sri Lankan skates and lobsters fisheries, scientific studies on tuna, conservation of spiny lobsters carrying eggs, assessment of algae constitution of Maduru-oya reservoir, water quality assessments, disease management, and implementation of information sharing projects are among the research projects launched by NAQDA. Research on marine mammals and studies on fish habitats such as coral reefs, sea grasses, and mangroves were conducted by NARA with relevant conservation recommendations. Government has taken actions in 2016 to promote high-seas fishing through the introduction of state-of-the-art fishing vessels and to encourage marine fish farming. Further, fundamental measures have been taken in promoting shrimp, crab, and other freshwater fish species cultivation and thereby developing the freshwater fishes to commercial levels with private sector contribution.

Sri Lanka, being an island nation and having eight times higher marine water resources compared to its land area, possess enormous natural resources including fish. Sri Lanka also gets benefits from those global actions taken by the UN by being a signatory to practicing the code of conduct for sustainable fisheries and many more agreements. At present, 86% of the country's total fish production is received from the marine sector. The marine water area that belongs to Sri Lanka will further expand by another 1 million square km in the near future according to the new maritime mapping system, thus resulting in an increase in both marine fish production and available fishing area of the country. Further, Sri Lanka is one of the countries having the world's highest freshwater resources with 4% of freshwater bodies compared to its land area. At present, only 10% of it is used for fish production. Availability of these natural aquatic resources provides immense opportunities to achieve high economic growth with guaranteed food and nutritional security for its population in near future by blue economic revolution.

References

Annual Performance Report (2016) Ministry of Fisheries and Aquatic Resources Development. Government of Sri Lanka

Annual Report NAQDA (2017) National Aquaculture Development Authority, Pelawatte, Battaramulla

CBSL (2017) Annual Report of Central Bank of Sri Lanka. Central Bank of Sri Lanka, Colombo

De Silva RI (1984–1985) The Sharks of Sri Lanka – a key to the different species and a preliminary checklist. Ceylon J Sci Biol Sci 17–18:56–66

EDB (2017a) Developing the Sri Lanka's fishing sector to face the global demand, EDB Blog, Export Development Board of Sri Lanka

EDB (2017b) Sri Lanka fisheries sector. Industry Capability Report, Export Development Board of Sri Lanka

FAO (2014) Fish and Human Nutrition. Food and Agriculture Organization of the United Nations, Rome

FAO (2015a) *A project report on Sustainable Management of By-catch in Latin America and Caribbean Trawl Fisheries* (REBYC-II LAC), Rome: Food and Agriculture Organization of the United Nations

FAO (2015b) SAVE FOOD: global initiative on food loss and waste reduction. Food and Agriculture Organization of the United Nations, Rome

FAO (2017) Country programming framework-Sri Lanka 2013–2017. Food and Agriculture Organization of the United Nations, Rome

Jag Pal S, Maurya AK, Verma HO, Pandey G, Amitha (2018) A review on role of fish in human nutrition with special emphasis to essential fatty acid. Int J Fish Aquat Stud 6(2):427–430

Rajasuriya A (2014) Field guide to reef fishes of Sri Lanka, vol 2. IUCN Sri Lanka Office, Colombo. Colombo 07, Sri Lanka

Sivasubramaniam K and Maldeniya R (1985) A review of the results of demersal fish resources surveys around Sri Lanka. Development of Sustainable Fisheries; Bay of Bengal Programme

SOFIA (2016) Contributing to food security and nutrition for all. The state of world fisheries and aquaculture. Food and Agriculture Organization of the United Nations, Rome

SOFIA (2018) Meeting the sustainable development goals. The state of world fisheries and aquaculture. Food and Agriculture Organization of the United Nations, Rome

Evolution of Agricultural Extension System in Sri Lanka

15

C. Sivayoganathan

Abstract

Agricultural extension, though plays an important role in agricultural development, continues to be a neglected component of the agricultural knowledge and information system in Sri Lanka. Agricultural extension service in the food crop sector of Sri Lanka had been built-up over the years as an evolutionary process. Various extension approaches such as conventional technology transfer, training and visit, integrated agricultural extension, and block demonstration were implemented in this sector along with the introduction of information and communication technology interventions in the recent past. Similarly, different extension approaches were adopted over the years to suit the specific requirements of the main plantation crops, namely tea, rubber, and coconut as well as sugarcane and minor export agricultural crops. The livestock sector also has an island-wide extension service similar to the food crop sector. The agricultural extension services provided by the state departments are supplemented to a limited extent by the extension activities of the universities, private sector companies, nongovernment organizations, and international development agencies. The major challenges facing agricultural extension in Sri Lanka are (a) lack of a comprehensive national agricultural extension policy, (b) limited recognition of agricultural extension service at all levels, (c) poor linkages among research, extension, and other agri-support services, (d) shortage of competent extension professionals, and (e) inadequate research in extension.

Keywords

Agricultural extension · Food crops · Plantation crops · Livestock sector · Information and communication technologies

---

C. Sivayoganathan (✉)
Faculty of Agriculture, University of Peradeniya, Peradeniya, Sri Lanka

B. Marambe et al. (eds.), *Agricultural Research for Sustainable Food Systems in Sri Lanka*, https://doi.org/10.1007/978-981-15-2152-2_15

15.1 Origin of Extension

The first use of the term "extension" originated in England in 1866 with a system of university extension which was taken up first by Cambridge and Oxford Universities and later by other educational institutions in England and in other countries (Swanson 1984). "Agricultural Extension" came into common use in the United States of America early in this century when the co-operative extension services were formed in each state in association with the Land Grant Colleges (Van den Ban and Hawkins 1985). Although the meaning of the term extension is well known and accepted by people who work in extension organizations, it is not well understood in the wider community. Extension is perceived in different ways by different people. Given below are the meanings of some words which are used instead of extension to describe related processes. The Dutch use the word *Voorlichting,* which means lighting the pathway ahead to help people find their way. The British and the Germans refer to "advisory work" (Baratung), which implies that the expert can give advice on how to achieve a person's goal, but leaves him/her with the final responsibility of selecting the best way to do so. The Germans also speak of "education" (*Erziehung*), as in the USA, where it is stressed that the goal of agricultural extension is to help people learn to solve problems themselves. The Australians relate to "furthering" (*Forderung*) or stimulating one to go in a desirable direction, which is rather similar to the Korean term "rural guidance." The French refer to *vulgarization* which stresses the need to simplify the message for the "common man," while the Spanish sometimes use the word *capacitacion*, which indicates the intention to improve people's abilities, although normally it is used to mean "training" (Van den Ban and Hawkins 1985).

Most agricultural extension services use a mixture of all these processes which have some common features which help us to define extension. Extension is an ongoing process of getting useful information to people (communication dimension) and then in assisting those people to acquire the necessary knowledge, skills, and attitudes to utilize effectively this information or technology (education dimension). The ultimate goal of extension is to enable people to use these skills, knowledge, and information to improve their quality of life (Swanson 1984).

15.2 Importance of Extension

The demand for food is increasing rapidly, especially in developing countries like Sri Lanka, with comparatively high rates of population growth. These countries will not be able to import much of their food because they have few exports to generate funds to pay for food imports. The food production should, therefore, be increased rapidly to avoid serious shortages. This increased production has to come mainly from intensive cultivation, that is, from increased productivity of the existing cultivated lands rather than from increased land use, as in the past. This is because, little unused land is still available, and it requires comparatively more capital to bring new land into production than to increase yields.

Increasing yield per hectare (land productivity) implies the shift from traditional methods of production to new scientific methods that include new technological components such as new varieties, cultural practices, and new farming systems. In many countries there are wide gaps between the yields that could be obtained through the use of available production technologies and the yields obtained by the majority of farmers. This is due to the wide knowledge–practice gap. Research done sometime back has shown that the level of adoption by farmers of the available agricultural technologies is 26% in Bangladesh and about 30% in India (Karim 1999). In Sri Lanka, too, there is a wide knowledge–practice gap.

Agricultural extension agents can certainly help narrow down these gaps by helping farmers form sound opinions and to make the correct decisions with respect to the use of available resources and the adoption of improved technology in agricultural production. Complex technologies especially need in-depth educational and communication inputs. However, the impact of agricultural extension will be the greatest only in the immediate growth potential areas where the other agri-support services such as credit facilities, farm to market roads, retail outlets for farm inputs, local verification trials, and markets for farm products are already developed (Mosher 1978). Agricultural extension, of course, has an indirect role to play in developing some or all of the above services. Extension thus lies at the heart of agricultural development and a strong network of extension is a vital prerequisite to reap a good harvest of the scientific, technological, and educational advancements of a country (Kashem and Halim 1999).

Food systems comprise activities related to food production, processing, distribution, marketing and trade, preparation, consumption, and disposal. They capture the path of food from farm to the food plate meeting food and nutritional needs of a nation. Agricultural extension plays a vital role in sustaining food systems. It builds capacity of farm-level producers and serves as the bridge between the farmer and the government sector agents and ultimately forms the foundation for applied research in national agricultural research system. The role of the extension worker in the present day goes beyond provision of skills and knowledge in farming. It now requires provision of information on best agricultural practices to produce high-quality products, value addition of the farm produce, culinary methods, nutritional knowledge, and women empowerment to make the food systems sustainable. This chapter provides an overview of the extension services provided by the government of Sri Lanka by subsector, presents the key challenges, and provides suggestions for improvement of the extension service provision.

15.3 Food Crop Sector

Agricultural extension service in the food crop sector of Sri Lanka had been built-up over the years as an evolutionary process (Sivayoganathan and Wirasinghe 1992). With the breakdown of vibrant agricultural economy, after series of foreign invasions starting from 1640 to 1812, the earliest attempt by the British to build up the food cop sector commenced after 1880s. Historical records reveal that from about 1880,

Agricultural Instructors (AIs) with two-year training in agriculture were appointed to work directly under the Government Agents who were vested with the responsibility of increasing food production in the country (Emitiyagoda 2009). In 1904, the Ceylon Agricultural Society (CAS) was established by the wealthier planters, land owners, and agriculturists with the objective of helping the native farmers. The extension work was carried out by the CAS in collaboration with the Government Agent. Food production thereafter was undertaken by the Ceylon Agricultural Society in association with the Government Agents. In 1921 the staff of the Ceylon Agricultural Society was absorbed into the Department of Agriculture (DOA) which was established in 1912. Though initially the DOA concentrated on the plantation sector, by around 1932 the extension staff was also entrusted with the responsibility of delivering new information and technology and supplying seeds and planting materials to the rural farming community. During this period the Agriculture Officers (AOs) were in charge of divisions while AIs were in charge of ranges delivering the services required by the crop and livestock farmers as well as the planters.

It was only in 1922 when the country faced severe food shortages as an aftermath of the First World War, agricultural extension service was developed as a part of the DOA. There were three divisions in 1923, which was increased to four in 1927, six in 1932, and nine in 1939, which remained the same until 1957. In 1939, the total number of AIs in the country was 74, which was too small to offer an effective agricultural extension service (Emitiyagoda 2009). Immediately after independence the government paid greater attention to rice production. In 1952, food production was considered to be of very high priority and a separate Department of Food Production (DFP) was created in the Ministry of Agriculture. In 1957, however, the DFP was disbanded and the staff was absorbed into the DOA. About 500 Food Production Overseers (FPOs) came into the DOA. Many of the FPOs did not have any agricultural training whatsoever. The designation of FPO was later changed to *Krushikarma Viyapthi Sevaka* (KVS) and the subsequent recruitments to this cadre were from persons with 1 year training in Practical Farm Schools. Thus, emerging below administrative hierarchy of the AIs, a new cadre of extension workers came in direct contact with the farmers.

In 1957, the AO grade was abolished and was replaced by a grade of District Agricultural Extension Officer (DAEO). The AO was in charge of a province whereas a DAEO was appointed to each district. Apart from the small increase in the number of AIs and appointment of large number of KVSs, 70 field demonstrators were appointed in 1950s to assist AIs in organizing field days and cinema shows. During the same period, a few female demonstrators were appointed to work with farm women on food preparation, food preservation, needle work, and handicrafts (Emitiyagoda 2009).

Over the years the agricultural extension service expanded, and in 1963 a separate Division of Agricultural Extension was created in the DOA. The extension service, which previously concentrated on rice, was expanded to include a few other crops such as chillies, onions, potatoes, and vegetables. The noteworthy developments up to 1979 were the introduction of annual agricultural implementation program, appointment of Government Agents as Deputy Directors of Agriculture to coordinate the District Agricultural Committees, and the establishment of about 500 Agricultural Service Centres along with the outlets for other agri-support services.

In 1980, recognizing the need to improve the extension service, the training and visit (T&V) system developed by Benor and Harrison (1977) was introduced through the World Bank assisted Agricultural Extension and Adaptive Research Project. The T&V system was, however, modified to suit the local conditions. Under the T&V system, there was one village-level agricultural extension worker (KVS) for every 750 farm families. These extension workers received regular fortnightly training by Subject Matter Officers (SMOs), visited the contact farmers and the farmer groups regularly, and advised them regarding the adoption of new farming techniques. Although the T&V system had some positive results, after the termination of the World Bank project, the government was unwilling to sustain the required level of recurrent funding, which was prohibitively high.

In 1989, under the 13th amendment to the constitution, the extension function of the DOA and Department of Animal Production and Health (DAPH) was largely devolved to the provinces. The public sector agricultural extension had further setbacks in the same year. The total cadre of 2400 KVSs were transferred to the Ministry of Public Administration as *Grama Niladharis* (GNs), thus creating a vacuum at the field level. The transfer of these village extension workers to the Ministry of Public Administration as GNs cut off the front line of the agricultural extension service resulting in a virtual collapse of the service at the village level. The AIs, who operated hitherto at the supervisory level, were thus left as the extension contact with the farmers and had to cover nearly 2500–3500 farm families. Moreover, the GNs who were multipurpose officers engaged in more pressing regulatory duties could not be used for agricultural extension work as they did not have the necessary knowledge, attitude, and skills to carry out this work. This situation had even led 71% of the 240 vegetable farmers interviewed to report that there was no agricultural extension service in their area (Hettige and Senanayaka 1992).

In order to address some of the new extension challenges, viz. cost effectiveness, disruption in the technical line of command, absence of grassroot-level extension workers, and the need for a farming systems approach, an approach based on integration of the discrete extension efforts of four line agencies, namely DOA, DAPH, Department of Export Agriculture (DEA), and the Coconut Cultivation Board (CCB), was implemented from 1993 to 1998 with assistance from the World Bank under the Second Agricultural Extension Project (SAEP). This integrated agricultural extension service had shown only limited success in bringing the extension officers of the four participant line agencies to work together and establish an integrated extension system at the field level (Sivayoganathan and Kotagama 1999). Although there were some professional advantages of the four agencies being integrated, the main driving force keeping them together was the receipt of project resources. Consequently, the integration was only temporary and the integrated approach could not be sustained after the end of the project.

The individual and group extension methods were intensified in a special program called *Yaya* (block), implemented since 2000, to increase the national rice production. The AIs selected paddy tracts having 20–25 ha of land area with 15–25 farmers, with higher productivity potential, better irrigation, and free of soil and land ownership problems. The farmer groups were provided with a package of integrated crop management practices. The DOA acted as the mediator between relevant organizations such as Fertilizer Companies, Banks, and other private organizations to supply all the inputs needed for the cultivation and for signing forward contract agreements to assure better market facility.

At the same time, in 2002, the government took some steps to improve the extension setup in the field to a certain extent by appointing nearly 9500 field-level workers as Agricultural Research and Production Assistants who were expected to devote 2 days a week for extension work. Those officers were, however, incompetent to advise farmers. They were trained by the DOA to enhance their technical knowledge and develop the necessary technical and organizational skills. These officers were expected to play an important role by helping farmers to organize themselves effectively.

In 2004, as an Information Communication Technology (ICT) initiative, for the first time in Sri Lanka "Cyber Extension" was implemented by the Audio Visual Centre of the DOA as an appropriate information exchange mechanism, which seemed affordable and convenient to rural farmers in satisfying their information needs (Wijekoon and Rizwan 2009). The project established 45 Cyber Extension Units (CEUs) at 45 *Govijana Kendra* offices (Agrarian Service Centres) during the period 2004–2006. Interactive Multimedia-based digital extension strategies were used in these units. Continuous monitoring and evaluation of digital extension mechanism were done and improvements made. After considering the rapid development of e-governance situation in Sri Lanka, internet connections were provided to CEUs to enhance national agricultural research and extension system by improving the generation and collaborative use of agriculture knowledge and information system. Among the ICT initiatives for agriculture in Sri Lanka, establishment of National Agricultural Information and Communication Centre (NAIC) with the vision of achieving excellence in ICT in agriculture for national prosperity, crop forecasting information system (cropnet.lk), management information system for capturing real-time data on seed and planting material availability, phytosanitary certificate issuing system for DOA, mobile apps of the DOA, *Govi mithiru* program with Dialog Axiata PLC, toll-free agricultural advisory service (Hotline no. 1920) which has become popular among the farming community and general public, and Agriculture *Wikipedia,* a participatory and interactive web (Wijekoon and Sisira Kumara 2018), are noteworthy.

The *Yaya* program which was initiated in 2000 was further intensified as the Granary Area Program (GAP) in some locations with the involvement of other major stakeholders. Some of the innovative extension tools such as Cyber extension service, rice knowledge bank web site, media campaign, interactive multimedia CDs, crop clinics, picketing campaigns, and cultivation and yield competitions were also introduced and implemented to educate farmers. The outcome of these intensified extension efforts for 3 years was the increase of national rice yield up to 4.3 t/ha (Emitiyagoda 2009).

Encouraged by the success of the *Yaya* program, the second phase was launched under the title *Yaya 2* with the aim of increasing the national productivity further. The main interventions included environment friendly Good Agricultural Practices (GAP) such as judicious use of agrochemicals, use of Integrated Plant Nutrition System (IPNS), Integrated Pest Management (IPM), use of mechanized farming techniques preferred by the young farmers, and increasing cropping intensity with the introduction of crop diversification to enhance the overall farm profitability. Appropriate technology packages for the main rice-growing agro-ecological regions were proposed. The extension approach was designed to increase the frequency of contacts, use of ICT, and frequent monitoring of the seasonal activities with the help of relevant stakeholders.

Twenty-five District Directors were appointed in 2012 covering all districts and they were attached to the District Secretariats. Their main responsibilities were coordination of agricultural programs of the particular district, rendering technical assistance to the District Secretary who is the administrative head of the district, rendering technical leadership to the District Agricultural Committee, and facilitating the activities of all government, nongovernment, and farmer organizations with respect to agricultural development in the district. These officers also helped strengthen the linkage between the central government and the provincial system (Arunapriya et al. 2018). Instead of the KVSs who were removed in 1989, a new cadre of Technical Assistants (TAs) was created in the DOA in 2015 and diploma holders in agriculture were recruited for this cadre. Recruiting and attaching these TAs to the AI's division was another crucial turning point in the agricultural extension system in Sri Lanka. These TAs carried out their duties under the direct supervision of AIs of the respective divisions. At present there is no uniform agricultural extension system common to all inter-provincial and provincial areas in the country.

15.4 Major Plantation Crops

The Ministry of Plantation Industries is the main governing body of the institutions responsible for the development and promotion of the main plantation crops, namely tea, rubber, coconut, and sugarcane.

15.4.1 Tea

In early 1860s, James Taylor established the first commercial tea plantation in Sri Lanka at Loolecondera estate, Hewaheta in the Kandy district (Rajasinghe and Samansiri 2014). Export of tea contributes substantially to the foreign exchange earnings of the country. The Tea Research Institute (TRI), which was established in 1925 to conduct scientific research into the problems faced by the tea industry, continues to be the main organization dealing with developing and disseminating technologies of tea cultivation and processing to the tea industry. Until late 1950s, the scientists provided advisory services to the managers of the tea plantations by visiting them when they faced problems. In order to cater for the increasing demand for advisory services, a separate Advisory Division was established in the TRI in 1959. It was then transformed into an "advisory and extension" service. The services that were limited to on-call estate visits to solve field problems were expanded to facilitate wider dissemination of technical information. The functions of the advisory staff included routine advisory work, study group, field days, discussions, publications, training, etc. These services were further supplemented by the senior research staff of the TRI. The TRI liaises with the Tea Small Holdings Development Authority (TSHDA), which was established in 1977 to cater to the needs of the tea small holdings sector, in disseminating technological information to tea small holders.

In early 1960s, the TRI realizing the need for decentralizing the advisory services opened Advisory and Extension Units at the TRI regional stations in Passara (1963) for Uva, Ratnapura (1964) for low country, and Hantana (1966) for the mid-country tea-growing regions. With the expansion of the tea small holder sector in the low country areas, the TRI realized the need for more stations in these areas and opened regional stations at Kottawa in Galle district (1980), Deniyaya in Matara district (1984), and at Matugama in Kalutara district (2008) (Wanigasundera 2015).

With the establishment of the TSHDA, a concerted effort was made to promote the development of small holder tea in Sri Lanka. There are 157 Tea Inspectors/Extension Officers operating at the field level under 26 suboffices, coming under eight Assistant Regional Managers (Extension) and supervised by six Regional Managers (TSHDA 2013). The extension services of the TSHDA follow the commodity-based production-oriented approach dealing with only tea cultivation.

15.4.2 Rubber

Rubber was first introduced to Sri Lanka in 1876 by British colonialists. The Rubber Development Department (RDD) was established in 1994 by amalgamating the Rubber Control Department (established in 1930s for the regulatory functions of the rubber industry) with the Advisory Services Department (ASD) of the Rubber Research Institute (RRI). The general extension service for the rubber growers is mandated to the RDD, while the RRI also provides technical advisory services through its research divisions and the ASD. The RRI is vested with the statutory responsibility for research and development activities. In addition to performing extension activities, RDD is also responsible for issuing permits for rubber planting, subsidy disbursement, and providing planting materials to small holders. According to the cadre positions given by the RDD, 182 RDOs are expected to serve as field extension officers catering to the rubber small holders whereas the actual number in service is only 98.

The ASD of the RRI, which was amalgamated to form the RDD in 1994, was again re-formed with 35 extension officers reallocated from RDD. These officers are engaged in the advisory and training functions of the ASD. In addition, the respective research divisions of the RRI also provide technical services to the rubber plantations, plant nurseries, and processors. The RDD extension officer has to serve an average of 815 growers and the extension officer of ASD has to serve nearly 2980 growers. It appears that there is a lack of coordination among the actors who serve the rubber small holders (Wanigasundera 2015).

15.4.3 Coconut

The Coconut Research Institute (CRI) established in 1929 was responsible for the development of the coconut sector. The development of technologies through research in various disciplines and extension activities in the sector were the major activities of the institute (Appuhamy 2009). The Coconut Cultivation Board (CCB) was established in 1971 to provide subsidies to island-wide coconut growers and assist them to get the best benefits from them. The extension services of the CCB are implemented through 17 regional stations located in major coconut-growing areas. The Coconut Development Officers (CDOs) are the field-level extension officers operating from Agrarian Service Centres, who not only administer the subsidy scheme but also provide technical guidance and inputs to small- and medium-scale coconut growers. There are 17 Regional Managers and 162 CDOs covering the coconut-growing regions in the country. The average land area to be covered by a CDO is about 3000 ha and the average number of holdings to be served is 10,362 (Wanigasundera 2015). This clearly shows that the CDO is unable to attend to the needs of the coconut growers on an individual basis. The main extension methodologies used by the CCB include individual field visit by CDO, group discussions, field days, crop clinics, exhibitions, etc.

In addition to the technology transfer to the coconut plantations, the CRI conducts technical training programs to field extension officers of CCB and many other extension personnel from the government as well as NGOs and private sector organizations. Also, the extension training for farmers and in-service training for officers of various agencies are given by two Coconut Development Training Centres located at Lunuwila and Medamulana.

15.4.4 Sugarcane

Sugar is a basic food commodity, next in importance to rice and wheat. The extension service of the sugar sector was initiated with the establishment of the Sri Lanka Sugar Corporation under the State Corporation Act No. 37 of 1957 (Perera 2009). The extension activities of the sugar industry are primarily under private ownership based on commodity development approach. The field extension officers (Agricultural Assistants and Field Assistants) monitor the crops in their assigned zones, and provide both technical advice and inputs at the same time. The extension service is provided by the sugar companies and the farmers have to bear the cost of extension service indirectly as the present system of payment for cane by the company includes the cost of extension service also. However, the current extension approach could also be viewed as a pluralistic approach as part of the extension service is provided by the Sugarcane Research Institute (SRI), apart from the extension activities conducted by the private companies.

15.5 Export Agricultural Crops

The export agricultural crops (excluding the main plantation crops, namely, tea, rubber, and coconut) represent the export-oriented perennials including spices, beverage crops, essential-oil-producing crops and stimulants where more than half of the produce is exported. The Department of Export Agriculture (DEA) established in 1972 is the formal government body responsible for developing the export agricultural crop sector in the country (Seneviratne and Gunasinghe 2009). The extension service is operating in 14 districts where suitable climatic conditions exist for the cultivation of export agricultural crops and implemented by 14 district-level Assistant Directors, under whom about 150 field Extension Officers provide services to the growers. Another 195 graduates who were appointed to the department also undertake some development services to the growers.

The extension interventions practiced by the DEA until the recent past consisted of the commodity-specialized approach combined with general extension approach, and training and visit approach. The export crops assistance scheme was incorporated with these approaches from the inception of the DEA. The Unified Extension and Training System (UETS), which operated with FAO assistance, commenced in 1986. The objective of the UETS approach was to integrate the extension activities of the DEA with the DOA in order to reach more farmers as a solution to the shortage of extension workers in the DEA. From 1993 to 1998 the DEA participated in the integrated agricultural extension service which was implemented under the SAEP with assistance from the World Bank. The *Ama* program implemented by the Ministry of Agriculture in 1990s came into effect with the termination of SAEP, which was another integration strategy based on the collective efforts of 18 agencies under the ministry with the participation of the beneficiaries. The extension efforts of the DEA in the recent years include crop zoning, plant certification, integrated nutrient management, market-oriented extension, post-harvest advisory service, and creation of export agricultural promotional societies for better coordination with farmers.

15.6 Livestock Sector

Livestock is an integral part of agro-production system in the country. In Sri Lanka dairy, poultry, goat, and swine are the main components in the livestock sector. The public livestock extension is a decentralized service similar to DOA and is mainly implemented by 287 government veterinary offices established in all parts of the country under the nine Provincial DAPH. Each veterinary office is manned by one Veterinary Surgeon and two Livestock Development Instructors (LDIs) (Wickramasooriya et al. 2009).

Most of the conventional extension approaches such as general agricultural extension approach, project approach, farming systems development approach, commodity-specialized approach, cost sharing approach, and participatory approach have been adopted by the Sri Lankan livestock sector to transfer technology and train farmers. Furthermore, relatively new approaches such as farmer field schools and livestock production villages have also been implemented in the livestock sector in the recent past. However, compared to the other sectors in Sri Lankan agriculture, the livestock extension service shows slow progress in adopting modern ICT for technology dissemination to farming communities.

15.7 Outreach Extension Services by the Universities

Agriculture universities in India and elsewhere are engaged significantly in extension services for farming communities. In Sri Lanka, there are no such agricultural universities or services. However, the Faculties of Agriculture and allied faculties of the state universities conduct outreach programs targeting farmer training and creating awareness among school children on an ad hoc basis.

The Agricultural Education Unit (AEU) of the Faculty of Agriculture at University of Peradeniya, Sri Lanka, organizes various training programs to farmers and rural youth in collaboration with national and international organizations. Another noteworthy extension initiative could be found in University of Rhuna in the Southern Province of Sri Lanka. In this program called "Grow more mushroom to enhance food security," different agricultural extension efforts have been implemented to uplift the living standards of the mushroom farmers, while increasing the productivity. The students following the Advanced Module of the Department of Agricultural Extension of the Faculty of Agriculture at University of Peradeniya conduct a series of comprehensive productivity enhancement training programs annually with the guidance of experts for a selected community as a part of their training. The Institute for Agro-Technology and Rural Sciences of the University of Colombo located in remote location of Hambantota district (Southern Province) conducts courses for farmers from diploma level up to PhD, provides tissue-cultured banana plants, and conducts farmer training. Further, the Open University of Sri Lanka offers some courses to individuals who intend to choose agriculture as their future career and the courses also enhance the working capacity of those in the field by targeting high levels of productivity and sustainability. The courses are offered under two distinct streams of studies—industrial studies in agriculture and engineering technology.

15.8 Non-State Sector Extension Services

According to Mahaliyanaarachchi and Bandara (2006) the structural change in agricultural extension could take two approaches: commercialization and privatization. Privatization is mainly changing the ownership of the extension service to private sector from public sector, which has been mainly funded and delivered by government agencies free of charge for decades. People in most of the developing countries have unpleasant experiences of privatization. Commercialization is not merely privatization. It does not need a change of ownership under commercialization. Ownership can be kept with the government or semigovernment organization, but the service is provided on a commercial basis. The strategies of commercialization include decentralization, public cost recovery, contracting of services, etc.

The private sector involvement in the food crop sector extension formally began in 1998, when the Asian Development Bank supported launching a pilot study to promote private sector participation in agricultural extension. This pilot study was done under the Second Perennial Crops Development Project (PeCroDeP). This study was implemented from 1998 to 2007. In 2001, three agribusiness firms were selected covering specialized lines of business, viz. input supplies, output procurement, and crop advisory and farmer services. The fee-based activities focused on loan processing for the establishment of spice, fruit, and floricultural crops as well as farm advisory services. The fees charged from farmers were set below the actual cost with the remainder covered by the government through project funds.

The experience of the pilot study showed that there is potential in Sri Lanka for a gradual introduction of a fee-levying private extension service among commercial farmers with a better ability to pay, leaving scarce government resources to serve the poor and subsistence-level groups of farmers more effectively. However, to be profitable, private sector extension services must be integrated into other commercial operations, such as the sale of farm inputs.

Input linked commercial extension services were in operation from early days with the introduction of plantation agriculture related to tea, rubber, and coconut by the British. Several firms, which supplied agro-inputs and processing machinery, and those engaged in produce marketing, provided advisory services and recovered the cost of the service by including a marginal proportion of the income or a commission in their selling or buying.

Since 2000, there has been a rapid growth in non-state actors providing extension alongside other farm services. For these organizations, agricultural extension remains a function that contributes to realizing the larger goal of agri-business development or overall rural development. Two types of such entities could be identified: (i) those dealing with the supply of agricultural inputs and/or procuring of agricultural products and (ii) development agencies.

The CIC Agri-Businesses (Pvt.) Ltd. is the leading private sector organization providing agricultural advisory services in Sri Lanka. It is the leading supplier of certified quality paddy seed in the country through its own farms and a network of contract growers. In 2009, the organization had 110 extension staff that worked directly with over 20,000 farmers who produced a variety of agricultural and livestock products like seed paddy, rice, fruits, vegetables, eggs, and yoghurt (De Zoysa 2014). The A. Baur and Company (Pvt.) Ltd., Hayleys Agriculture Holdings Ltd., Brown & Company PLC., Lankem Ceylon PLC, and Unipower (Pvt.) Ltd. are some of the other private sector organizations providing extension support to the farming community in Sri Lanka.

Furthermore, some international development agencies such as Food and Agriculture Organization (FAO), Canadian International Development Agency (CIDA), Japanese International Co-operation Agency (JICA), and German International Cooperation Agency (GIZ) as well as local, national, and international nongovernment organizations have also been engaged in agricultural extension activities in implementing their development project activities.

15.9 Present Challenges and Future Directions

15.9.1 National Agricultural Extension Policy

The National Agricultural Extension Committee (NAEC), which was constituted in 1992 as a Standing Committee under the Sri Lanka Council for Agricultural Research Policy (SLCARP) under sect. 19(4) of the SLCARP Act No. 47 of 1987, was entrusted with the responsibility to develop a policy for agricultural extension service. Committee membership included representatives from the government departments, universities, and the private sector organizations engaged in the provision of agricultural extension service. The role of the committee as stated was to advise CARP on policy matters regarding, (a) agricultural extension and training, (b) organization, coordination, planning, execution, and funding of agricultural extension services, extension pilot programs, and training, and (c) research-extension-farmer linkages. Although the committee met a few times between 1992 and 1995 and discussed some policy issues related to agricultural extension, the support envisaged was not forthcoming from the authorities and it could not serve the purpose for which it was constituted.

Attaching the NAEC as an appendage to SLCARP did not seem to be a satisfactory arrangement for the efficient functioning of the former. A separate body could be established with the necessary authority to handle policy issues regarding agricultural extension. A somewhat better arrangement suggested, but not implemented, was to change the name of SLCARP to SLCAREP (Sri Lanka Council for Agricultural Research and Extension Policy) and to have two committees on equal footing with the necessary authority and support to allocate resources and handle policy issues regarding agricultural research and agricultural extension. Without a comprehensive policy, agricultural extension is unlikely to get the priority and attention it deserves. Formulation and implementation of an extension policy should, therefore, be the first priority for improving extension delivery in Sri Lanka (Sivayoganathan 2014).

15.9.2 Recognition of Extension Service

Agricultural extension dealing with transfer of technology is a neglected component in the agricultural knowledge and information system, which is composed of research, extension, and the user. The role of extension should not be viewed as mere transfer of technological information to farmers but as encompassing partnerships, cooperation, and interdependence as its integral components resulting in improved interactions among the actors in the agricultural knowledge and information system. Agricultural extension should be duly recognized as a profession. In this context, the Sri Lanka Agricultural Extension Association (SLAEA), a professional body of agricultural extension workers, academics, and managers of extension services has a major role to play.

The mindset that "anybody can do extension" should be changed. The decision makers should be convinced by showing the real social and economic impacts of extension scientifically by using a comprehensive set of indicators in the monitoring and evaluation of extension projects. Of rather inadequate budgetary allocation for agricultural extension in the state institutions, nearly 85% of the recurrent budget is accounted for staff salaries and allowances. However, research conducted on extension system worldwide indicates that an extension system needs to allocate about 30% of its recurrent budget to program and operational costs to be effective in carrying out its functions and responsibilities (World Bank 2007). The budgetary allocation for extension should be increased and the field extension workers should be motivated by giving the necessary incentives such as greater opportunities for career advancement and program support to perform their duties effectively.

15.9.3 Research-Extension and Other Linkages

Agricultural research should be well focused on the needs of farmers and consumers, rather than being governed by the availability of research facilities and funds. Also, the outcome of existing research should be satisfactorily communicated to the extension service. The present mechanism for research–extension linkage in the agricultural sector is rather weak and should, therefore, be strengthened.

Further, there is very limited coordination among the state, private sector, and nongovernment organizations in the provision of agricultural extension services. Similarly, the coordination between the extension service organizations and input suppliers as well as the central and provincial extension services is far from satisfactory. It is, therefore, necessary to have a national-level dialogue with wide participation of all stakeholders engaged in the provision of agri-support services to farmers and develop policy guidelines to integrate agricultural extension as an effective partnership model. The state controlled largely top-down extension should be changed to a collaborative service where the state, the private sector, the nongovernment organizations, and the international development agencies work together bringing more benefits to the rural farming community.

15.9.4 Human Resources: Availability, Deployment, and Capacity Building

Agricultural extension work, to be effective, should be undertaken by adequate number of qualified, trained, and competent extension professionals. At present, there is a shortage of extension manpower especially at the field level. This is further aggravated by the improper deployment of the available extension workers by the different agencies and the lack of coordination among these workers. Suitable persons should be recruited without any further delay to fill the existing vacancies in the planned cadres especially in the provincial extension service.

The extension workers require not only technical skills in agriculture but also the necessary skills in dealing with farmers and other stakeholders. They have to undergo preservice training, induction or orientation training, and various in-service training programs. Although presently various in-service training programs are being conducted, it should be further intensified and better training facilities and support provided. Also, the profession of extension should be recognized in par with research and equal opportunities should be given to extension staff to pursue higher studies.

15.9.5 Use of Diversified Communication Media

The conventional face-to-face individual extension methods cannot be continued due to the shortage of field extension workers and the higher costs involved except for cases where individual attention is absolutely necessary. The extension service should, therefore, rely on greater use of mass media and group extension methods such as farmer training classes and block demonstrations. Furthermore, strengthening the ICT-based interventions will help effectively link the farming communities with other stakeholders in the value chain. Thus, every effort should be made to popularize the already developed ICT-based innovative applications among the farming communities.

15.9.6 Research in Extension

Agricultural extension is a science—a complex social science. Findings of agricultural extension research are needed to understand the problems of technology transfer, research–extension gap, farmer attitudes towards and adoption of agricultural innovations, and other such issues. Gains and losses of farmers are not necessarily the same as perceived by the farmers themselves, the crop scientists, extension officers, and policy makers. These issues are to be examined through extension research. Crop scientists, extension officers, and the farming community at large can benefit from the findings of such research. However, extension research conducted in Sri Lanka leaves much to be desired. It is largely restricted to the research studies conducted by the university staff and students both at undergraduate and postgraduate levels and some evaluation studies conducted by selected state institutions, international development agencies and nongovernment organizations. This kind of research should not be confined to the above organizations. The agricultural research institutes should have provision to conduct both crop and extension research. At the same time, a research division could be established in the extension organization to conduct extension research. Although extension research is rather difficult, the indicators are now developed to evaluate the extension service by examining both its effectiveness and efficiency. The necessary support and recognition should be given by the state and other funding agencies for such research.

References

Appuhamy PAHN (2009) Experience and challenges in technology transfer in the coconut sector of Sri Lanka. In: Sivayoganathan C (ed) Proceedings of the agricultural extension conference on experiences and challenges in agricultural extension: meeting farmer needs. Sri Lanka Agricultural Extension Association, Gannoruwa, pp 213–225

Arunapriya SA, Sudasinghe JR, Weerakoon WRWMSNP, Wijayathunga RSCWMABM (2018) Historical evolution of agricultural extension system in Sri Lanka. In: Jayawardena LNAC, Suryamani M, Sivayoganathan C (eds) Proceedings of the International Conference on Transforming Agricultural Extension Systems: Towards Achieving Relevant Sustainable Development Goals for Global Impact. PRDIS and Sri Lanka Agricultural Extension Association, Kandy, pp 1–6

Benor D, Harrison JQ (1977) Agricultural extension: the training and visit system. World Bank, Washington, DC

De Zoysa S (2014) Private sector involvement in agricultural extension in Sri Lanka. In: Wijeratne M, Mahaliyanaarachchi RP (eds) Agricultural Extension: Recent Interventions for Development in Sri Lanka. Sri Lanka Agricultural Extension Association, Peradeniya, pp 87–98

Emitiyagoda SS (2009) Extension services with special reference to the functions of the Department of Agriculture, Sri Lanka: historical development, current position and future directions. In: Sivayoganathan C (ed) Proceedings of the Agricultural Extension Conference on Experiences and Challenges in Agricultural Extension: Meeting Farmer Needs. Sri Lanka Agricultural Extension Association, Gannoruwa, pp 77–97

Hettige ST, Senanayaka SMP (1992) Highland vegetable production and marketing systems. Agrarian Research and Training Institute, Colombo

Karim Z (1999) Agro-technology transfer: the quest for a better design of systems. In: Proceedings of the Symposium on Agricultural Extension Services in the 21st Century: Needs and Challenges. Department of Agricultural Extension, Bangladesh Agricultural University, Mymensingh, Bangladesh, pp 7–19

Kashem MA, Halim A (1999) Foreword. In: *Proceedings of the Symposium on Agricultural Extension Services in the 21st Century: Needs and Challenges*. Department of Agricultural Extension, Bangladesh Agricultural University, Mymensingh, Bangladesh

Mahaliyanaarachchi RP, Bandara RMAS (2006) Commercialization of agriculture and the role of agricultural extension. Sabaragamuwa University Journal 6(1):13–22

Mosher AT (1978) An introduction to agricultural extension. Agricultural Development Council, New York

Perera MS (2009) Extension service of the Sri Lankan sugar industry. In: Sivayoganathan C (ed) Proceedings of the Agricultural Extension Conference on Experiences and Challenges in Agricultural Extension: Meeting Farmer Needs. Sri Lanka Agricultural Extension Association, Gannoruwa, pp 235–251

Rajasinghe JCK, Samansiri BAD (2014) Evolvement of agricultural extension in tea research institute and present applications. In: Wijeratne M, Mahaliyanaarachchi RP (eds) *Agricultural extension: recent interventions for development in Sri* Lanka. Sri Lanka Agricultural Extension Association, Peradeniya, pp 111–128

Seneviratne MAPK, Gunasinghe UM (2009) Extension approaches adopted in the export agricultural sector in Sri Lanka. In: Sivayoganathan C (ed) Proceedings of the Agricultural Extension Conference on Experiences and Challenges in Agricultural Extension: Meeting Farmer Needs. Sri Lanka Agricultural Extension Association, Gannoruwa, pp 122–138

Sivayoganathan C (2014) Face to Face, Agricultural Extension in South Asia (AESA)

Sivayoganathan C, Kotagama HB (1999) Agricultural extension at crossroads in Sri Lanka. In: *Proceedings of the Symposium on Agricultural Extension Services in the 21st Century: Needs and Challenges*. Department of Agricultural Extension, Bangladesh Agricultural University, Mymensingh, Bangladesh, pp 59–64

Sivayoganathan C & Wirasinghe S (1992) Agricultural extension in the food crop sector of Sri Lanka: past, present and future. Jubilee year publication, Faculty of Agriculture and Postgraduate Institute of Agriculture, University of Peradeniya, Sri Lanka

Swanson BE (1984) Agricultural extension: a reference manual. FAO, Rome

TSHDA (2013) Annual report of the tea small holdings development authority (2012)

Van den Ban AW, Hawkins HS (1985) Agricultural extension. John Wiley and Sons, New York

Wanigasundera WADP (2015) Status of extension and advisory services in Sri Lanka. Working Paper No. 1, Agricultural Extension in South Asia (AESA)

Wickramasooriya BC, Kendaragama KMT, Wijayathilaka TP (2009) Approaches adopted in rural transformation related to livestock sector. In: Sivayoganathan C (ed) *Proceedings of the Agricultural Extension Conference on Experiences and Challenges in Agricultural Extension: Meeting Farmer Needs*, 27th to 28th August. Sri Lanka Agricultural Extension Association, Gannoruwa, pp 98–109

Wijekoon R, Rizwan MFM (2009) ICT initiative or agricultural extension of the Department of Agriculture. In: Sivayoganathan C (ed) Proceedings of the Agricultural Extension Conference on Experiences and Challenges in Agricultural Extension: Meeting Farmer Needs, 27th to 28th August. Sri Lanka Agricultural Extension Association, Gannoruwa, pp 297–329

Wijekoon R, Sisira Kumara G (2018) An overview of education and learning to meet the SDGs with special reference to recent innovations of the agriculture sector in Sri Lånka. In: Jayawardena LNAC, Suryamani M, Sivayoganathan C (eds) Proceedings of the International Conference on Transforming Agricultural Extension Systems: Towards Achieving Relevant Sustainable Development Goals for Global Impact, 10th to 11th May. PRDIS and Sri Lanka Agricultural Extension Association, Kandy, pp 20–26

World Bank (2007) Reviving Sri Lanka's agricultural research and extension system: towards more innovation and market orientation. World Bank, Colombo

Performance and Potential of Agricultural Insurance: Global and Sri Lankan Perspectives

16

D. V. Pahan Prasada

Abstract

Agriculture insurance is still taking root, though not new, to mainstream agriculture in Sri Lanka. With increasing market-based interventions in agriculture such as transformation of traditional supply chains to agribusinesses using partnership ventures, forward contract agreements, movement away from input subsidies to cash transfers, the importance and demand for insurance tools have surfaced. This chapter discusses the evolving need for agricultural insurance and addresses implications of concurrent regional and global developments in agricultural insurance on local farming. Performance of public sector and private sector agricultural insurance schemes are discussed. The financial implications of implementing agricultural insurance are illustrated mainly with respect to the paddy insurance scheme. We find that the feasibility of the insurance scheme has been marginally better in *Yala* season overall as seen in lower loss ratios. The paddy insurance scheme has failed to differentiate the variation of risk by the season and assessed damages and compensations following an alternative logic.

Keywords

Agriculture Insurance · Private sector · Public sector · Crop · Livestock

16.1 Risk and Insurance in Agriculture

Agriculture is fraught with risk at multiple points of the value chain. The risks at production are numerous and well known. Processing, storage, and retailing stages are vulnerable to both natural hazards and market threats. The ability to mitigate

D. V. P. Prasada (✉)
Department of Agricultural Economics and Business Management, Faculty of Agriculture, University of Peradeniya, Peradeniya, Sri Lanka

B. Marambe et al. (eds.), *Agricultural Research for Sustainable Food Systems in Sri Lanka*, https://doi.org/10.1007/978-981-15-2152-2_16

risks and to smoothen the shock of the realized risks is the key to conducting agriculture as a business venture.

Management of risk through different strategies is an area that has been researched and understood to a greater degree both in high-income farming and low-income settings. The coping strategies of the farmers as individuals, households, and communities are contextual and numerous. Such strategies vary widely in terms of their effectiveness. Still, the exposure to risk even at the subsistence to low-income smallholder level (setting aside the agribusiness operator's case) is too large to ignore. In a classical analysis, Rosenzweig and Binswanger (1993) found that smaller and poorer farmers in a semiarid region in India sacrificed 27% of their expected income to reduce risk. Further evidence on the limitations of traditional risk management tools can be witnessed in the long-standing poverty traps in smallholder agriculture. While poverty traps are not created entirely due to risk in farming, they are indirectly perpetuated by the inadequacy of traditional risk mitigation methods.

The other side of unmitigated risk in agriculture falls on humanity at large through the shocks to food supply. As population continues to grow in many vulnerable areas, the social and economic cost of a malfunctioning and inefficient food supply chain could expose economies and nations to other types of risks such as political and environmental (culminating in catastrophic outcomes). The planetary level shocks will also continue to deteriorate the conditions further since climate change is predicted to increase the frequency and severity of many weather-related disasters (World Bank 2011).

Integration into global agri-food markets has made the capacity of traditional risk-management strategies in many agricultural settings redundant. To protect their farmers from market forces originating from international prices and the demand uncertainties of the globalized food markets, many low-income country governments have implemented trade protections, price supports, and other protective measures which provide a limited cushion to external risks. However, such external pressures along with the simplification of landscapes (i.e., mono-cropping) due to agricultural modernization of agriculture systems have made farmers more vulnerable to economic stress and environmental hazards (Clapp 2012).

Given the inadequacy of traditional tools and clear evidence of gaps, one wonders as to why insurance is not adopted in agricultural sector in parallel to its extensive use in other areas of economics activity. Insurance is a market-based instrument with important economic and behavioral properties necessary to a modernized agricultural sector. For long years, the lack of adoption of insurance, especially in smallholder agriculture, has been discussed as a concern which needs attention form policy and business community alike. To understand the role insurance can potentially play in smallholder agriculture, it is crucial to consider insurance in broader context of financial management in smallholder agriculture. The insurance markets are not functioning in isolation in agriculture. The role of insurance is closely linked to the function of credit. Farm households, especially smallholders, do not make adequate investments due to difficult to access credit from formal institutions (i.e., banks). This in turn limits the ability of the smallholder to link to other financial tools such as insurance. Those who do have access to insurance are unwilling to seek credit because the collateral requirements would expose them to further risk. The end result of underdeveloped insurance markets is a vicious cycle of under-investment, poor productivity, and rural poverty (Boucher et al. 2008).

The argument for creating a conducive framework for insurance is multifaceted. In the absence of moral hazard (the insured farmer behaves more riskily due to the availability of insurance) and adverse selection (more risky farmers seek insurance coverage while less risk-prone farmers ignore the coverage), insurance schemes provide a useful economic function. Insurance is based on the concept of risk pooling and acts as a market-based social safety net. The fiscal burden of publically funded social safety nets and public relief programs make public insurance a nonviable option for most low-income governments. An added complication with universal public insurance programs is the free-riding problem and disincentive to innovation.

Among the indirect benefits of agriculture insurance to the system at large, insurance programs help transform agriculture by facilitating the adoption of improved technology by farmers, as risks attendant on advanced production methods and higher investments are shared. Insurance is also a protection to the financial institutions that give credit to farmers, and should therefore lead to reduction in the interest rates charged (UNCTAD 1994).

16.2 Penetration of Insurance in Agriculture

History of risk mitigation instruments in agriculture is limited by the presence of documentation. Earliest recorded incidences relate to Western Europe. Crop-hail insurance was offered in Germany as early as the 1700s and livestock insurance was offered in Germany, Sweden, and Switzerland by 1900. Early implementations were single peril type products. Federal crop insurance in the USA was first authorized in Agricultural Adjustment Act of 1938. The Federal Crop Insurance Act of 1980 (USA) made crop insurance the primary form of catastrophic protection available for farmers (Glauber et al. 2002). Japan implemented a multiple peril crop insurance program in the early twentieth century that provided nationwide coverage for paddy rice, wheat, barley, and mulberries (Yamauchi 1986). A number of public sector multi-peril crop insurance (MPCI) schemes were established in Latin America (for example, Brazil, Costa Rica, and Mexico) and Asia (India, the Philippines), often linked to seasonal production credit programs for small farmers (Kerer 2013). To date the highest penetration rates in user-subscribed agricultural insurance are reported in European countries (Fig. 16.1).

Over 100 countries in the world have some form of documented agricultural insurance. A World Bank assessment (Mahul and Stutley 2010) regarding the extent of usage of agricultural insurance around the world from 2007 estimated that 104 countries had some form of agricultural insurance in place that year. However, 88% of the value of premiums was collected in high-income countries (mostly in North America and Europe), while lower-middle-income and low-income countries

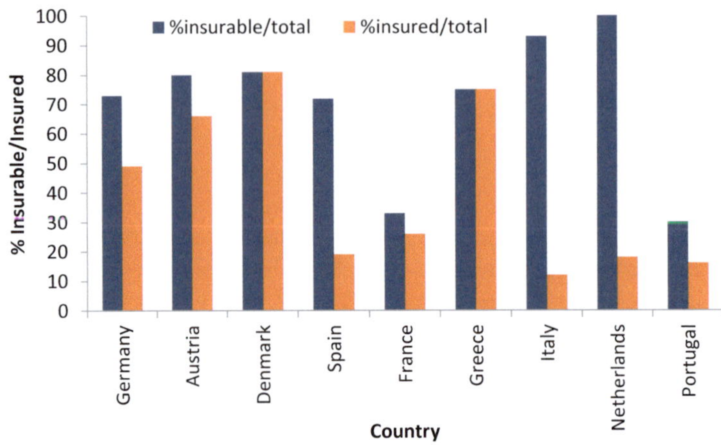

Fig. 16.1 The level of insurance penetration in EU agriculture (based on Eurostat data)

accounted for only 7.5%. This statistic refers to realized premiums and includes the subsidy element which is widespread regardless of the income status of the country. In terms of premium value by each country, the USA, Canada, Mexico, and Spain dominate the world crop insurance market. In terms of premium value to farm gate value, there is wide variation across different regions of the world: North America (69%), Western Europe (21%), Latin America (5%), Asia (3%), Australia (1%), and Africa (1%) (Roberts 2005).

In the low- and middle-income countries, improved forms of agricultural insurance have not yet achieved sufficient scales, except in Brazil, China, India, and Mexico. In the other low-income countries, the evolution of agricultural insurance instruments follows sector-wide initiatives (such as public sector restructuring) and has remained within the public sector orbit in most instances. The absence of private sector in agricultural insurance is widely observable in low-income countries (Binswanger and Deininger 1999).

16.2.1 Evolution of Agricultural Insurance Instruments in Sri Lanka

In Sri Lanka, the first experimental crop insurance scheme was established in 1958 as a pilot project covering rice cultivation only. The scheme was legislated under the Crop Insurance Act No. 13 of 1961. The scheme covered 26,000 acres (approx. 10,522 ha) of paddy in five districts. The experience during the first 15-year period was quite favorable. By 1973, 16% of the total area cultivated with paddy was insured by this scheme in both seasons (Sandaratne 1974).

Marking the second stage of agricultural insurance, the crop insurance board was established in 1973 under the Parliamentary Act No. 27 of 1973 (Agricultural Insurance Law No. 27 of 1973) to operate a comprehensive agricultural crop insurance scheme, covering all major crops and livestock. In case of rice and other crops, insurance protection was provided against lack of water, drought, excessive water, floods, diseases, insect infestation, and damage by wild animals and losses due to nonadherence to approved methods of farming (i.e., Department of Agriculture recommendations). This scheme was partially subsidized with the administration costs being borne by the state. Other objectives were to stabilize farm incomes, thereby promoting agricultural production, and also to undertake research for the promotion and development of Agricultural Insurance. Other crops like green gram, cowpea, chili, soya bean, and even livestock, especially cattle and poultry, came under the program by 1985. The total area under insurance cover increased to 200,000 acres (approx. 80,937 ha). A large percentage (85%) of the total acreage insured is paddy and other crops that received agricultural credit. The institutional framework was further expanded by the Agricultural and Agrarian Insurance Board (AAIB) Act No. 20 of 1999 (Agricultural and Agrarian Insurance Act No. 20 of 1999) which operated under the objective of running the scheme on a self-financed basis (AAIB annual reports/AAIB reports to the Parliament of Sri Lanka, various years).

With the revision of the act in 1999, the scope of the Insurance Board was widened extensively. The legislative expansion included "insurance for agricultural and horticultural crops and medicinal plants, livestock, fisheries and forestry, agricultural equipment and implements, the storage and preservation of agricultural and horticultural produce and produces of medicinal plants and fisheries and forest products." Provision of medicinal benefits and social security schemes was also mandated. Thus, the newly formed AAIB was effectively taking on the scope from other statutes such as Farmers Pension and Social Security Benefit Scheme Act No. 12 of 1987 and the Fisherman's Pension and Social Security Benefit Scheme Act No. 23 of 1990. While the scope was consolidated, the 1999 modification paved way for a more liberal insurance sector by opening the sector to private insurers. The export crop insurance scheme was implemented by the AAIB in collaboration with the Department of Export Agriculture in 2002. Pepper, coffee, cocoa, cinnamon, and areca nut are the main export crops insured by the AAIB.

Although paddy insurance is compulsory under sect. 11 of the Agricultural Insurance Law, insurance subscription remained effectively voluntary and participation by farmers has ranged around 10%. Further changes followed suit. Area under crop insurance was only about 1% of the total paddy lands in 2005 as reported by the (AAIB, various years). The direct focus of the Amendment to the act in 1999 Agricultural Insurance Act was making provisions for the private sector involvement in crop insurance. The results, however, have not been satisfactory. By 2005, only the Ceylinco Insurance Company Limited (CICL) has entered the field. To date, there had been no new entrants. However, the agricultural credit services by private actors have expanded. Most credit services have quasi insurance elements embedded. Another development that has taken root extensively is forward contract agreements initiated by agribusinesses. Both credit and contract agreements have reduced the need for explicit private insurance programs for farmers. The Agribusiness companies are insured for their projects through corporate/business insurance.

A notable addition to private providers is SANASA insurance, which is a subsidiary of cooperative and development finance network spread across the country. In 2015, SANASA unveiled an index-based insurance under a development grant from the International Finance Corporation (IFC) and the World Bank. The cover was extended to paddy and tea under a pilot program.

16.2.2 Paddy Insurance Scheme

Paddy insurance covers total or partial loss of yields due to flood, drought, plant diseases, pest damages, and damages by wild animals. Period of insurance cover spans from planting up to the date of harvest of the crop. Coverage and premium vary depending on the risk level and the land class (as under major irrigation, minor irrigation, and rainfed). Three tiers of risk exposure are classified and premium rate of low risk is set to 5% of the coverage, medium risk as 7.5% of the coverage, and high risk as 10% of the coverage. Calculation of indemnity was based on cost of cultivation under each land class. Under the major irrigation schemes, insurance coverage ranged between LKR 10,000 and LKR 15,000 per acre[1] (approx. USD 140 to 210 per ha) and premium rates varied between LKR 500 and LKR 750 per acre (approx. USD 7–10.5 per ha; AAIB, various years).

16.2.3 Compulsory Insurance Tied to Fertilizer Subsidy (i.e., Kethata Aruna Pohora Diriya Insurance)

This scheme was established for farmers who receive subsidized fertilizer. Damages caused to cultivations due to droughts, flood, and wild elephants were covered. Farmers contribute at the time of purchasing fertilizer during each cultivation season. The insurance premium of LKR 450 (approx. USD 2.6) was charged at the time of purchasing fertilizer (for every 50 kg) at the Agrarian Services Centers (ASCs). The coverage in case of total loss is set to a maximum of LKR 25000 (approx. USD 142) per ha for damages due to floods, drought, and wild elephants (Table 16.1). However, with the introduction of fertilizer cash subsidy in 2016, the collection of insurance premiums was suspended. Despite no premium collection from the farmers, government continues LKR 10,000 per acre (approx. USD 140 per ha) indemnity for paddy cultivation using a crop levy fund.

16.2.4 "Kethata Aruna" Insurance Scheme of 2017

Initiated in the parliamentary budget of 2017, this insurance scheme covers a minimum of LKR 40,000 per acre (approx. USD 560 per ha) for paddy and five other field crops, namely, maize, soya bean, big onion, potato, and chili. The premium is

[1] 1 USD = approx. LKR 178

Table 16.1 Scheme of compensation under the '*kethata aruna*' program until 2017 (Source: AAIB website)

	Damages in first 30 days (LKR)	Damages from 30 days to flowering (LKR)	Damages from Flowering to harvest (LKR)
Acre	4000.00	6000.00	10,000.00
Ha	10,000.00	15,000.00	25,000.00

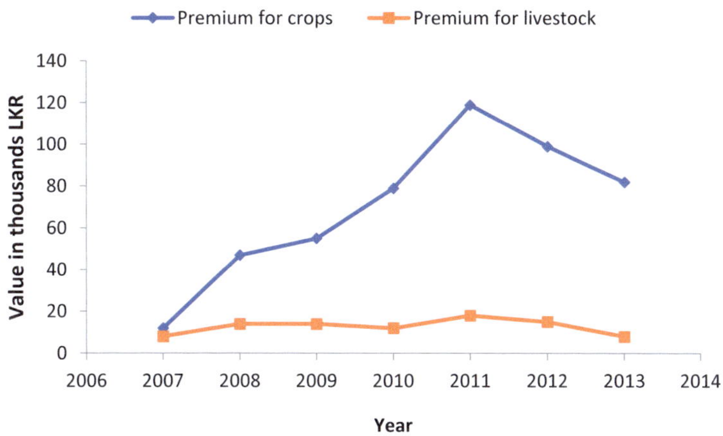

Fig. 16.2 Recent trends in premium collection (in thousands LKR) in crop vs. livestock insurance (Source: AAIB data). Note: 1 USD = approx. LKR 178

set to LKR 675 (with 80% subsidy) per acre (approx. USD 9.5 per ha) at the start and planned to be increased over a 5-year period to reach a 50% subsidized rate. The expectation is to progressively remove the subsidy while increasing the compensation. There is no published evidence of the implementation and the performance of the new scheme yet.

16.2.5 Livestock Insurance

In the dairy sector, the introduction of imported foreign breeds, particularly the European breeds such as Friesian, Jersey, Brown Swiss, and Ayrshire, for breeding purposes was coincidental to the intensification of the sector in Sri Lanka and other low-income countries. For example, the livestock insurance scheme in Sri Lanka commenced in 1975 to provide insurance for imported high-value animals and insurance penetration has remained stable (Fig. 16.2). Based on an estimate of life expectancy and incidence of disease, the insurable age of the animal is determined for the purposes of premium setting. Sri Lankan guidelines for insurable age are as follows: dairy cows 2–12 years; dairy buffalo 4–14 years; stud bulls 2–8 years; stud buffalo 3–8 years; draught bulls 3–8 years; and draught and ploughing buffalo 3–10 years

(AAIB, various years). Permanent total disability is covered with an indemnity of up to 90% of the cover. The value of the cover was set initially to LKR 10,000 per animal for dairy cattle and buffalo and stud bulls and buffalo, LKR 6000 for draught bulls, and LKR 4000 for draught and plough buffalo. For fully financed animals, insurance is set equal to the loan. All animals financed were to be insured. The premium rate is 4% for coverage up to LKR 8000 and 4.5% for LKR 10,000. Group discounts of 10%, 15%, 20%, and 30% are allowed for 2–5, 6–10, 10–50, and more than 51 animals. A no-claim refund of 25% of the total premium is allowed on 5-year basis. Approximately 25% of the total animals insured are non-financed property.

16.3 Financial Performance of Agricultural Insurance in Sri Lanka

16.3.1 Financial Performance of Crop Insurance

The financing of the crop insurance scheme was a burden on the treasury throughout. The difficulty emerged pointedly at the conversion of fertilizer subsidy to a cash transfer. Concurrently, the then government introduced a third party funded insurance fund by legislating a Crop Insurance Levy as 1% of the profit after tax as applicable to all institutions under Banking Act No. 30 of 1988; Finance Companies Act No.78 of 1988; or Regulation of Insurance Industry Act No. 43 of 2000.

Agricultural Loan Protection Insurance Scheme was established to benefit the Financial Institutions that contribute to the crop insurance levy (in order to help manage the risks incurred in lending for paddy cultivation). Under this program, claims are paid for LKR 10,000 per acre if the damage to crop takes places close to harvesting; the amount payable will be 40% (LKR 4000) if damage occurs within 30 days from the date of sowing and 60% (LKR 6000) if damage occurs thereafter up to blooming. The Financial Institutions should apply claims through relevant divisional secretariat with proper assessment.

The statistics of compulsory paddy insurance display a declining trend of both premium value and indemnity value over a decade starting from mid-1990s. While it is clear that insurance uptake is falling, the financial performance of insurance is stable. There is a modality across *Yala* and *Maha* seasons. While the total premium collected has tallied with the total indemnities paid, there is a strong divergence between the total claims and indemnities. Observing the data, it is clear that AAIB has determined indemnities mainly based on the premium income and the correlation to total claims appears to be weak (Fig. 16.3). Figure 16.4 disaggregates the insurance performance (measured in loss ratios) by season and the type of insurer.

Indemnity as percentage of insured value is an indicator which can reveal the insurance performance both to the insurer and insured. There is no design through which this ratio can be maintained as a constant. Performance of the indicator over the years for the AAIB insurance reveals that the government insurance is not necessarily infeasible as a financial instrument. Figure 16.5 shows a stable pattern of standardized premiums and percentage indemnified.

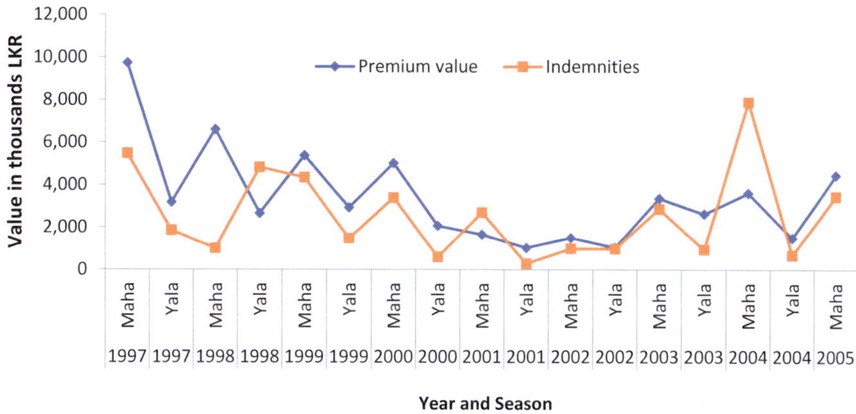

Fig. 16.3 Premium and indemnities on paddy insurance (Source: AAIB data)

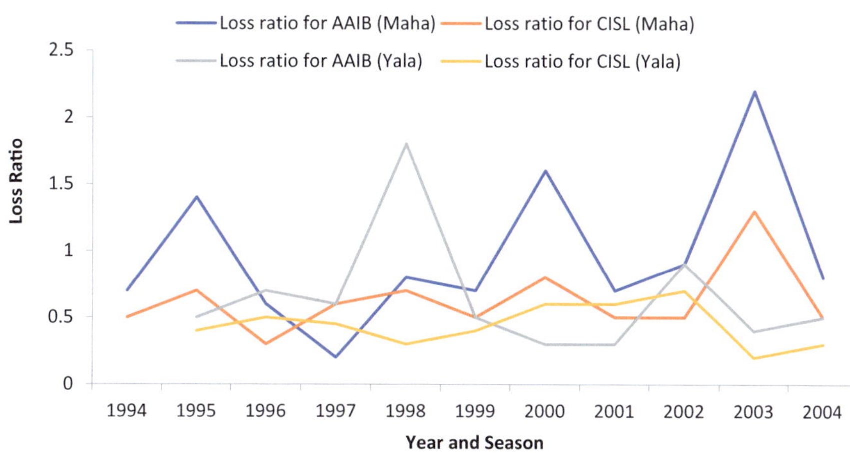

Fig. 16.4 Private and public loss ratios by cropping seasons *Yala* and *Maha*. *Note:* loss ratio is the ratio of total losses incurred in claims divided by the total premiums earned; AAIB: Agricultural and Agrarian Insurance Board; CISL: Ceylinco Insurance and Securities (Pvt.) Ltd.

Involvement of private actors in insurance was facilitated by the Finance Companies act of 1988 and subsequent reforms in the AAIB Act. Ceylinco Insurance and Securities Ltd. (CISL) was the leading private actor to break into agricultural insurance. Their scheme targeted mainly paddy farmers and performed profitably compared to the government equivalent (Fig. 16.6[2]). However, there were widespread complaints on irregularities on how private insurers acted

[2] A key indicator of insurance programs is the Loss ratio. It refers to ratio between the sum of total indemnities and costs of insurance and the total premium collected.

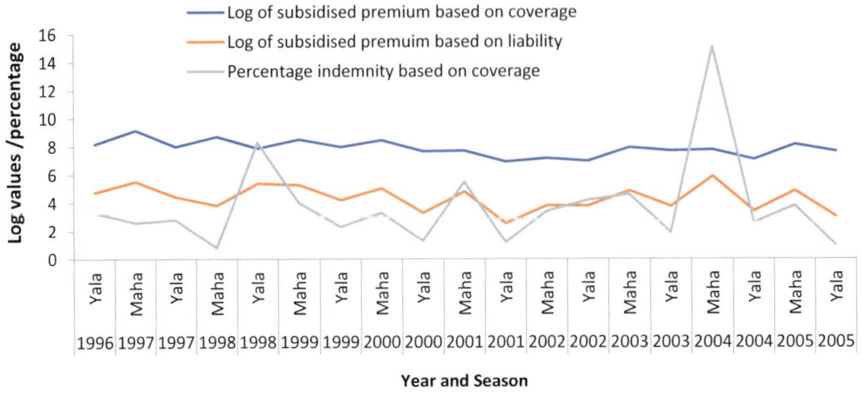

Fig. 16.5 Variation of insurance indicators in terms of ratios over the years (Source data: AAIB)

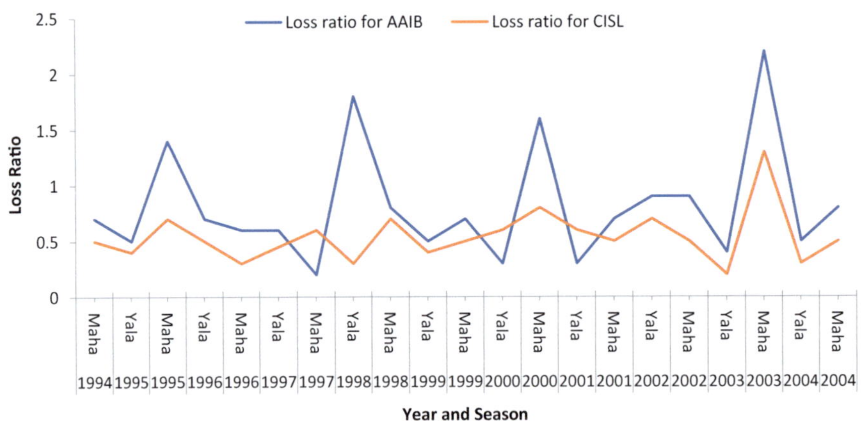

Fig. 16.6 Public vs. private company performance (source: AAIB data) Note: CISL is the private insurance provider

in assessing damages and paying compensation. However, over the years, the involvement of private sector in crop insurance has dwindled in the portfolio of the above company.

16.3.2 Financial Performance of Livestock Insurance

Livestock insurance has performed better than the crop insurance from a financial perspective. The average loss ratio is 0.44 in Sri Lanka (over the period 1996–2007) indicating that premiums collected have surpassed the indemnified liability (Fig. 16.7). This is a unique condition indicating that livestock production indeed has lower risks than could be actuarially possible. Alternatively, the premiums may be artificially set at a high value or the indemnity listing of damages may be undervalued.

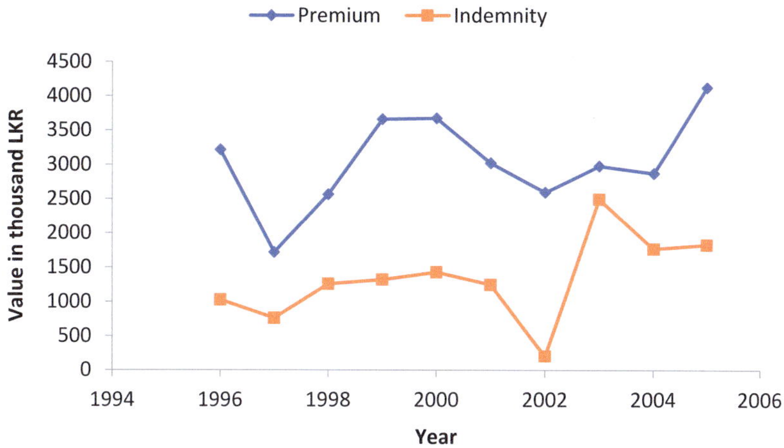

Fig. 16.7 Livestock insurance premium and indemnity (Source of data: AAIB)

16.4 Seasonality and Insurance Performance: Evidence from Paddy Insurance

The differences in cropping patterns from *Yala* to *Maha* seasons pose a unique problem to insurance programs. Since the cropping decisions vary by rainfall expectation and area cultivated is structurally different between the two seasons, there are difficulties in targeting insurance. While the acreage cultivated (paddy) in *Maha* is significantly higher to that of *Yala*, the penetration of insurance remains constant over the years (Fig. 16.8). The ratio of insured land to total cultivated area diverged in early years, but the rate has reached to 1% in both *Yala* and *Maha*. While data disaggregated by cropping season is not available for the period from 2005 to 2013, the most recent data also shows the divergence between *Yala* and *Maha* patterns as discussed later in this section. In the recent years, the financial performance of the insurance scheme measured by the loss ratio[3] has slightly deviated between the seasons (Fig. 16.9). Feasibility of the insurance scheme has been marginally better in *Yala* season overall as seen in lower loss ratios.

However, the data on indemnity payments made in the last 3 years show that even though extent under insurance under the compulsory scheme is starkly different across seasons (extent under insurance in *Maha* is double that of *Yala*), the losses indemnified are not different. This observation highlights one of the key issues of the performance of insurance in paddy sector. In effect, the insurance scheme has failed to differentiate the variation of risk by the season and assessed damages and compensations following an alternative logic.

[3]This is the ratio between total indemnity and collected premium.

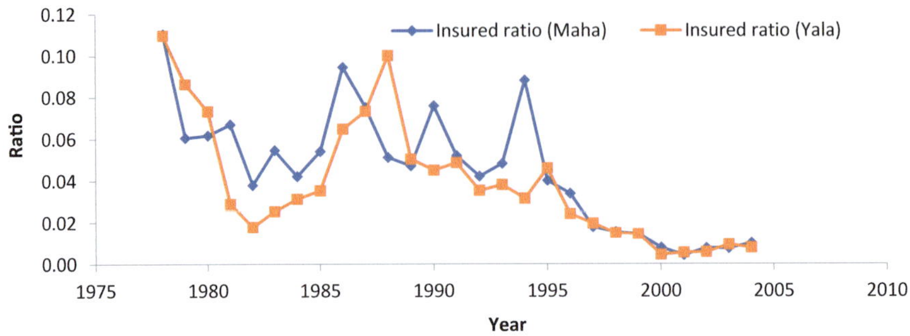

Fig. 16.8 Insurance penetration, reported as a ratio of insured land to cultivated land, between *Yala* and *Maha* seasons (note: AAIB has recorded data by season only till 2005. Data separated by cropping season is not available for recent years)

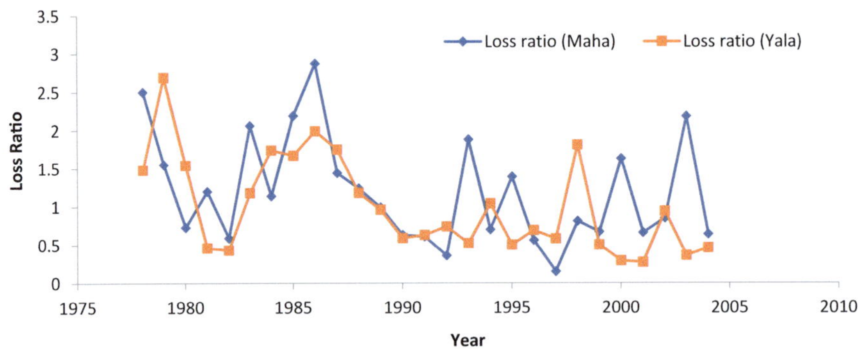

Fig. 16.9 Insurance performance, reported as the ratio between indemnity and premium, between *Yala* and *Maha* (AAIB data; (note: AAIB has recorded data by season only till 2005. Data separated by cropping season is not available for recent years)

To investigate the performance of insurance from the farmer's perspective, we take the data from 2013[4] to 2016 for both seasons to develop a comparison of the two seasons in terms of premium value per acre and indemnity paid per acre. For the six seasons under consideration, the extent insured and extent indemnified are displayed by season in Fig. 16.10. The *Yala* and *Maha* data clearly show a disproportionality of indemnification by season for the 3 years considered. In particular, in terms of likelihood of compensation, farmer has higher benefit in *Yala* than *Maha*. This observation can be further illustrated by considering the per acre values of premium and indemnity by season (Fig. 16.11). The difference of value to the farmer for an acre of land insured clearly is higher for *Yala* in terms of likelihood of being compensated.

[4] Data by cropping season is not available for the period 2005–2013, according to AAIB.

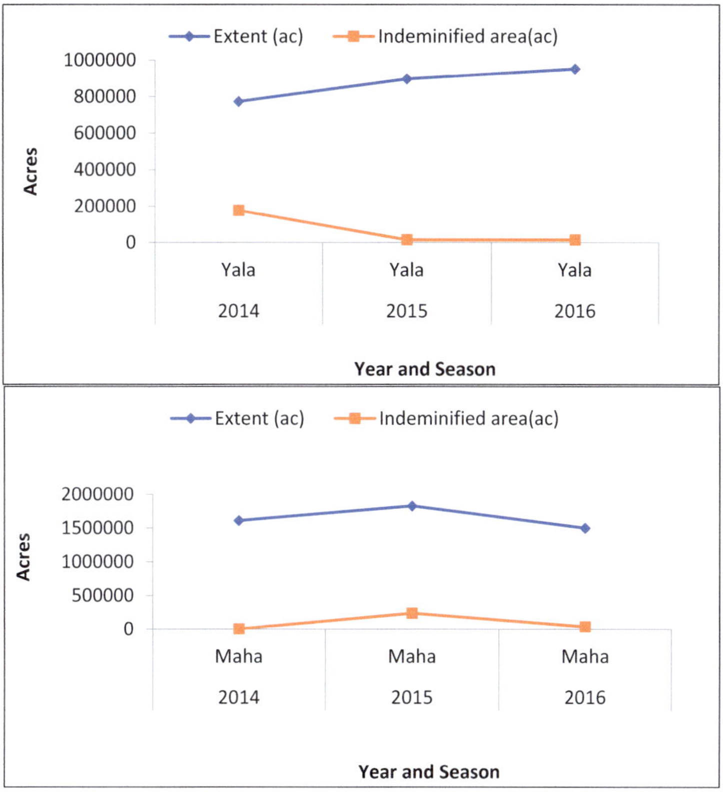

Fig. 16.10 Insurance performance in *Yala* (top panel) and *Maha* (bottom panel) season in terms of land subscribed and land indemnified (in acres)

16.5 Financing Insurance: Global Experience

While farmers use insurance as a risk coping mechanism, the same uncertainty is an opportunity for arbitrage. Thus, the development of a variety of agricultural insurance/derivative products follows the business motive of making money from the arbitrage opportunity. This is quite similar to other insurance products in the financial markets which are designed to generate returns while serving a socially important purpose (Martin and Clapp 2015). While this logic is true about all private insurance schemes, public agricultural insurance schemes in low-income countries operate as compensation schemes with subsidized premium. The state holds a fund to disburse claims largely funded by taxpayer contributions. This format of insurance administration carries a compulsory subsidy for the premium. Often, the premiums are collected as part of payment for inputs. In essence, the insurance program runs without a conscious motive to be insured for an identified risk on the part of the farmer. Given this context, there are no financial incentives for the farmer to adopt other risk mitigation strategies and often ends up creating more space for moral hazard.

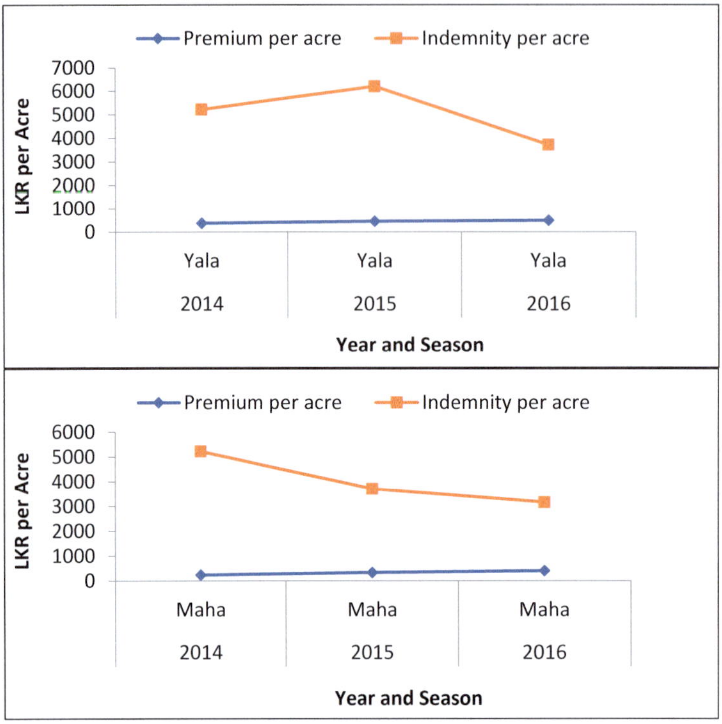

Fig. 16.11 Premium value (in LKR) per acre and indemnity (in LKR) per acre for *Yala* (top panel) and *Maha* (bottom panel)

In order to learn the lessons of running large insurance programs in farming sector, it is important to look at high-income countries that have run programs for decades. As per global data, nearly 57% of agricultural insurance schemes are underwritten by private insurers and nearly all of these come from high-income countries. European countries have adopted insurance quite substantially via financing models which involve substantive state subsidies. According to the Eurostat data (Fig. 16.12), the largest subsidies in Europe are operational in Italy. In all indemnity-based insurance schemes, there is however a deductible falling between 10% and 20% of the payout. The presence of deductible alleviates the adverse selection and moral hazard issues to a certain degree. The premium as a percentage of payout varies from 3% to 10% approximately. In most low-income to lower-middle-income countries, since the government has taken the role of insurer, the deductible and the subsidy has been bundled together, effectively charging a premium under 10% of the payout.

Loss ratios (loss paid as a ratio of premium) of the crop insurance programs in Sri Lanka was as high as 1.74 in 1986 when a comparative study of agricultural insurance was performed first (FAO 1991). This was only lower than that of India for the same season but exceeded all low-income countries in Asia and Africa which had state-run agricultural insurance. The loss ratio has remained at these values

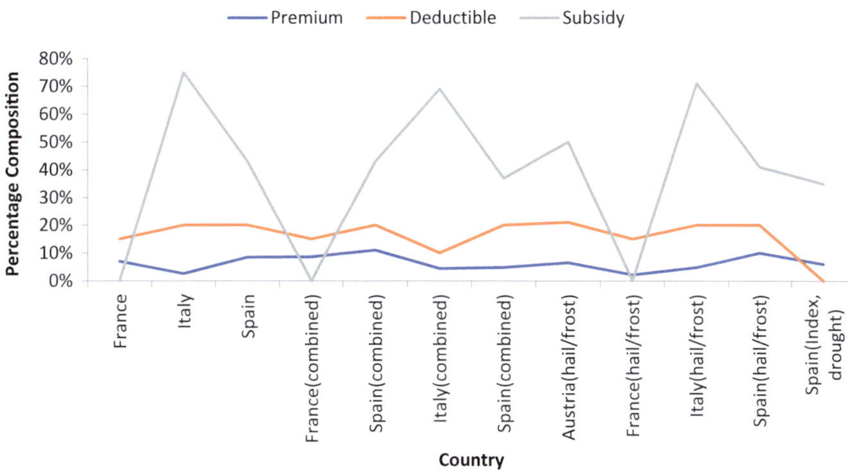

Fig. 16.12 Composition of total premium by percentage component and by country. *Note:* The percentages are averaged values of those documented for wide variety or perils and agricultural activities (Source: Eurostat data – multiple years)

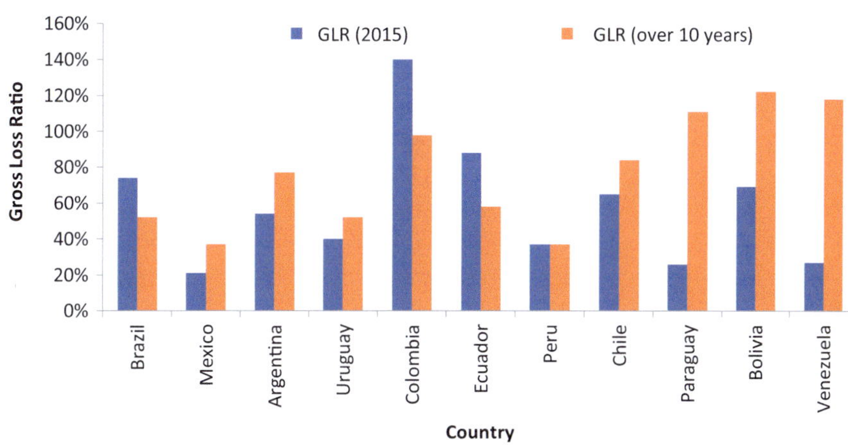

Fig. 16.13 Short term and long term Gross Loss Ratios (GLR) as percentages (Source: SwissRe 2016)

even to date, according to AAIB. Further, according to AAIB sources, drought years and political initiatives have worsened the loss ratios in recent years. The comparative experience from South America is useful for establishing the magnitude of loss ratios in countries where agro insurance have taken root. Figure 16.13 displays the calculated loss ratios for year 2015 and the average loss ratios for 10-year period. In the more politically stable countries of South America, the difference between short-term and long-term loss ratios is smaller indicating that year-to-year variation in agricultural activity is not volatile.

16.6 Challenges and Unresolved Issues in Implementing Insurance

The most classical type of insurance is fixed premiums and fixed compensation models. The premium per acre is fixed on the basis of average yields. The rates of premia are often set up at two levels depending on the credit exposure of the farmer. In instances where cropping practices and management practices are not uniform, the fixed premium and fixed compensation model is not as appealing. The usual practice is to benchmark on a unit land basis, a single crop, and single peril basis. Even for a situation as uniform as a single crop and single peril, a higher premium rate with higher coverage is often operated targeting the farmers who take agricultural credit while a lower premium rate and lower coverage level for the farmers who have not obtained agricultural credit.

An affiliated contentious issue in agro-insurance is the determination of indemnity during a loss. Calculation of indemnity varies across countries and historically followed two methodologies: yield based and cost of input based (there may be mechanisms for revenue forgone as well but often not practiced). In the initial state-run scheme, Sri Lanka (and many other countries) adopted a yield-based assessment of indemnity (verified by the mandated officer at the time of harvest). There were instances where input-based indemnity was calculated. Philippines in Asia and Zimbabwe in Africa are notable examples. However, input-based methodology depends heavily on existence of reliable farm records which is usually lacking in low-income smallholder farming.

Agricultural insurance is usually benchmarked on a geographical area and, in certain instances, a crop type. Usually area is an administrative boundary or an agro ecological region. Crop insurance in Canada and India are examples that were implemented through an area approach. However, the area approach in Canada proved to be inequitable, as it did not ensure a fair distribution of benefits among the farmers (Turvey et al. 2002). Farmers with yields closest to the average would be the ones to get the most benefits. Turvey et al. (2002) point out that the area approach was not only unbalanced but also ineffective. They suggest that individual approach to crop insurance is better for reducing risk, but it also implies the use of higher premiums. Another modality is offering insurance on group basis. This is quite similar to the concept of group-based credit/microfinance. Group insurance has its own merits. Delivery and servicing become easier and administrative costs can be kept low. If the group is sufficiently large and homogeneous, problems of adverse-selection, and to some extent the problem of moral hazard, can also be mitigated. Insurance on a group basis is therefore an important strategy for extending insurance to riskier and remote areas where insurer outreach is poor.

One of the factors inhibiting the growth of agricultural insurance in low-income countries is the lack of facilities for assessment of loss. While in high-income settings third party officers are involved, most countries adopt a system of committee-based assessment. The committee usually comprises agriculture sector key officials with field-level exposure. For instance, in Sri Lanka, during the first implementation of agriculture insurance, the agronomist, the village level officer of the Agrarian

Services Committee, and the Extension Officer of the Agricultural Department were the members of the committee (AAIB, various years).

It is difficult to reach farmers individually, since the cost of delivery and servicing is likely to be prohibitive. Furthermore, problems of adverse-selection and a limited spread of risks can be mitigated if all are insured. Hence, practical considerations may require the program to be automatic or obligatory, either by mandating concessionary credit or input subsidy recipients to take insurance as in India and Sri Lanka. In the recent implementation of crop insurance in Sri Lanka, the conditionality appeared as mandatory for recipients of fertilizer subsidy.

16.7 Indexation of Insurance: Recent Developments

There are a number of inherent limitations to the conventional insurance schemes. These limitations and drawbacks include the difficulty of controlling moral hazard, inability to control adverse selection, inadequate and inefficient institutions which make monitoring and assessment of risks of many farmers more costly, and many weather-related hazards are often correlated with each other which make assessment of losses and damages more challenging for insurance providers. As an alternative to conventional insurance, Weather Index Based Insurance (WIBI) products have been promoted by international development agencies with local insurance providers in developing countries in Africa and Asia. WIBI relies on the objective measure or indicator of risk due to weather such as rainfall and temperature. An example of a commonly used weather index related to crop yield is the cumulative rainfall for each of the crop growth and development stages. Another example is the number of days with temperatures above or below a certain threshold temperature (Banerjee and Berg 2012).

Four countries in the ASEAN have either known successful commercial index insurance products (Philippines), small crop weather index insurance pilot programs (Indonesia and Thailand), or crop weather index insurance pilots designed and are awaiting implementation (Vietnam) (FAO 2011). However, the feasibility of WIBI depends much on a number of factors including the development of the appropriate weather index as well as the provision of enabling institutional framework for its implementation and operations. Weather-indexed insurance design is not practically perfect. For instance, Munich-Re Inc. failed to achieve commercial scale in introducing a flood index insurance product in Jakarta due to low value proposition, high premium charge, failure to adjust to the preference of the market, and an incomplete understanding of the hydrological context of the area covered (Bhat and Mukherjee 2013). In other ASEAN countries, efforts using river gauge data as a proxy for flood damage has been piloted.

The WIBI can also address moral hazard and adverse selection. Moreover, since pay-outs or payments of indemnities are based on observed weather data from accredited weather gauging stations, costs of monitoring and assessment are minimized (Barnett et al. 2008). Thus, operational cost is low and rapid pay-offs are expected. Nevertheless, WIBI has also some disadvantages related to basis risk among farmers, limited targeting of hazards, the need to replicate and adjust the index, and inadequacy of available weather data for the cropped areas (IFAD 2011).

The WIBI also appear prominently in the FAO's call for "Climate-Smart Agriculture" (FAO 2013). It has also caught the interest of major agro-input suppliers. For instance, Syngenta Foundation launched a weather-based product in Kenya in 2009, which was later converted to a private business in 2014. Monsanto Inc. purchased The Climate Corporation in 2013, a weather insurance underwriter, and is now planning to develop index insurance products for Indian and South American farmers (Gilbert 2014). Karlan et al. (2012) find that a rainfall-based index insurance contract has stronger effects than direct cash grants on farmers' investment levels in Ghana. However, Binswager-Mkhize (2012) questions the general availability of information and provides a cautionary critique of the recent shift towards index-based insurance.

16.8 Summary and Conclusion

Performance of the agricultural insurance is a function of multiple overlapping variables, such as the degree of state support/intervention, the depth and breadth of distribution channels, retention levels, and others. Evidence suggests that performance varies widely from country to country. Even so, there are certain common characteristics—relatively high loss ratios, high operating cost structures, high volatility of losses and premiums, and high price competition. Agriculture loss ratios generally exceed non-life market averages largely because of the segment's higher exposure to catastrophe risks compared to other lines of business. Claims are subject to the vagaries of weather and climate, and hence loss experience tends to be volatile and uncorrelated with trends in the rest of the insurance industry.

There is a potential role that insurance can play in organized and modern agriculture. If farmers wish to integrate with the global value chains, the need for risk management through insurance may surface as important. Given that state features heavily in the financing of insurance in the low-income countries, with more integration to the international value chains, the market-based insurance operated through private insurers is the key to developing the agricultural insurance sector. Developments in indexed products and other innovative developments will generate the alternatives needed for localized and crop-specific conditions.

References

Banerjee C, Berg E (2012) Policy for implementation of Index Based Weather Insurance revisited: the case of Nicaragua (No. 122448). European Association of Agricultural Economists

Barnett BJ, Barrett CB, Skees JR (2008) Poverty traps and index-based risk transfer products. World Dev 36(10):1766–1785

Bhat S, & Mukherjee P (2013) Towards de-risking disasters: taking stock of microinsurance for disaster risk reduction – index based microinsurance in South and South East Asia. Uttar Pradesh: MicroSave

Binswager-Mkhize HP (2012) Is there too much hype about index-based agricultural insurance? J Dev Stud 48:187–200

Binswanger HP, Deininger K (1999) Explaining agricultural and agrarian policies in developing countries. World Bank, Washington, DC

Boucher S, Carter M, Guirkinger C (2008) Risk rationing and wealth effects in credit markets: implications for agricultural development. Am J Agric Econ 90(2):409–423

Clapp J (2012) Food. Polity Press, Malden, MA

FAO (2011) Agricultural Insurance in Asia and the Pacific region. Food and Agriculture Organization of the United Nations, Regional Office for Asia and the Pacific, Bangkok. http://www.fao.org/3/i2344e/i2344e00.htm. Accessed 24 June 2019

FAO (2013) Climate-smart agriculture: sourcebook. Food and Agriculture Organization of the United Nations, Rome

Food and Agriculture Organization (FAO) (1991) A compendium of Crop Insurance Programmes. Food and Agriculture Organization of the United Nations, Rome

Gilbert K (2014) Index insurance take root as climate change stings agriculture. Institutional Investor. https://www.institutionalinvestor.com/article/b14zbjncgzd730/index-insurance-takes-root-as-climate-change-stings-agriculture. Accessed 24 June 2019

Glauber JW, Collins KJ, Barry PJ (2002) Crop insurance, disaster assistance, and the role of the federal government in providing catastrophic risk protection. Agric Finance Rev 62(2):81–101

International Fund for Agricultural Development (IFAD) (2011) Weather Index-based Insurance in Agricultural Development. ISBN 9789290722762

Karlan D, Osei R, Osei-Akoto I, Udry C (2012) Agricultural decisions after relaxing credit and risk constraints. Working Paper (18463), National Bureau of Economic Research

Kerer J (2013) Situation of agricultural insurance in Kenya with reference to international best practices. GIZ/MoA (ACCI), Nairobi. http://www.ruralfinanceandinvestment.org/sites/default/files/ACCI_Insurance-Background-Kenya_6-2013.pdf. Accessed 24 June 2019

Mahul O, Stutley CJ (2010) Government support to agricultural insurance: challenges and options for developing countries. World Bank, Washington, DC

Martin SJ, Clapp J (2015) Finance for agriculture or agriculture for finance? J Agrar Chang 15(4):549–559

Roberts RA (2005) Insurance of crops in developing countries, vol 159. Food & Agriculture Organization, Rome

Rosenzweig MR, Binswanger HP (1993) Wealth, weather risk and the profitability of agricultural investment. Econ J 103:56–78

Sandaratne N (1974) Using insurance to reduce risks in peasant agriculture, The Agricultural Development Council, USA

SwissRe (2016) Agricultural insurance in Latin America: taking root. http://agroinsurance.com/wp-content/uploads/2016/08/Agricultural-Insurance-in-Latin-America-en.pdf. Accessed 24 June 2019

Turvey CG, Hoy M, Islam Z (2002) The role of ex ante regulations in addressing problems of moral hazard in agricultural insurance. Agric Finance Rev 62(2):103–116

UNCTAD (1994) Agricultural insurance in developing countries. https://unctad.org/en/PublicationsLibrary/unctadsddins1Rev.1_en.pdf. Accessed 24 June 2019

World Bank (2011) World development report 2010: development and climate change. World Bank, Washington, DC

Yamauchi T (1986) Evolution of the crop insurance program in Japan. In: Hazell P, Pomerada C, Valdez A (eds) Crop insurance for agricultural development: issues and experience. Johns Hopkins University Press, Baltimore, pp 223–239

Agricultural Research for Sustainable Food Systems: Recommendations with Special Reference to Sri Lanka

17

Buddhi Marambe, Warshi S. Dandeniya, and Jeevika Weerahewa

Abstract

This chapter summarizes the recommendations and conclusions presented in 16 other chapters of the book. Scientific research conducted on different sub-components of food systems in Sri Lanka reveals that (i) a balanced approach that effectively utilizes the merits of traditional technologies with proper blend of modern agro-inputs is required to make the food systems sustainable, (ii) agricultural production and land policies are to be reoriented to diversify farm-level production, (iii) food systems are to be made more nutritionally sensitive through appropriate policies, institutions and technologies to promote production and consumption of nutritious and healthy foods, (iv) a well-coordinated agricultural policies, food policies and nutrition policies is to be implemented to address the challenges faced by the food systems and (v) building of human resource capital, institutional infrastructure and stakeholder links are great assets in the transition process of current food systems to sustainable food systems, and furthering economic development while achieving SDGs.

Keywords

Food systems · Sustainability · Agriculture · Nutrition · Sri Lanka

B. Marambe (✉)
Department of Crop Science, Faculty of Agriculture, University of Peradeniya, Peradeniya, Sri Lanka
e-mail: bmarambe@pdn.ac.lk

W. S. Dandeniya
Department of Soil Science, Faculty of Agriculture, University of Peradeniya, Peradeniya, Sri Lanka

J. Weerahewa
Department of Agricultural Economics and Business Management, Faculty of Agriculture, University of Peradeniya, Peradeniya, Sri Lanka

B. Marambe et al. (eds.), *Agricultural Research for Sustainable Food Systems in Sri Lanka*, https://doi.org/10.1007/978-981-15-2152-2_17

17.1 Introduction

Sustainable food systems have become the major focus of global efforts to ensure food supply to growing populations. The Sustainable Development Goals (SDGs; UN 2015), with more intrinsic relationship with Goal 2 (end hunger, achieve food security and improved nutrition and promote sustainable agriculture), Goal 12 (ensure sustainable consumption and production patterns) and Goal 13 (take urgent action to combat climate change and its impacts), facilitated the transition of global food systems to become more sustainable. The FAO (2014) has identified five principles to support a common vision for sustainable agriculture and food, namely (1) resource efficiency, (2) action to conserve, protect and enhance natural resources, (3) rural livelihood protection and social well-being, (4) enhanced resilience of people, communities and ecosystems and (5) responsible governance. However, the environmental, social and economic dimensions that lead to sustainability issues of the modern food systems have drawn the attention of many researchers in the recent past (Béné et al. 2019), while recognizing the need for a systematic transformation. Different chapters of this book addresses various aspects of the major components of food systems with a special reference to Sri Lanka. The recommendations and the conclusions of these chapters are summarized here to provide concluding remarks on historical development of the agricultural research for sustainable food systems.

17.2 Overall Conclusions and Recommendations

Sustainable agriculture in the global arena has progressed as a production system to achieve self-reliance in food, a concept of stewardship, and a system for nourishing rural communities (MacRae 1990). Béné et al. (2019) reported some key challenges for policy-making and research on sustainable food systems focusing around (1) a clear and urgent need to provide precise meaning of a sustainable food system, (2) a need to more explicitly acknowledge the local-specific nature of food systems and (3) the thinking on sustainable food system should be cognizant of the presence of trade-offs. A historical perspective plays an important role in understanding the structure and function of food systems at present and plan and develop a more sustainable system in the future. This volume thus has focused on the historical development of the agricultural research for sustainable food systems with a special focus on Sri Lanka. Following section summarizes the overall synthesis on the outcome of the analysis.

- The food systems are influenced by interactions between many factors, i.e. technical, economic, social and cultural, and such dynamics should be considered in designing appropriate policies as they could confound research, analysis and development of sustainable operations. Transformative changes of agriculture and food systems are the current needs in all countries, but with different priorities to align with different economies.

- The shifts in food systems cause profound changes in dietary habits and hence the status of nutrition. Though declining trends in many forms of undernutrition, particularly among females, over time has been observed in Sri Lanka, it still faces the challenges in undernutrition, particularly among the estate sector and rising level of over-nutrition across all age cohorts particularly among the urban residents. The society is on the verge of entering into the fourth stage of nutrition transition which is characterized by consumption of energy dense food, overweight and obesity, and rising non-communicable diseases. The food systems are to be made more nutritionally sensitive through appropriate research, pricing, trade, food and nutrition policies to promote production and consumption of nutritious and healthy foods.
- The necessity has arisen to restore and conserve the natural ecosystems surrounding the agricultural ecosystems to sustain the food production in the long run, especially in providing nutrient-rich food through ecologically sound agricultural technologies. Furthermore, the cascaded tank village system in Sri Lanka was designated as a Globally Important Agricultural Heritage System (GIAHS) by the Food and Agriculture Organization of the United Nations (FAO 2014). The tank cascades will continue to play a crucial role at global, regional and national level for food security and sustainable development of food systems in the context of climate change.
- Sri Lanka hosts a very high soil diversity; thus, systematically documenting and using soil information is crucial for sustainable management of the land resource. Detailed soil information has emerged as a prerequisite for addressing environmental issues, food security, energy security, water security, and human health, and ensuring sustainable food systems in global, regional and national scale. Over the past few decades inventorying soil information gradually advanced with a consistent development of expertise on advances in soil survey, classification and mapping. The digital soil mapping techniques have been used in Sri Lanka mostly to generate detailed soil information required at a small scale for land resource management related to agricultural purposes. With necessary government interventions and multidisciplinary team work, finer resolution digital soil information can be generated aiming a multitude of applications to support decision-making in land resource management for strengthening sustainability of food systems.
- Conventional breeding techniques practised in food crops over decades through hybridization have faced many obstacles, where some crop wild relatives do not show incompatibility for cross pollination with other varieties. Innovative breeding techniques such as bridging varieties, embryo rescue, double haploids, chromosomal substitution, protoplast fusion and genetic transformation need to be explored, though they are costly operations with no assured success. Enriching the gene pool with trait-specific germplasm will help in varietal development in perennial food crops that would lead to sustainability of food systems as well as the industrial development. An integrated approach using the best of conventional and modern technologies would help developing countries in their path for development and sustenance in the food systems.

- Methodologies adopted for research and extension for varietal testing and release of selected food crops, especially rice, in Sri Lanka have made a significant progress in their adoption at the field level. More than 98% of the rice varieties cultivated by the farming community in Sri Lanka comprise new high-yielding varieties, showing a significant shift from traditional varieties cultivated since 1959. Such coordinated varietal improvement programmes through participatory approaches for other food crops involving a multidisciplinary team of experts and field practitioners using a blend of technologies developed by various research institutes and universities, with the adoption of appropriate regulations for plant varietal protection and intellectual property management, will no doubt enhance the varietal development with private sector participation and varietal adoption rates by the farming community. Further, well-directed scientific explorations to assess the pharmaceutical and nutraceutical values of the local germplasm of food crops will help attracting global demand and open up new market opportunities. These aspects will immensely contribute to the sustenance and resilience of the food systems.
- Progress has been shown in the recent past in the development of technological packages with more labour and cost-saving techniques in food crop production. Agricultural input intensification has taken place over the years with farming communities shifting away from the traditional but low-yielding technologies supported by a well-coordinated research and extension programmes owing to the green revolution. The positive impact of such technologies in providing food for growing populations has been witnessed globally. Approximately 90% of the global growth in crop production and 80% growth in developing countries were a result of higher yields and increased cropping intensity, while the remainder came from land expansion. However, the negative impacts of such input intensified-agriculture especially on the human and environmental health are currently being debated, as the pressure on renewable water resources and energy sources would continue and need urgent attention of the scientists and practitioners focusing on more judicious use of resources considering the sustainability of the future food systems.
- Pests (insects, weeds and pathogens) have played a critical limiting role in the history of agriculture, and their impacts have been aggravated since the later part of the twentieth century. Their management has been difficult owing to the need of biological and ecological information prior to planning and adoption of effective control measures. The work towards integrated pest management (IPM) has yielded some success stories, with more attention on the non-chemical pest control methods. The growing demand for pesticide residue-free food products and the use of plant-based crop protection methodologies need further attention of the scientists and researchers in developing a sustainable system for pest management. Here, new product development should be ensured through proper test protocols while also identifying their impacts on non-target organisms, fate in the environment, etc. These would not only overcome the technical barriers for product registration and use in Sri Lanka but also make the marketed food products safe for consumption.

- An increase in global and local demand for agri-produce from eco-friendly farming systems, as evidenced in the recent past, has provided an guide for fine-tuning the technologies that are specific for futuristic sustainable cropping systems (e.g. ecological agriculture, organic farming) and protected culture, which needs attention of all stakeholders in the food crop production sector. A balanced approach for effectively utilizing the merits of traditional technologies with proper blend of modern agro-inputs would no doubt benefit the food systems in terms of efficient use of resources and their sustainable use.
- The global demand for animal-based food is predicted to increase by 70% by 2050 with poultry sector being the main contributor. Animal feed industry thus plays an important role in determining the much-needed products to the society. However, with the growing population and limited land available due to urbanization and other agriculture ventures, the feed crop production faces enormous challenges to meet the demand. Use of alternate feed resources (non-traditional feeds and by-products) and increase in feed utilization efficiency through feeding management, including precise feed formulation, have been suggested to support the food systems at global and national scale.
- Sri Lanka provides a rich ground for expansion of the livestock and poultry industries to cater to the ever-growing consumer demand and to maximize its contrition to food and nutritional security. Limited resources coupled with weak strategic approaches have made industry to lose the opportune advantages. Mixed production system of animals could be a driving force and indeed a strength bringing in solutions to many challenges faced by the industry. Furthermore, being an island nation with an eight times higher marine water resources compared to its land mass, the country possess enormous natural resources including fish. About 86% of the country's total fish production is received from the marine sector and the country has the world's highest freshwater resources, i.e. 4% of freshwater bodies compared to its land area, but only 10% of it is used for fish production. Availability of these natural aquatic resources provides immense opportunities to achieve high economic growth with guaranteed food and nutritional security for its population in near future by blue economic revolution. Thus, better utilization of the industry dichotomy and supportive policy orientation are required for the sustainability of livestock food systems in the country.
- Extension is a component that has received a stepmotherly in Sri Lankan agriculture and has been viewed as a process than mere transfer of technological information to farmers. The analysis clearly indicates that a comprehensive agricultural extension policy is required for this important sub-sector to have the priority attention that it deserves. Limited co-ordination among the state, private sector and non-government organizations in providing agricultural extension services and weak research-extension dialogue have been the major impediments for the disarray experienced in the agricultural extension system in the country. Use of diversified communication media and strengthening ICT-based interventions are expected to help linking farming communities effectively with other stakeholders in the agricultural value chain.

- Performance of agricultural insurance systems varies widely from country to country with a relatively high loss ratio, operating costs, high volatility of losses and premiums, and high price competition. The potential role of insurance in organized and modern agriculture food systems should never be undermined but be considered as a strength that integrate rural farmers with global value chains, and an important risk management tool. In Sri Lanka, state intervenes heavily in financing agricultural insurance as in many other low-income countries. Market-based insurance operated through private insurers with indexed products and other innovative developments is the key to the development of the sector.

The review of historical developments in the agricultural research and sustainable food systems in Sri Lanka revealed that well-focused research to generate scientifically valid information is required for the in-depth understanding of the pathways to achieve sustainable food systems in the country. Agricultural policies, food policies and nutrition policies should be integrated to combat the challenges faced by the food systems. Reorientation of agricultural and land policies to diversify farm-level production is essential to explore the development paths in the country's agricultural economy. The long felt need of building human resource capital, institutional infrastructure and stakeholder links are great assets in the transition process of current food systems to sustainable food systems, and furthering economic development while achieving SDGs.

References

Béné C, Oosterveer P, Lamotte L, Brouwer ID, de Haan S, Prager SD et al (2019) When food systems meet sustainability – current narratives and implications for actions. World Dev 113:116–130

FAO (2014) Building a common vision for sustainable food and agriculture: principles and approaches. Food and Agriculture Organization of the United Nations, Rome

MacRae R (1990) Strategies to overcome institutional barriers to the transition from conventional to sustainable agriculture in Canada: the role of government, research institutions and agribusiness. PhD thesis. McGill University, Canada

UN (2015) Transforming our world: the 2030 agenda for sustainable development. United Nations, New York, pp 39